WITHDRAWN

D1391341

ANIMAL HEALTH AND WELFARE IN ORGANIC AGRICULTURE

Animal Health and Welfare in Organic Agriculture

Edited by

M. Vaarst
Danish Institute of Agricultural Sciences
Research Centre Foulum, Tjele, Denmark

S. Roderick
Duchy College
Camborne, Cornwall, UK

V. Lund
Swedish University of Agricultural Sciences
Skara, Sweden

and

W. Lockeretz
Tufts University
Boston, Massachusetts, USA

Askham Bryan College
LIBRARY BOOK

CABI Publishing

CABI Publishing is a division of CAB International

CABI Publishing
CAB International
Wallingford
Oxon OX10 8DE
UK

Tel: +44 (0)1491 832111
Fax: +44 (0)1491 833508
Email: cabi@cabi.org
Web site: www.cabi-publishing.org

CABI Publishing
875 Massachusetts Avenue
7th Floor
Cambridge, MA 02139
USA

Tel: +1 617 395 4056
Fax: +1 617 354 6875
Email: cabi-nao@cabi.org

©CAB International 2004. All rights reserved. No part of this publication may be reproduced in any form or by any means, electronically, mechanically, by photocopying, recording or otherwise, without the prior permission of the copyright owners.

A catalogue record for this book is available from the British Library, London, UK.

Library of Congress Cataloging-in-Publication Data
Animal health and welfare in organic agriculture / edited by M. Vaarst ...[et al.]
p. cm.
Includes bibliographical references.
ISBN 0-85199-668-X (alk. paper)
1. Organic farming-Europe. 2. Livestock-Europe. 3. Animal health-Europe. 4. Animal welfare-Europe. I. Vaarst, M. (Mette) II. Title.
S605.5.A55 2004
630.5'84'094-dc21 2003011238

ISBN 0 85199 668 X

Typeset by Wyvern 21 Ltd, Bristol.
Printed and bound in the UK by Cromwell Press, Trowbridge.

Contents

Contributors

Alrøe, H.F., *Danish Research Centre for Organic Farming, Danish Institute of Agricultural Sciences, Research Centre Foulum, PO Box 50, DK-8830 Tjele, Denmark.*

Ambrosini, F., *Dipartimento di Scienze Zootechniche dell'Universitá di Firenze, Via del Casine 5, I-50144 Florence, Italy.*

Baars, T., *Louis Bolk Institute, Department of Organic Animal Husbandry, Hoofdstraat 24, 3972 LA Driebergen, The Netherlands.*

Baumgartner, J., *Institute of Animal Husbandry and Animal Welfare, University of Veterinary Medicine Vienna, Veterinarplatz 1, A-1210 Vienna, Austria.*

Bennedsgaard, T., *Department of Animal Health & Welfare, Danish Institute of Agricultural Sciences, PO Box 50, DK-8830 Tjele, Denmark.*

Bestman, M., *Louis Bolk Institute, Hoofdstraat 24, NL-3972 LA Driebergen, The Netherlands.*

Boivin, X., *INRA, Theix, 63122 St Genes Champanelle, France.*

Conington, J., *Scottish Agricultural College, West Mains Road, Edinburgh EH9 3JG, UK.*

Gray, D., *Scottish Agricultural College, Veterinary Science Division, Mill of Craibstone, Bucksburn, Aberdeen AB21 9TB, UK.*

Hektoen, L., *Department of Large Animal Clinical Sciences, The Norwegian School of Veterinary Science, Pb 8146 Dep., N-0033 Oslo, Norway.*

Henriksen, B., *Norwegian Centre for Ecological Agriculture (NORSØK), N-6630 Tingvoll, Norway.*

Hovi, M., *Veterinary Epidemiology & Economics Research Unit, University of Reading, PO Box 236, Reading RG6 6AT, UK.*

Idel, A., *League for Pastoral Peoples, Pragelatostr. 20, 64372 Ober-Ramstadt, Germany.*

Johnsen, P.F., *Department of Animal Health & Welfare, Danish Institute of Agricultural Sciences, PO Box 50, DK-8830 Tjele, Denmark.*

Kelly, H.R., *School of Agriculture, Food and Rural Development, University of Newcastle, Newcastle NE1 7RU, UK.*

Kiley-Worthington, M., *ECO Research Centre, Little Ash Eco Farm, Throwleigh, Okehampton, Devon EX20 4QJ, UK.*

Knierim, U., *Institute of Animal Hygiene, Welfare & Farm Animal Behaviour, School of Veterinary Medicine Hannover, Buenteweg 17p, 30559 Hannover, Germany.*

Kristensen, T., *Department of Agricultural Systems, Danish Institute of Agricultural Sciences, Research Centre Foulum, PO Box 50, DK-8830 Tjele, Denmark.*

Krutzinna, C., *Department of Animal Health, The University of Kassel, Nordbahnhofstrasse 1a, D-37213 Witzenhausen, Germany.*

Lockeretz, W., *Friedman School of Nutrition Science and Policy, Tufts University, Boston, MA 02111, USA.*

Lund, V., *Department of Animal Welfare & Health, Swedish University of Agricultural Sciences, PO Box 234, SE-532 23 Skara, Sweden.*

MacNaeihde, F., *Teagasc, Johnstown Castle Research & Development Centre, Wexford, Ireland.*

Martini, A., *Dipartimento di Scienze Zootechniche dell'Université di Firenze, Via del Cascine 5, I-50144 Florence, Italy.*

Menke, C., *Institute of Animal Husbandry and Welfare, Veterinarian University of Vienna, 1040 Vienna, Austria.*

Niebuhr, K., *Institute of Animal Husbandry & Animal Welfare, University of Veterinary Medicine, Vienna, Veterinarplatz 1, A-1210 Vienna, Austria.*

Padel, S., *Institute of Rural Studies, University of Wales, Aberystwyth SY23 3AL, UK.*

Pryce, J.E., *Livestock Improvement Corporation Ltd, Private Bag 3016, Hamilton, New Zealand.*

Roderick, S., *Organic Studies Centre, Duchy College, Rosewarne, Camborne, Cornwall TR14 0AB, UK.*

Roiha, U., *University of Helsinki, Mikkeli Institute for Rural Research and Training, Lonnrotinkatu 3–5, 50100 Mikkeli, Finland.*

Rydhmer, L., *Department of Animal Breeding & Genetics, Swedish University of Agricultural Sciences, Funbo-Lovsta, S-755 97 Uppsala, Sweden.*

Seabrook, M., *School of Biosciences, The University of Nottingham, Sutton Bonington Campus, Loughborough LE12 5RD, UK.*

Schmid, O., *Forschungsinstitut für biologischen Landbau, FiBL (Research Institute of Organic Agriculture), Ackerstrasse, CH-5070 Frick, Switzerland.*

Sørensen, P., *Danish Institute of Agricultural Sciences, Research Centre Foulum, PO Box 50, DK-8830 Tjele, Denmark.*

Striezel, A., *Atzelsberger Str. 10, D-91094 Bräuningshof, Germany.*

Studnitz, M., *Department of Animal Health & Welfare, Research Centre Foulum, PO Box 50, DK-8830 Tjele, Denmark.*

Sundrum, A., *Department of Animal Nutrition & Animal Health, University of Kassel, Nordbahnhofstra. 1a, 37213 Witzenhausen, Germany.*

Thamsborg, S.M., *Centre for Experimental Parasitology, Royal Veterinary & Agricultural College, Bulowsvej 13, DK-1879 Frederiksberg C, Denmark.*

Trujillo, R.G., *Instituto de Sociología y Estudios Campesinos, Universidad de Córdiba, Ave. Menendez Pidal s/n, 14004 Córdoba, Spain.*

Vaarst, M., *Department of Animal Health & Welfare, Research Centre Foulum, PO Box 50, DK-8830 Tjele, Denmark.*

Verhoog, H., *Louis Bolk Institute, Department of Organic Animal Husbandry, Hoofdstraat 24, 3972 LA Driebergen, The Netherlands.*

Wagenaar, J.P., *Louis Bolk Institute, Department of Organic Animal Husbandry, Hoofdstraat 24, 3972 LA Driebergen, The Netherlands.*

Waiblinger, S., *Institute of Animal Husbandry & Animal Welfare, University of Veterinary Medicine, Vienna, Veterinarplatz 1, A-1210 Vienna, Austria.*

Walkenhorst, M., *Forchungsinstitut für biologischen Landbau, FiBL (Research Institute of Organic Agriculture), Aclerstrasse, CH-5070 Frick, Switzerland.*

Weary, D., *Animal Welfare Program, Faculty of Agriculture, The University of British Columbia, 2357 Main Mall, Vancouver, BC, V6T 1Z4 Canada.*

Wemelsfelder, F., *Animal Biology Division, Scottish Agricultural College, Bush Estate, Penicuik, Edinburgh EH26 0PH, UK.*

Younie, D., *Scottish Agricultural College, Craibstone Estate, Bucksburn, Aberdeen AB21 9YA, UK.*

Zollitsch, W., *Department of Animal Science, University of Agricultural Sciences Vienna, Gregor Mendelstrasse 33, A-1180 Vienna, Austria.*

Preface

The rapid growth of organic farming has been among the most remarkable changes in European agriculture in the past several decades. Its significance extends beyond what can be stated numerically, such as the number of holdings that have been converted to organic management. Perhaps even more significant is the influence that it has had on European agriculture as a whole. By offering an alternative vision of agriculture based on a more balanced and mutually supportive relationship among the crops, soils and animals on a farm and between the farm and both its biological and social environment, organic farmers have profoundly affected the debates about many areas of European agricultural policy, especially regarding food quality and safety, biodiversity, environmental protection, and resource consumption.

Animals have been important in organic farming systems from the beginning, and animal health and welfare has always been a goal. However, until recently, more attention was paid to the crop side of organic systems, and animals were not a high priority in formal research and development of organic farming. But that has now changed. There is now recognition of the need to understand animal health and welfare better. What is now required is to begin to understand this in the context of organic farming principles and put this into practice. Because of the wide range of production systems and growing conditions throughout Europe, this is a very difficult task that presents many exciting challenges.

This book is a contribution towards that task. Its purpose is to advance the understanding of organic animal husbandry, drawing mainly on research and practical experience with organic farming in Europe. It takes the goals and fundamental ideas of organic farming as a given. The purpose is not to debate the validity of these goals, nor to compare the performance or societal consequences of organic and conventional systems. In saying this, however, we hope we are maintaining a (constructively) critical attitude in the sense of recognizing and acknowledging the problems and limitations of current organic systems, in the hope of contributing towards practical and effective solutions.

The book was conceived as an activity of the Network on Animal Health and Welfare in Organic Agriculture, an EU-funded network of some 17 researchers working at 17 universities and research institutions in 13 European countries. Between 1999 and 2001, the network held five meetings devoted to such topics as animal welfare, diversity, feeding, breeding and positive health measures. The final conclusions and recommendations from this Network is given in the Appendix of this book.

Most of the 48 authors (including the four editors) were drawn from the network, which meant that the network meetings provided a good opportunity for the editors and authors to work together. This also allowed the authors to benefit from having their topics discussed by the larger group. However, because the subject is so broad and because we wanted the book to reflect the wide range of European conditions and perspectives, we recruited additional authors as needed. This enabled us to draw on the best expertise available for each topic; in addition, it allowed us to have several authors per chapter, something we considered highly desirable because many of these chapters involve controversial issues on which there are many valid viewpoints. Most of the authors, like the four editors, have been significantly involved in organic farming for a considerable time, whether as researchers, advisers, educators or practitioners. However, because so much of the relevant expertise is not unique to organic farming, we did not impose this as a requirement, and enlisted several authors whose experience has mainly been in 'conventional' agriculture.

Because of the large number of authors involved, a word must be said about the role of the editors. We conceived this as a book, not as a collection of separate papers. However, it had to start out as separate papers, of course, since initially each author had only a general idea of the other chapters. To turn this material into a book, the editors took the liberty of moving material around among chapters for the sake of a more coherent structure. (It was fortunate that we could meet three times for several days each to carry out this task.) This means that many authors contributed beyond the individual chapters that bear their names. Moreover, several authors made valuable comments on their colleagues' drafts.

A problem in any book about organic farming is how to deal with diversity. A key principle in organic farming is that a farming system must be adapted to the local environment. We have confined ourselves to Europe, a region in which organic animal husbandry is particularly important and where much of the world's research on organic animal health and welfare has been done. Even dealing just with Europe, it was a formidable challenge to encompass the diversity of growing conditions and production systems; however much we would have liked to, it would have been out of the question to cover the rest of the world in comparable depth. However, we hope that this book will offer something to researchers and practitioners elsewhere. That is because this book is only in part about facts, which do not necessarily apply elsewhere. But in part it is about organic ideas and about thinking organically. Good organic farming involves principles, and these principles are broadly applicable, even though they must be applied one location at a time. We hope that beyond providing an up-to-date summary of issues important for animal health and welfare in European organic farming, this book will help others to understand the underlying principles better, and thereby to enhance the health and welfare of animals on organic farms in their parts of the world.

More broadly, we hope this book will stimulate constructive discussions among our readers, inspire them to take up some of the challenges in the field, and most of all, encourage them to develop visions of a future agriculture in which both animals and people are well served.

Acknowledgements

The editors would like to thank the following people for their time and effort in assisting in reviewing chapters in this book: Hugo F. Alrøe, Xavier Boivin, Niklas Cserhalmi, Douglas Gray, Per Jensen, Margit Bak Jensen, Lizzie Melby Jespersen, David Main, Helena Röcklingsberg, Eric Sideman, Dan Weary and Tony Wilsmore. A number of the authors also assisted in the review process, and their contributions are gratefully acknowledged. We would also like to thank Tim Hardwick at CAB International for his advice, patience and encouragement during the long process of writing and editing this book.

Foreword

Engelhard Boehncke

In its early days, the development of organic agriculture concentrated on such matters as soil fertility, plant health and the design of crop rotations. The development of animal husbandry lagged by 10 or 15 years, and still has not caught up. This is surprising, for two reasons. First, from the earliest days, a leading idea in organic farming was that healthy soil, healthy plants, healthy animals and healthy people are inseparably linked. Second, somewhat later, the negative side-effects of intensive animal production became a major driving force behind the advance of organic agriculture.

In any case, until now we have had no books about animal health and welfare in organic agriculture. What makes this book stand out is its extraordinary breadth and depth. Numerous European scientists from many countries present their knowledge and the results of their experience in the various fields of ecological animal science.

Animal husbandry systems vary greatly from region to region and from country to country. The regional conditions under which the animals live and the landscapes they have formed over the years are very different. This book does not attempt to even out this diversity of organic animal husbandry in different European countries. On the contrary, beyond merely being acknowledged, it is analysed thoroughly and examined from various directions.

Still, there are common basic principles of the how and why of keeping animals in organic agriculture. These are given ample attention as well, as is appropriate, given the European standards that set up a uniform framework for the role of animals in organic farming systems. Independent of country or region, they aim at an optimal integration of animals into the nutrient cycle of the farm organism. Furthermore, they help to minimize the ecological damage potentially caused by animal production.

Animal health is given particularly high priority on all kinds of organic farms. Therefore, the reader of this book will find extended chapters dealing with many aspects of animal health and diseases in organic agriculture. Throughout, it is made clear that not only human health, but also animal health is more than the mere absence of diseases. The authors also point out how animal diseases are influenced by feeding, breeding and handling, as well as veterinary health care and treatments. In other words, animal health has its physical, physiological and psychological sides.

The authors of this book could not work just on animal health; rather, they also have dealt very successfully with animal welfare. This is because of the development of two young disciplines: farm animal ethology and animal ethics. During the past 30 years, ethology was able to demonstrate how appropriate management systems can be designed. Again there is a variety of possibilities and solutions. However, they all have one thing in common: they give the animals the chance to realize their natural, inborn and learned behaviour, and that, as we know, supports their health and welfare.

Animal ethics also plays an important role in organic agriculture, representing the humane basis of animal welfare. Modern animal ethics has taught us that we should treat animals as a part of creation and that they have their own dignity and value.

It is to be hoped that the authors of this book have broken a path that others will follow. May it help all farmers, veterinarians, consumers, students and scientists to contribute to the improvement of animal health and welfare, not just in organic but in all of agriculture.

1

Organic Principles and Values: the Framework for Organic Animal Husbandry

Mette Vaarst,[1] Stephen Roderick,[2] Vonne Lund,[3] Willie Lockeretz[4] and Malla Hovi[5]

[1]Department of Animal Health and Welfare, Research Centre Foulum, Danish Institute of Agricultural Sciences, PO Box 50, DK-8830 Tjele, Denmark; [2]Organic Studies Centre, Duchy College, Rosewarne, Camborne, Cornwall TR14 0AB, UK; [3]Department of Animal Welfare and Health, Swedish University of Agricultural Sciences, PO Box 234, SE-532 23 Skara, Sweden; [4]Friedman School of Nutrition Science and Policy, Tufts University, Boston, MA 02111, USA; [5]Veterinary Epidemiology and Economics Research Unit, University of Reading, PO Box 236, Reading RG6 6AT, UK

Animals in Organic Farming

Animal husbandry has been a substantial part of the growth in organic farming in the past several decades. In Europe, a common set of standards has been established for organic farming. However, organic farming is not just a niche production sector, based on rules that allow its products to carry a particular label. More than that, it is a concept with a history, built on a set of coherent values. Standards are established to allow these values to be embodied in practical farming methods and actual products.

Animals have been important in organic farming systems from the outset, both conceptually and in practice, and high animal health and welfare status of the animals has always been an important goal for organic husbandry. Animals should be an integral component of the organic farm, part of a system in which all parts interact to their mutual benefit, and where a harmony is created between the land, the animals and the people. The farm strives to be a closed system, producing feed for its own animals and incorporating their manure into crop production.

© CAB International 2004. *Animal Health and Welfare in Organic Agriculture* (eds M. Vaarst, S. Roderick, V. Lund and W. Lockeretz)

Nevertheless, as opposed to the crops, animals are not just parts of this system; they are also sentient creatures and as such they deserve special moral considerations. This makes their management fundamentally different from that of crops. When a farmer is learning to manage a farm with no chemical inputs, sometimes a field may become overgrown with weeds, for example. The farmer might not be able to do much about it, and may simply accept the loss as an inevitable part of learning how to farm organically.

However, it is not acceptable to let animals suffer or die. This moral aspect of dealing with sentient beings gives animals a special status on the farm. They are individuals that need to be looked after, that can suffer, and that can interact with each other and with the humans around them. Animal welfare – the notion that animals have experiences and are sentient beings – gives humans a moral obligation to treat animals well and to intervene before they suffer or die. This obligation allows us to use synthetic medicines for treating sick animals – the only circumstance where use of 'chemicals' is allowed and recommended in organic farming in some cases and by some people to avoid suffering.

Whilst the avoidance of suffering is important in both organic and conventional animal husbandry, the organic farming principles go much further than that in pursuit of animal welfare. One of the basic principles of organic farming refers to access to natural behaviour for organically managed animals, which substantially broadens the concept of 'welfare'. Moreover, this heightened understanding of the lives of animals needs to be put into practice. Because of the wide range of production systems and growing conditions throughout Europe, this is a very difficult task that presents many exciting challenges.

Discussion is needed not only of how organic farming influences the lives of the animals, but also of how our awareness of animals and their welfare can influence the development of organic farming. The special situation of animals – as both sentient individuals and parts of a farming system – raises interesting questions. Because 'individuals' must be handled differently from 'parts of the farm', the animal herd in some cases seems rather disconnected from the organic farm.

The visionary ideas of organic farming are well developed in regard to crop production, where the farmer must rely on non-chemical techniques. On livestock farms, there often is not much difference between organic and conventional management, for example in regard to disease treatment or prevention. In interview studies of organic farmers in the Scandinavian countries, farmers expressed frustration that no really 'new' or 'alternative' ways to manage the animals have emerged (Vaarst, 2000a; Vaarst *et al.*, 2001; Lund, 2002). Some farmers considered the use of alternative medicine as something new, but had not undertaken other, more fundamental changes in their husbandry methods, such as breeding for increased disease resistance or introduction of more species-appropriate housing. Farmers' advisers and veterinarians have also noted that they cannot really see any clear goals that make organic animal husbandry very different from conventional (Vaarst, 2000b). Both groups (conventional and organic farmers) agree on good animal welfare as a goal. Whilst restrictions on the use of conventional medicine may change farmers' perspectives somewhat, no radical alternatives or specifically 'organic' characteristics seem to guide the development of organic animal husbandry. One of the aims of this book is to present 'real' alternatives and to establish characteristics that may guide this development in the future.

A further complexity is that differences in climate, traditions, farm characteristics and farming conditions among European countries make it a challenge to state a common understanding of what European organic animal husbandry is. Nevertheless, since August 2000, uniform standards are supposed to guide all these different systems.

These dilemmas form a background for asking about the identity of organic animal husbandry. On the basis of its current status, what can we point to as relevant and realistic possibilities for its future development in theory and practice? For example, in recent years the understanding of animal welfare in organic farming has increasingly been discussed as including functional integrity and a striving towards 'naturalness'; how can this be integrated into daily practice? This book is an attempt to explore questions such as these, and to see organic animal husbandry within the broad concept of organic farming. The core values of organic farming and its development over the past several decades will be the first step in understanding where organic animal husbandry is today – as well as possible future developments.

Development of Organic Farming and its Core Values

The principles of organic farming have been stated in different ways that vary somewhat in what they emphasize, but agree on the fundamental issues. The Scandinavian organic farming associations have accepted the following statement:

> Organic farming describes a self-sustaining and persistent agro-ecosystem in good balance. As far as possible, the system is based on local and renewable resources. It builds on a holistic view that incorporates the ecological, economical and social aspects of agricultural production in both the local and global perspectives. In organic farming Nature is considered as a whole with its own innate value, and Man has a moral obligation to farm in such a way that cultivated landscape constitutes a positive aspect of Nature.
>
> (DARCOF, 2000)

Niggli (2000) emphasizes the following core values:

- Respecting and enhancing production processes in closed cycles.

Askham Bryan College
LIBRARY BOOK

- Stimulating and enhancing self-regulatory processes through system or habitat diversity.
- Using strictly naturally derived compounds, renewable resources and physical methods for direct interventions and control (e.g. of weed, insects, etc., with only a few specifically listed exceptions).
- Considering the wider social, ethical and ecological impacts of farming.

In a discussion report from the Danish Research Centre for Organic Farming (DARCOF, 2000), three basic principles are identified for organic farming:

- The *cyclical* principle, where collaboration with Nature is promoted, and through this, versatility, diversity and harmony are all ensured. Recycling and use of renewable resources are included in the practice of this principle.
- The *precautionary* principle, which is partly tied to the concept of functional integrity, in contrast to resource sufficiency as the underlying the concept of sustainability.
- The *nearness* principle, promoting transparency and cooperation in food production.

It is notable that very few of these principles directly relate to animal husbandry. Whilst this lack of definition is partly addressed by the existing standards and regulations, the underlying understanding of animals' role in organic agriculture is often not easily deduced from the basic principles, as shown by the Scandinavian surveys mentioned above in this chapter. This book hopes to address these issues further, aiming to strengthen both theoretical and practical understanding of organic animal production.

Sources of the Principles and Values of Organic Farming

The history of organic farming consists of several tracks going in the same general direction, but diverging in their specifics. Understanding both the common principles and the diversity of practices is important in developing the organic approach further.

In the German-speaking countries (Germany, Austria and Switzerland), organic farming arose from five sources (Vogt, 2000):

1. The so-called 'life reform' movement of around the turn of the last century, which opposed the industrialization of farming, urbanization and domination of the modern world by technology, and called for a more natural way of living.
2. A crisis in agriculture and agricultural sciences between the two World Wars, with decreasing yields, soil exhaustion, plant diseases and pest infestations.

3. Biologically oriented agricultural sciences, developing at the turn of last century.
4. Holistic views of nature; a romantic natural philosophy of how to live, and scientific–ecological principles of how to farm.
5. East Asian farming cultures with emphasis on sustainability over centuries, featuring techniques such as composting, transplanting cereals and recycling of human wastes, as publicized very widely in the West in the influential US book, *Farmers of Forty Centuries* (King, 1911).

Some of these ideas create paradoxes in organic standards of today. An example is municipal sewage sludge, which is not permitted as a fertilizer because it is too contaminated, even though using it on cropland would fit very well with the basic organic ideas of nutrient recycling and application of organic matter to the soil.

In the UK and other English-speaking countries, the idea of 'healthy food from healthy soil' was a major motivation behind organic farming. ('Is our health related to the soil?' is how the earliest advocate of organic farming in the USA, J.I. Rodale, opened his book, *The Organic Front* (Rodale, 1949), a question he answered with an unambiguous yes.) Another powerful influence was the extreme soil erosion in the USA. This was especially severe during the 'Dirty Thirties', when large parts of the country were ravaged by wind erosion (in the region known as the 'Dust Bowl') or by water erosion (which formed gullies up to tens of metres wide in the parts of the South dominated by cotton monoculture).

In the German-speaking countries, two main lines of organic farming were established early in the 20th century: 'natural agriculture' (based on both science and a return to a more 'natural' lifestyle) and 'bio-dynamic agriculture' (based on anthroposophy, from which Rudolf Steiner derived his ideas on farming that he first presented in 1924) (Vogt, 2000). Three lines were added later: organic–biological agriculture and biological agriculture (both in the 1950s–1960s), and ecological agriculture during the 1980s (Vogt, 2000). The first of these was motivated by the desire to preserve rural life against industrialization; the latter two kept the scientific orientation of natural agriculture but abandoned its lifestyle elements (Vogt, 2000).

In the UK, important early efforts were inspired by the work of Sir Albert Howard in India concerning compost, plant breeding and health, recycling of organic municipal wastes, and the idea of looking at whole farming systems, not just parts (Howard, 1943). In 1946, Lady Eve Balfour founded the Soil Association to promote organic principles. The first standards were published by the Soil Association in 1967 (Rundgren, 2002). In France, organic farming was promoted in the 1960s by Claude Aubert and the association 'Nature et Progrés', who published the first set of standards in 1972. In the Scandinavian countries, the first bio-dynamic farms were established in the 1930s, and organizations for organic farming were formed first in Norway (1971) and most recently in Iceland (1993). The first governmental standards were put forth in Denmark in 1987 and in the rest of Scandinavia by 1995 (Lund, 2002).

The most common development pattern in Europe has been stepwise, following the historical trends just described. This involves an early establishment phase, mainly of bio-dynamic farmers (1920s and 1930s), followed by some decades of little or no growth, then a sometimes dramatic increase in the number of organic farmers, typically in the late 1980s and the 1990s. Box 1.1 gives examples of different rates and paths of development of organic farming in Europe. The fraction of each country's agricultural land converted to organic farming varies from about 1% to over 10%. There are also significant variations among countries and

Box 1.1. Development of organic farming in various European countries. Source: historical background primarily from Organic Farming in Europe – Country Reports (http://www.organic-europe.net/country_reports). Statistics on organic area and number of farms from survey conducted in February 2003 (Yussefi and Willer, 2003), except Swedish figure from Lis-Britt Carlsson (Jordbruksverket, Sweden, 2003, private communication). Figures are for 'land under organic management', including 'in conversion' and certified land.

In *France,* 1.4% of agricultural land is organic, with 10,364 farms, but in contrast to most European countries, the major rise of organic production happened during the 1980s. It stagnated in the mid-1990s, mainly because of the implementation of Council Regulation 2092/91 establishing organic standards, which apparently were too binding to appeal to farmers, with the whole organic sector getting reorganized.

Belgium's development was similar, with 1.6% of all agricultural land organic (694 farms). The first farms converted to organic farming in the 1960s, but the number of organic farms is decreasing now. Consistent with farming traditions in Belgium, there is wide difference between regions (e.g. Walloon region with intensive farming, and the more hilly Flanders with grassland farms).

In the *UK,* 4.0% of all agricultural land is organic, but only a small portion of this is arable, with the rest being grassland. In parts of the UK, such as the southeast, there is intensive organic farming, whereas Scotland has large areas, but with little possibility of intensive farming.

In *Denmark,* organic accounts for 6.5% of all agricultural land after a particularly strong rise in the mid-1990s. More recently, the situation has stagnated, with some organic farmers reverting to conventional management or getting out of farming because of poor market conditions. About four-fifths of all registered organic farms have animals, with dairy farms being the most prominent type.

In *Sweden*, where only 7% of all land is agricultural, 15.8% of the agricultural land is organic, and the organic sector is still growing slowly but steadily. (This figure is for farms receiving for EU agrienvironmental payments for organic management; about 7% of agricultural land is certified organic.) The average size of Swedish organic farms is 54 ha.

In *The Netherlands*, 1528 organic farms occupy 1.9% of all agricultural land. Animals of some kind are kept on 45% of these farms. In the late 1990s, organic farming grew at more than 25% per year, but this dropped to 14% in 2000 and 8% in 2001.

In *Germany*, 3.7% of all agricultural land is organic, with 632,000 ha (10,400 farms). German organic farming has had three growth phases: (i) 1968–1988, when the negative consequences of industrialized farming and general pollution became obvious, and the organization 'Bioland' was formed (1971), which set standards and created a forum for development and guidance; (ii) 1988–2000, when organic farming spread to East Germany after the reunification; (iii) since 2000, characterized by a strong governmental emphasis on sustainability.

In *Austria*, 11.3% of all agricultural land is organic, with 18,292 farms on 285,000 ha. Growth was particularly strong in the mid-1990s. The main factors were the federal subsidies introduced in 1991, an agrienvironmental programme introduced in 1995 and favourable conditions for conversion of many grassland farms.

In *Spain*, 1.7% of all agricultural land is organic, with 15,607 farms on 485,000 ha.The average size of an organic farm is 31 ha, which is higher than conventional farms (18 ha). Spain is rich in diverse ecosystems with natural conditions covering big land areas. Fewer than 1 in 10 organic farms has animal husbandry, primarily cattle and sheep.

In *Italy*, 7.9% of all agricultural land is organic; Italy's 1.23 million ha is by far the largest organic area in Europe. The private organization Italian Association for Organic Agriculture (AIAB) presented the first regulations for Italian organic agriculture in 1988. Today, it has 14,000 members – farmers and consumers. Organic animal production started in 1994 in regions such as Tuscany and a few others, and since implementation of the EU regulation, organic animal production has become more widespread.

In *Switzerland*, development has been very much influenced by early pioneers in bio-dynamic farming, and research has been carried on since the 1940s. The rate of conversion was steady for 40 years, but increased dramatically since 1990, from about 800 to 6169 organic farms today on 9.7% of the total agricultural land. There is great diversity, from mountain areas, where 10–30% of all land is organic, to the west, where stockless arable farms and horticulture dominate, and where 1–3% of the land is organic. Organic milk accounts for 30% of the total market, and organic vegetables and fruit 24%.

regions regarding living conditions for humans as well as animals, and the importance of different products and elements within a farm.

Development of organic animal husbandry and the organic farming sector over the past few decades has been stimulated by governmental initiatives such as subsidies, research programmes and legislation. Consumers have been involved to varying degrees in the development and formulation of goals, and private organizations have often been a driving force in promoting and consolidating the organic movement in different countries. Organic farming and bio-dynamic farming have developed in Europe in parallel tracks, partially influenced by developments in other countries, but strongly influenced also by local conditions.

An important recent influence has been the increased demand for organic animal products among European consumers concerned about several developments in modern animal production, exemplified by the outbreaks of BSE and foot-and-mouth disease. (In theory, organic meat is free of BSE, because feeding animal products to ruminants is prohibited in organic farming, although there still is a risk from meat from newly converted organic farms or farms where prohibited materials were fed deliberately or accidentally.) Beyond the specific fears they aroused, these outbreaks made consumers sceptical about today's farming practices and the structure of modern agriculture, with its heavy concentration and its emphasis on specialization and intensive production. For many, this has made alternatives such as organic farming look more attractive.

Finally, a very important recent development, sparked in large part by the growing consumer interest in organic foods, has been the EU regulations on organic farming practice (EC Regulations 2092/91 and 1804/1999), described in detail in Chapter 4. The European standards are intended to form a bridge between principles and practices in a way that clearly indicates to the consumers what the product they buy stands for. The challenge arises because the standards must accommodate very diverse living conditions, farming traditions, preferences among farmers as well as consumers and social structures.

The Outlook for European Organic Farming and the Role of Animals in it

Organic farming was developed and driven by people searching for a sustainable way of farming. They often sold their products directly to consumers, bypassing the many marketing steps between the farm and the consumer.

Today the situation is very different. European organic farmers are helped by premium prices and, in some cases, public subsidies. They are subject to regulations and inspections and, to an ever greater extent, sell their produce through supermarket chains and multinational companies.

Organic products are transported long distances, and marketing through supermarket chains often means introducing new – and not necessarily 'organic' – criteria of quality, e.g. that the vegetables must be of uniform size, shape and colour.

Under such circumstances, the principles of organic farming, with their emphasis on recycling, local communities and agroecosystems, harmony, and precaution, could easily be overwhelmed. In an age of large-scale marketing strategies and growing international trade, what will happen to principles such as closed nutrient cycles and connections between consumers and farmers in a mutual understanding of the foundations of the farming system?

The explosive growth of organic farming requires us to pause and think about the future. In the light of recent developments, there is an obvious need to identify, clarify and make all the stakeholders (farmers, advisers and other professionals connected to the farm, and consumers) more aware of the values in the products themselves and in the way they are grown, processed and marketed.

One way for organic farming to thrive while remaining true to its core values may be to maintain and strengthen the diversity within organic systems. Diversity may even be an innate requirement of organic farming systems. Although these systems are based on common principles, they are also based on the idea of site-specific adaptation to local circumstances. This in turn allows for diverse strategies for dealing with issues within the animal production system, such as feeding strategies over changing seasons or access to range.

Organic Farming: a Potential Influence on Mainstream Agriculture?

Organic farming has clearly established itself as a significant part of the European food market, and one can reasonably expect that it will continue to grow. But can it do more than that? Can it also exert a positive influence on the rest of agriculture? Whilst rejection of chemical use has often been seen as the main defining feature of organic farming by the conventional sector, organic farming differs from the mainstream in many other and perhaps more significant respects. It is driven by its own positive goals and ideals, and is not only a challenge to what its supporters find objectionable in mainstream agriculture. It represents a philosophy for production, distribution and consumption of food in a responsible way that offers a sustainable framework for the environment and all living organisms, on both the local and global level. Its core values are interrelated and form a broad ecological notion of sustainability.

Therefore, beyond offering an alternative to those dissatisfied with conventional farming, organic farming can also be a positive force for the development of all farming. This can be seen in Denmark, where the influence of organic farming has inspired the development of farming policy as a whole, especially regarding the environment. In Sweden, the (conventional) Farmers' Association has explicitly stated that organic farmer is a forerunner and 'spearhead' for conventional agriculture.

At the conference 'Organic Food and Farming – Towards Partnership and Action in Europe', organized by the Danish Ministry of Food, Agriculture and Fisheries in May 2001, organic farming was analysed in relation to the Common Agricultural Policy in Europe (CAP). The conference concluded that organic farming offered measurable benefits for the environment, rural development, food quality and farm income. It was seen as contributing to the objectives of the CAP, and as a catalyst for reaching these objectives in the larger European agricultural system.

The currently growing emphasis on animal welfare in agriculture in general is likely to be an area in which the organic farming can lead the rest of the agriculture. Based on an alternative world view, it can also stimulate and challenge the academic as well as the practical debate regarding what animal welfare should entail for the animals.

To Conclude

What are the consequences of taking the concept of organic farming as the framework for keeping animals? To fulfil the challenge of developing animal farming practices that are in accordance with both local conditions and the basic principles and ideas of organic farming, everyone involved must have a good understanding of both animal husbandry and organic farming. This book is intended to contribute to a better understanding of both fields, and to stimulate discussion on how the two can best be combined for the future development of organic animal husbandry.

Having these intentions means that we will initiate a discussion of how the farming environment influences the management of animals, and how a philosophy of farming involving respect for nature, harmony and farmers' responsibilities regarding the environment and society may interact with good welfare. One major consequence of these interactions is that the solutions are complex and involve several levels of the farm. The use of antibiotics and other artificial inputs creates a distinction between the organic herd and other areas of organic farming, because such inputs are being legally applied in one part of the the farm, whilst it is not acceptable to use any chemical inputs on the crops. This difference also underlines the fact that animals are sentient beings.

On the other hand, the aim for naturalness and natural behaviour as a goal for organic animals links the goals of organic farming and animal welfare. Viewing the animal as a whole organism, whose health should be supported and promoted, and viewing the whole herd in the context of the farm further add to a complex and holistic picture of organic animal husbandry. The animals and organic farming are not separate elements that must be forcibly combined, with many inevitable compromises. Gradually, and with help from innovative, theoretical thinking and the development of daily practice, a more balanced concept of organic animal husbandry will emerge. We hope that this book will contribute to a fruitful dialogue leading to this result, knowing all the while that the process will need to go on as long as there is organic farming.

References

DARCOF (2000) *Principles of Organic Farming. Discussion document prepared for the DARCOF Users Committee.* Danish Research Centre for Organic Farming, Tjele.

Howard, A. (1943) *An Agricultural Testament.* Oxford University Press, New York and London. Reprinted by Rodale Press, Emmaus, Pennsylvania.

King, F.H. (1911) *Farmers of Forty Centuries, or Permanent Agriculture in China, Korea and Japan.* Reprinted by Rodale Press, Emmaus, Pennsylvania.

Köpke, U. (2000) The evolution of environmentally sound sustainable farming systems: perspectives and visions. In: Alföldi, T., Lockeretz, W. and Niggli, U. (eds) *IFOAM 2000 – The World Grows Organic. Proceedings of the 13th International IFOAM Scientific Conference, Basel, 28–31 August 2000*, pp. 714–717.

Lund, V. (2002) Ethics and animal welfare in organic animal husbandry. An interdisciplinary approach. Ph.D. thesis. *Acta Universitatis Agriculturae Suecia, Veterinaria* 137. Swedish University of Agricultural Sciences, Skara.

Niggli, U. (2000) Ethical values and historical dogmas of organic farming: promise or peril for its future development. In: Robinson, P. (ed.) *Two Systems – One World.* EurSafe 2000, 2nd Congress of the European Society for Agricultural and Food Ethics, 24–26 August 2000, Royal Veterinary and Agricultural University, Copenhagen, pp. 23–27.

Rodale, J.I. (1949) *The Organic Front.* Rodale Press, Emmaus, Pennsylvania.

Vaarst, M. (2000a) Landmændenes oplevelse af omlægning til økologisk drift. In: Kristensen, E.S and Thamsborg, S.M. (eds) *Vidensyntese om sundhed, velfærd og medicinanvendelse ved omlægning til økologisk mælkeproduktion.* DARCOF Report no. 6. Danish Research Centre for Organic Farming, Tjele, pp. 47–64.

Vaarst, M. (2000b) Omlægning til økologisk drift set fra dyrlægers og konsulenters synsvinkel. In: Kristensen, E.S and Thamsborg, S.M. (eds) *Vidensyntese om sundhed, velfærd og medicinanvendelse ved omlægning til økologisk mælkeproduktion.* DARCOF Report no. 6. Danish Research Centre for Organic Farming, Tjele, pp. 15–46.

Vaarst, M., Alban, L., Mogensen, L., Thamsborg, S.M. and Kristensen, E.S. (2001) Health and welfare in Danish dairy cattle in the transition to organic production: problems, priorities and perspectives. *Journal of Environmental and Agricultural Ethics* 14, 367–390.

Vogt, G. (2000) Origins, development, and future challenges of organic farming. In: Alföldi, T., Lockeretz, W. and Niggli, U. (eds) *IFOAM 2000: The World*

Askham Bryan College
LIBRARY BOOK

Grows Organic. Proceedings of the 13th International IFOAM Scientific Conference, Basel, 28–31 August 2000, pp. 708–711.

Yussefi, M. and Willer, H. (eds) (2003) *The World of Organic Agriculture – Statistics and Future Prospects – 2003*. International Federation of Organic Agriculture Movements, Tholey-Theley, Germany. http://www.soel.de/oekolandbau/weltweit.html

2

The Role of Animals in Farming Systems: a Historical Perspective

Ton Baars,[1] Jan Paul Wagenaar,[1] Susanne Padel[2] and Willie Lockeretz[3]

[1]Louis Bolk Institute, Department of Organic Animal Husbandry, Driebergen, The Netherlands; [2]Institute of Rural Studies University of Wales, Aberystwyth SY23 3AL, UK; [3]Friedman School of Nutrition Science and Policy, Tufts University, Boston, MA 02111, USA

Editors' comments

An important goal of organic farming is to create harmony between all the living organisms and all levels on the farm, and between the farm and its surroundings. Therefore, a major theme in this book is how to better integrate the animals into the agroecosystem of an organic farm. A relevant question to open such a discussion concerns the roles of the animals on an organic farm – why do we keep them at all, and are they needed to create a 'harmonious' farm and a sustainable agroecosystem? The process of domestication ought to shed light on this question, because it is a story of a symbiotic development in which the relation between humans and animals developed – ideally! – to the benefit of both. The authors of this chapter open this discussion by giving us a historical perspective on how husbandry of several important species has developed, with emphasis on Western Europe.

Introduction

In contrast to the type of intensive, specialized agriculture prevailing in the industrialized countries of the world today, organic agriculture is based on the idea of an integrated agricultural system bound to a particular location, as described in Chapter 1. The analogy is often used of the farm as an organism, a coherent, harmonious unit tied to the land. Organic farming's distinctive philosophy and goals enter into discussions of issues such as resource use, animal welfare and environmental damage from farming systems.

© CAB International 2004. *Animal Health and Welfare in Organic Agriculture* (eds M. Vaarst, S. Roderick, V. Lund and W. Lockeretz)

To understand how organic animal husbandry can fulfil the organic principles, it is useful to look at the historical development of agricultural systems, the process of domestication and the ecological roles of different species of farm animals. Knowing the origin of farm animals and their original environment also is important from an ethological point of view, because it helps us to provide them with living conditions compatible with their natural behaviour and the needs of their species. In this chapter, we are concerned with the ecological aspect, whereas the ethological aspects are discussed in Chapters 6 and 7.

Phases in European Agricultural Development

Broadly speaking, three phases can be distinguished in the development of today's agriculture: (i) a pre-agricultural hunter–gatherer society with shifting cultivation; (ii) the development of a diverse *agri-cultura* ('caring for the field') based on local circumstances; and (iii) an industrialized farming system that is much less restricted by its ecological context or impacts (Klett, 1985).

During the first phase, agriculture was practised at the expense of the continent's natural forests through shifting cultivation. Farmers would burn small areas of forest and use them for several years as arable land, until the soil was exhausted.

In the second phase, forced by a growing population to use scarce available resources in the most efficient way, each region of Europe developed its own locally adapted agricultural system (of which several modern derivations are described in Chapter 3). Soil fertility and climate determined the relationship between crops and livestock in each landscape, and different crop varieties and animal breeds developed. These 'landraces' were thus an expression of the landscapes and the agricultural systems of which they were a key element (Hengeveld, 1865; Hagedoorn, 1934; Bottema and Clason, 1979; Naaktgeboren, 1984).

Only comparatively recently, at the end of the 19th century, did agriculture gradually move into its current, partially industrialized, phase. This phase ended the strong interdependence of crops and animals within each cultural landscape. The availability of cheap energy and the introduction of artificial fertilizers were important reasons for a shift towards farming without animals and increased meat consumption in the West (Edelman, 1974; De Smidt, 1979). Artificial fertilizers diminished an important ecological functions of animals – supplying manure – and cheap transport made it possible for animals to be fed independently of the farm's agricultural cycle. With animals no longer needed to supply nutrients, parts of their grazing areas could be opened up and, with the help of modern and efficient machinery and artificial fertilizers, be transformed into arable land. Farm animals were reduced to the level of mere objects, where high production level was the sole focus, pushed by modern breeding and farming methods.

Partly as a reaction against the negative effects of the industrialization of agriculture, organic agriculture started to develop in the 1920s and 1930s, as described in Chapter 1 (Vogt, 2000). Within the movement, there has been an ongoing debate over which elements of traditional and industrialized agriculture should be incorporated into modern organic farming. Among the issues being discussed are whether indigenous breeds should be preferred or if artificial insemination can be accepted (Chapter 16), as well as the appropriate size of farms, including the size of livestock herds, the level of specialization and whether some versions of high-tech agriculture can be considered organic.

A very difficult question is how far we must go in fulfilling organic farming's emphasis on loose integration between animal and cropping systems. Must all organic animal production be based only on feeds and forages grown on the same farm? Do all organic farms need to keep animals?

To provide background to such discussions, in the following sections we first discuss the process of domestication of farm animals and changes to agricultural systems in the last century. This is followed by more detailed discussions of this topic for the three most important animal species in European organic agriculture: pigs, chickens and cattle.

Domestication in Relation to Ecology and Adaptation

In the past 10,000 years, only some 50 animal species were domesticated (Hale, 1962; Van Putten, 2002). An important element for animals to be domesticated at all is their ability to live in small groups with clear social dominance. Humans had to be accepted in this social structure as a leader, at least with larger animals (Naaktgeboren, 1984; Van Putten, 2002), which had important implications for the man–animal relationship. Humans could control breeding – that is, select the animals' mates – only if sexual behaviour was not too complicated (Van Putten, 2002). Although the Greeks and Romans were already breeding domestic animals, in Western Europe, animal breeding is not recorded until the Middle Ages (Idel, 1993).

The human–farm animal relation can also be perceived as a form of symbiosis (Zeuner, 1963, Chapter 2; Bökönyi, 1986; Rollin, 1995) or as incorporating strong symbiotic elements (Jarman *et al.*, 1982, p. 59). According to this view, domestication evolved through a form of behavioural co-evolution. Domestication of the major species probably took place separately and at different times by different processes. Similar theories of development of mutual dependence and co-evolution between agricultural plants and humans have been suggested as the origin of agriculture (Anderson, 1956; Rindos, 1980; Mannion, 1999). The domestication of animals and their roles in the development of agriculture cannot be seen in isolation from human interpretations or perceptions on the one hand, and the animal's environment on the other. The human perception is translated into the relation between man and animal. However, animal husbandry strongly influenced not only the domesticated animal populations but also the human societies into which they were assimilated (Clutton-Brock, 1981; Meadow, 1989). Domestication has been described as a cultural as well as a biological process (Clutton-Brock, 1989), and it has even been suggested to have influenced human identity (Hastorf, 1998).

Domesticated animals have fulfilled various functions in the service of man in the course of human development: sacrificial beast, draught animal, producer of hides, wool, meat, eggs, milk and manure. Man's relationship to animals has varied according to the function.

Two main factors have been involved in these developments: human demands and the ecological role of the animal in the agricultural system as a whole. Out of necessity, to make optimal use of resources, each species had important functions in pre-industrial agricultural systems. Furthermore, breeding and selection produced native breeds adapted to specific man-made cultural surroundings or cultural landscapes. Thus, in Europe alone, 180 local breeds of cattle were developed (Feelius, 1985). It is likely that the creation of the contemporary cultural landscapes was influenced by the various domesticated animals in the agricultural ecosystem through their physiology, food requirements and behaviour.

Historically, production results achieved in animal husbandry were strongly affected by the opportunities that specific surroundings offered to the animals and the role the animals played in agriculture. The first change took place in the second half of the 19th century, when the introduction of artificial fertilizers revolutionized farming, reducing the importance of the manure supply, as mentioned. The second change took place mainly since the 1950s, when animal breeding was largely separated from any specific environment (Baars and Nauta, 2001; Nauta *et al.*, 2002). Landraces as geographic varieties were replaced by commercial production breeds (Naaktgeboren, 1984).

Consequences of the Transformation of Agriculture in the 20th Century

Human activity has greatly changed the cultivated landscape, particularly in the 20th century, when it became more uniform and monotonous to make it more efficient for agricultural use. Consolidation of small parcels, straightening streams, clearing uncultivated headlands, and installation of drainage all contributed to decreasing the variety of ecological niches in the agricultural landscape. In the second half of the 20th century, widespread use of herbicides led to further uniformity.

The role of animals also changed in this period. Some animal production moved to 'landless' systems, with animals having no essential role within the farm. Mechanical traction ended the subsidiary role of cattle and horses as draught animals. The loss of the mixed role of cattle as providers of meat, milk and draught power led to the rise of the specialized beef and dairy breeds, although dual breeds remained dominant in many less intensively farmed areas. With the introduction of more widespread cotton production and man-made fibres, sheep lost their main role as fibre providers, instead being used mainly for their meat, and in less intensively farm areas for landscape conservation. The introduction of therapeutic use of pharmaceuticals meant that animals could be kept in less space and in an increasingly artificial environment that did not correspond to their physiological and ethological needs (Engel, 2002).

The increased control over the agricultural environment reduced the need for native breeds adapted to specific local environments (Nauta *et al.*, 2002). High short-term production could become the overall goal instead of longevity and lifetime production (Bakels and Postler, 1986; Haiger *et al.*, 1988) and there were no longer constraints associated with the animal's need for a natural environment with a finite carrying capacity. With farmers not having to worry about whether their own farms could provide feed for their animals, the different species lost their specific ecological significance.

The separation since the 1950s between animal breeding and specific agricultural environments was increased further by the separation of animal feeding from feed production. In Europe, an animal feed industry was founded, delivering all kinds of feedstuffs anywhere, and with feed imports, the farm's 'environment' has become global. Housing and feeding became more and more uniform, and farm animals were bred mainly for production purposes. This led to a drastic reduction in the number of landraces and the introduction of worldwide breeds like the Holstein–Friesian dairy cow and hybrid brands of poultry and pigs (Albers, 1998; Baars and Nauta, 2001; Nauta *et al.*, 2002). Instead of multi- or dual-purpose animals, animals were bred for specialized production goals: meat or milk; eggs or meat.

Whilst industrial agriculture implied certain advantages for the animals compared with earlier agrarian systems, such as secure access to feed, balanced rations and better knowledge regarding health care and the nutritional value of animal feeds, there also was a depreciation of the intrinsic value or integrity of the farm animals (Verhoog, 2000). As a result, adverse effects of breeding and the associated intensive farming methods were solved by interfering with the animals themselves, such as by mutilation of beaks, tails and horns, or removal of toes from breeding hens (De Jonge and Goewie, 2000). Other interventions included artificial insemination and unnatural births (turkey and beef cattle with an extra rump muscle), routine preventive medication (inoculations and probiotics in pigs and poultry), and protection through feed additives (antibiotics). Alternatively, the suffering of animals to some degree was accepted as an integral part of the economic system (Visser and Verhoog, 1999), for example keeping breeding broiler parents in permanent hunger. High productivity per animal was achieved by keeping animals in an increasingly artificial and fully controlled environment (heating, high density, all-in, all-out). However, not all systems have gone this route; exceptions remain, such as pastoral systems that are self-sustaining and rely minimally on inputs. Nevertheless, mutilations may also occur in such systems, for cultural and other reasons.

As noted, organic farming arose in part as a reaction against this separation of agriculture from the local environment. On one level, organic livestock production should be guided by standards such as the EU Regulations (see Chapter 4). However, on another level, organic farming is based on insights concerning the integrity of the system and the animal (see Chapter 5). To put these insights into practice, we need to take account of how the different species of animals originally were domesticated and what ecological function they fulfilled on the farm. In this way we can avoid wasting time seeking solutions elsewhere for problems that originate within the agricultural system itself. Therefore, the rest of this chapter outlines the domestication of pigs, chickens and cattle, the most important farm animals in Western Europe, and highlights some implications for their ecological role in organic farming systems.

Pigs

The domestic pig descended from the Eurasian wild boar (*Sus scrofa*) about 9000 years ago. When agriculture spread to Europe around 5000 BC, the domesticated pig was brought along by the first farmers. The European and Asian subspecies were domesticated independently, but towards the end of the 18th and the beginning of the 19th century, Asian breeds were used to improve European pig breeds (Darwin, 1868; Jones, 1998; Giuffra *et al.*, 2000).

Wild boars still are common in many countries throughout Europe, Asia and North Africa, although in the 17th century, they nearly were wiped out by hunting in parts of Europe (Ten Cate, 1972; Naaktgeboren, 1984; Van Putten, 2000). Wild boars live in the forest in small family groups or packs. They need wetland areas both for the control of their body temperature and their skin condition (mud bathing). The design of their rooting snout explains their need for a soft soil condition, found in wetter areas (Van Putten, 2000).

Originally, tame pigs also were kept in the forest. Pigs remained in the pre-agricultural gatherer stage for a long time, eating plants and forest fruits from the natural environment. In the first millennium, they hardly were tamed, but instead roamed free in the woods all year round. Pigs scraped together their largely vegetarian diet by rooting around the forest floor, grazing young plants and eating carrion. Their diet consisted mainly of edible roots, ferns, dicotyledonous herbs and nuts. However, they also ate small amounts of larvae, insects and worms, and hunted young mammals and ground-dwelling birds (Van Putten, 2000).

Later, people began to tend them, and they spent less time in the forest. In the Middle Ages, large herds were fattened mainly on acorns, but also beech nuts and chestnuts; pigs still were partially pastured in Europe's remaining forest areas until the end of the 19th century (Ten Cate, 1972).

In the Middle Ages, large areas of forest were cleared for agriculture and to obtain timber and firewood. In place of the forest came moors and steppes, which were grazed by ruminants (cattle, goats and sheep) (Clark, 1947). This made it increasingly difficult to keep pigs exclusively in the forest. The loss of forest happened first in the Mediterranean, and later in other parts of Europe (Ten Cate, 1972). In the late 18th and early 19th century, the oak lost its role in pig fattening in Western Europe. However, we can still see a tree-based style of pig farming in Spain (Andalusia) and Portugal (see Chapter 3).

In North America, large areas of forest were cleared starting in the early 19th century. Pig keeping developed rapidly, using a system of pasturing and finishing in the forest, combined with maize to supplement the diet. With the rapid increase in maize production in the fertile 'Corn Belt' of the central USA, 'corn-fed pork' became one of the country's major agricultural products beginning in the mid-19th century (Ten Cate, 1972).

The Corn Belt exemplifies the potential of pigs to compete for food with humans, because their digestive systems and nutrient needs are similar. The competition was not strong when pigs lived in the forest or roamed freely through medieval towns, partly feeding themselves. Even when they began to be penned, they were raised mainly on wastes, with large numbers kept near places where cereal waste accumulated, including breweries, distilleries, mills and bakeries. In ancient Rome, bakers and brewers traditionally fattened pigs. In areas where dairy farming was practised, each farm kept a few pigs to consume the by-products of butter- and cheese-making (buttermilk, whey). Pigs could be kept alongside oxen that were fed on crushed barley, allowing the pigs to eat the undigested parts of the barley in the oxen dung (Ten Cate, 1972).

The situation changed when improvements in the agricultural system, namely the three-course rotation and the introduction of potato and leguminous crops, eventually made the pig more adapted to a cultural environment and less of a forest animal. Pigs were still fattened largely on wastes but also on cultivated crops such as cereals and boiled potatoes. The change in animal type also was partly due to cross-breeding between the European and Chinese pig at the end of the 18th century. The Chinese pig was early-maturing, fertile, finer of build, fatter, less hairy and better adapted to living in a pen. Many productive breeds, such as the Yorkshire, Berkshire and Pietrain, were developed from this pig by decades of breeding and selection. These replaced the old local breeds and were the basis for many continental breeds (Comberg *et al.*, 1978).

After a decline in the mid-19th century, pig farming saw a resurgence. This time, however, pigs were kept indoors. In the 20th century, the nature of pig farming changed considerably. Whereas pigs originally had been raised mainly for their fat, in the 20th century, farmers increasingly bred lean animals with little fat. From the 1950s onwards, pig farming became increasingly industrialized. The total pig population and the number of pigs per farm grew rapidly, and was concentrated in particular areas. The pigs were no longer let out, and the climate in the pens was artificially controlled. Sows were confined in farrowing pens. Piglets were weaned at only 3–4 weeks. The development of inorganic nitrogen fertilizers and the increased use of phosphorus and potassium fertilizers (natural forms at first, later in manufactured forms) made it possible to greatly increase the area given over to cereal production. In the 1950s, an outdoor system of pig farming was introduced in England and Denmark. Sows were kept in a rotation system with arable crops (see Chapter 3), in portable huts rotated around the farm.

Current feeding practices have brought pigs into direct competition with man for food. The German term *Veredlungswirtschaft* (literally, 'production of improvements') expresses the idea that using cereals to produce pork is a way of improving them compared with consuming them directly. The decision to produce meat instead of cereals for direct human consumption is a matter of political and moral choice. In an organic system, however, pigs should return more to their role as processors of waste products alongside the true roughage eaters, i.e. the ruminants, although roughage should also be part of the pig's diet. In addition to being important for their well-being, this also may reduce the competition for food between humans and pigs.

Chickens

The ancestor of the modern chicken is the red jungle fowl (*Gallus gallus*), which still lives in the jungles of Southeast Asia (Kruijt, 1964; Crawford, 1990; Fumihito *et al.*, 1994). Domestication took place about 9000–6000 years ago (Crawford, 1990), but the chicken did not reach the Mediterranean area until around 1500–1000 BC and Scandinavia until as late as 500–0 BC (West and Zhou, 1989).

Chickens have been involved in the lives of humans in various ways: as food; as a source of feathers for warmth, ornamentation and sleeping comfort; for religious rituals and sacrifices; and as entertainment (cock fighting). Cock fighting probably was more responsible than food production for the domestication and spread of the chicken. Conscious selection for good fighting ability probably started soon after domestication. In this selection, many other characteristics must have been affected, such as body size and conformation, shortening of the moulting period, and possibly heightened activity of certain endocrine factors, since they are important in determining a bird's chance of success in a fight. This could also have affected the agricultural potential of the species (Wood-Gush, 1959).

Several early Greek and Roman writers discussed poultry production, especially the process of hatching eggs and the development of the chick embryo. The Roman writer Varro elaborated on how to raise poultry. Long before scientific proof of the importance of ultraviolet light, vitamins and animal protein, Varro offered an interesting piece of advice: 'Chicks ought to be driven into the sunshine and on the dung heaps to wallow, for by doing so they grow stronger. This holds good not only for chicks, but for all poultry, both in summer and whenever the weather is mild' (Termohlen, 1968).

The Romans raised agriculture to the level of a science. The complexity of the poultry industry of the Greco-Roman world is impressive: they developed specialized breeds, paid attention to minute details of husbandry and recognized the problems of marketing and costs.

With the decline of the Roman Empire, large-scale poultry husbandry seems to have waned. The chicken became the scavenger of the barnyard, and there was little conscious effort to improve its production. Records from around the early 14th century suggest that poultry had only a minor economic role. Until the late 19th century, it seems to have been generally accepted that keeping poultry was economically feasible only on a small scale. The *Winkler Prins Encyclopaedia* (1885) states that only when eggs became expensive was it economical to buy feed for chickens. Otherwise they should be raised on what they scavenged in gardens around the house or on kitchen leftovers.

The development of modern poultry industry, especially in the USA, was fostered by industrial development. With the growth of cities, better markets became available for agricultural products. Industrial development contributed directly to an expanding poultry industry: mechanical inventions such as the steel plough, the reaper and the grain drill greatly increased the supply of feeds, and long-distance transportation became possible (Termohlen, 1968).

In Western Europe, entrepreneurs in the 1930s began keeping poultry on specialized farms, and management and housing changed. Hens were kept in roofed houses connected to spacious runs (15–20 m^2 per hen).

Because they kept flying over the fences, wing clipping came into practice. As production intensified, the problem of cannibalism increased (Ketelaars, 1992). In the mid-1950s, the average pen density in The Netherlands was 3–4 birds per m^2. The birds increasingly were kept inside to prevent dirty eggs. The complete balanced ration was developed, including grit, which is required for optimal digestion of grain. In the 1930s, battery cages were introduced in Western Europe and the USA. They offered an improvement in the hygienic conditions under which hens were raised: the hens were separated from their dung, and the risk of parasitic diseases and zoonoses was reduced. Also, the new system was thought to allow egg production inside cities and thus provide consumers with fresh eggs daily. Because of their high cost, battery cages did not immediately become popular. Only after extensive research and attention in the poultry press in the 1960s did they became more widespread (Ketelaars, 1992). However, free-range poultry systems are making a comeback in several countries, in response to either legislation or market demand.

In today's conventional agriculture, breeding, housing and management are highly industrialized, both for laying hens and table birds. By the end of the 20th century, modern hybrid hens could lay almost an egg per day all year round, and meat types reached a mature weight of over 2 kg within 6 weeks, with large proportions of leg and breast meat (Albers, 1998).

Some organic farms have specialized in egg or poultry-meat production. For environmental reasons, poultry systems are land-based, the chickens are kept in free-range conditions, and the number of hens per unit of barn area is restricted. Also, the standards state that at least a certain proportion of the feed should be grown on the farm itself. However, most organic poultry production is not necessarily integrated into a diverse farming system, although this is the ideal that the standards aim towards. Even on organic farms, chickens, like pigs, remain competitors for food that would be suitable for direct human consumption, such as cereals and high-value proteins. The question must be asked whether this is in line with organic principles. The competition also has direct economic implications (as Winckler Prins above had realized), because the price of organic cereal-based feeds is linked to the price of milling cereals, making it very difficult to meet the consumers' expectations of cheap eggs and poultry meat to which they grew accustomed through industrialized conventional poultry production. Free-range organic poultry systems also face disease challenges that were an important reason for the introduction of the battery cages. No ideal solutions have yet been found, which is probably why organic standards are debated more intensely for poultry than for most other species (see also Chapter 4). Thinking about the origins and nature of poultry may illuminate the limitations facing the development of organic poultry.

Cattle

Western European cattle breeds descend from the aurochs, *Bos primigenius*, a large, sturdy animal living in herds in transitional areas with woodland and open spaces (Bakker, 1948; Naaktgeboren, 1984; Fokkinga, 1995). It was widespread over most of the northern hemisphere with the exception of North America. The earliest sign of decreasing size of the aurochs, which is interpreted as a sign of being kept in captivity, is found around 6200 BC in Turkey. It probably was extinct on the British Islands already by the Bronze age (*c.* 1250 BC), but in central Europe the aurochs continued to live alongside domesticated cattle until the 17th century. The last individual is said to have been killed in Poland in 1627 AD.

Cattle have been kept in the Western world for their usefulness as dairy, beef and draught animals, and for a long time they often were regarded as fertility symbols. Together with sheep and goats, cattle have a central role in agriculture. This is related to their special digestive system: being ruminants they can use high-fibre feeds that cannot be directly consumed by humans. This contributed to the development of arable farming by making grazing land a source of soil fertility for arable land (Haiger, 1991). In parts of Europe, the animals were housed indoors at night, summer and winter, and their manure was collected. This practice had an important ecological role in perpetuating this interdependent agricultural system, in which everything in the landscape, including the animals, had a specific function. The Swedish saying 'The meadow is the mother of the field' shows the traditional farming community's insight into the relation between animals grazing meadows (and forests) and the fertility of the farmed land. In various areas of Europe, a system of grazing rights developed on common lands (also including pigs). In some places, this system continued until the beginning of the 20th century. In many hill farming areas of Wales, as well as in Norway, common grazing rights for ruminants remain an important element of the farming system.

Naaktgeboren (1984) mentions three types of cattle, related to historical and geographical origins: breeds kept on steppes; mountain or highland breeds; and lowland breeds. These three main areas existed until the end of the 18th century. Since the 17th century breeds were specialized for production purposes: either meat, milk or dual purpose. In the 18th century, England became an important country for cattle breeding, where breeds such as the Jersey and Hereford developed (Idel, 1993). In the late 1880s, a small selection of Dutch black and white cattle was imported into the USA and bred to create the Holstein–Friesian, a high-production breed that 100 years later dominated the rest of the world (Wiersema, 1989).

Cross-breeding and selection allowed milk production to be extended beyond calf-rearing, whereas originally, milk production in the first 6 months after calving was possible only if the cow was in constant contact with the calf (Fokkinga, 1985, 1995) More generally, this period saw the emergence of an industry based on pedigreed stock, and in England, 'by 1800 the importance of quality in livestock was becoming fully recognized by progressive farmers, and animal husbandry had made a giant stride away from the haphazard breeding and indiscriminate standards of former times' (Chambers and Mingay, 1966, p. 69).

Despite major changes in modern times – increasing yields, greater use of concentrates in the diet, and longer periods of housing – dairy farming in Europe has remained a more natural process than poultry and pig farming. In beef production, both for veal and finishing, feedlot systems based largely on milk or cereals in the diet are more common. In contrast, in many parts of northern Europe the dairy cow has largely remained a grazing or at least a grass-fed animal.

The introduction of skimmed milk powder in the mid-1950s led to a specialized calf-rearing industry for white veal, and the use of maternal and whole milk for calf rearing declined. The emphasis in dairy breeding on high milk production has also affected beef production, with a shift away from culled dairy animals towards specialized beef breeds, with animals slaughtered at an early age to improve the quality of their meat.

In many organic systems, cattle and other ruminants are kept because of their special ability to consume green forages, particularly legumes, with the resulting contribution to soil fertility as well as milk and meat production. Because conventional production systems for cattle are not as specialized as for other animal species, the changes needed during conversion may appear less far-reaching. However, even many organic milk producers are specialized and cannot fully take advantage of the benefits of ruminants in mixed farming systems. Therefore, challenges remain and compromises must be found to balance different aims.

The European organic legislation (see Chapter 4) regulates the space requirements of cattle and prescribes access to grazing, which helps to reduce the parasite challenge. However, under certain conditions it still allows cows to be kept tethered, with only limited outdoor access. Fodder can be distributed as silage inside the barn. The aim of increased production may lead to heavy reliance on concentrates, instead of coarse fodder in both in beef and milk production, even on organic farms. This is likely to be detrimental for the health of the animal and will mean the loss of one of the key benefits of keeping ruminants compared with other livestock, their ability to live on forage legumes and grasses, feeds that humans cannot consume directly but that have significant benefits for soil fertility.

Concluding Comment

In many situations, organic farmers have to make difficult choices, particularly when converting farms that previously were specialized, whether in animal or crop production. Fulfilling the organic principles for animal husbandry does not mean returning to the agriculture of previous centuries, but every animal species and its specific contribution to the whole farming system should be respected. Ruminants in organic systems use forage legumes and grasses that contribute to soil fertility. The omnivorous pigs can use wastes but require some animal or other high-quality protein in a balanced diet. Similarly, small numbers of poultry can live off wastes and by scratching around in the yard.

However, in many organic systems cattle are fed cereals and pulses, and pig and poultry diets also rely heavily on them. This puts the animals in direct competition with humans and often has ethical as well as ecological implications, because they consume the cereal production of 'borrowed' hectares in other parts of the country or even abroad. The current shortage and high price of organic cereals therefore has a direct effect on the economic viability of organic animal production. Premiums of more than 100% are not uncommon for whole chickens (Hamm *et al.*, 2002). An alternative is to introduce systems based on home-grown forages and waste products, for example from vegetable production and milk processing, assuming that sufficient care is taken to ensure adequate hygiene to prevent diseases.

At the same time, organic farming should be aware of the ecological role of the different farm animals in thinking about future breeding, feeding and housing strategies. It is a matter of debate whether conventional strategies fit into organic systems, with their emphasis on closed systems with recycling of nutrients. Organic animal husbandry should be developed out of a systemic approach based on an ecological relationship among the animal, its environmental needs and its integrity (Spranger and Walkenhorst, 2001).

References

Albers, G.A.A. (1998) Future trends in poultry breeding. *Proceedings of the 10th European Poultry Conference, Jerusalem*, Vol. 1, pp. 16–20.

Anderson, E. (1956) Man as maker of new plants and plant communities. In: Thomas, W.A., Jr (ed.) *Man's Role in Changing the Face of the Earth*. University of Chicago Press, Chicago, Illinois, pp. 363–377.

Baars, T. and Nauta, W. (2001) Breeding for race diversity, herd adaptation and harmony of animal build: a breeding concept in organic farming. In: Hovi, M. and Baars, T. (eds) *Breeding and Feeding for Animal Health and Welfare in Organic Livestock Systems. Proceedings of the Fourth NAHWOA Workshop, Wageningen, 24–27 March 2001*. University of Reading, Reading, UK, pp. 107–113.

Bakels, F. and Postler, G. (1986) Grundlagen einer Rinderzucht auf

Lebensleistung. In: Sambraus, H.H. and Beohncke, E. (eds) *Ökologische Tierhaltung. Alternative Konzepte* 53, Verlag C.F. Müller, Karlsruhe, Germany, pp. 81–88.

Bakker, D.L. (1948) De geschiedenis van het rund. In: Maliepaard C.H.J. (ed.) *Rundvee*. Uitgeverij Contact, Amsterdam, pp. 21–40.

Bökönyi, S. (1986) Definitions of animal domestication. In: Clutton-Brock J. (ed.) *The Walking Larder: Patterns of Domestication, Pastoralism and Predation*. Unwin Hyman, London, pp. 22–27.

Bottema, S. and Clason, A.T. (1979) *Het schaap in Nederland*. Thieme, Zutphen, The Netherlands.

Chambers, J.D. and Mingay, G.E. (1966) *The Agricultural Revolution, 1750–1880*. Schocken Books, New York.

Clark, G. (1947) Sheep and swine husbandry in prehistoric Europe. *Antiquity* 21, 122–132.

Clutton-Brock, J. (1981) *Domesticated Animals from Early Times*. Heineman & British Museum, Natural History, London.

Clutton-Brock, J. (1989) Introduction in domestication. In: Clutton-Brock J. (ed.) *The Walking Larder: Patterns of Domestication, Pastoralism and Predation*. Unwin Hyman, London, pp. 7–9.

Comberg, G., Behrens, H., Bollwahn, W., Fiedler, E., Glodek, P., Kallweit, E., Meyer, H. and Stephan, E. (1978) *Schweinezucht*. Verlag Eugen Ulmer, Stuttgart, Germany.

Crawford, R.D. (1990) Origin and history of poultry species. In: Crawford, R.D. *Poultry Breeding and Genetics*. Elsevier, Amsterdam.

Darwin, C. (1868) *The Variation of Animals and Plants under Domestication*. John Murray, London.

De Jonge, F.H. and Goewie, E.A. (2000) *In het belang van het dier. Over het welzijn van dieren in de veehouderij*. Van Gorcum, Assen.

De Smidt, J.T. (1979) Een nutriëntenbalans voor de potstal-landbouw. *Contactblad voor oecologen, mededelingenblad ten behoeve van het oecologisch onderzoek in Nederland* 4, 149–161.

Edelman, C.H. (1974) *Harm Tiesing over landbouw en volksleven in Drenthe, deel 1*. Van Gorcum, Assen.

Engel, C. (2002) *Wild Health. How Animals Keep Themselves Well and What We Can Learn from Them*. Weidenfeld & Nicolson, London.

Feelius, M. (1985) *Genus Bos: Cattle Breeds of the World*. MSD AGVET, Merck & Co, Rahway, New Jersey.

Fokkinga, A. (1985) *Koeboek*. Educaboek, Culemborg, The Netherlands.

Fokkinga, A. (1995) *Een land vol vee*. Misset, Doetinchem, The Netherlands.

Fumihito, A., Miyake, T., Sumi, S.-I., Tadaka, M., Ohno, S. and Kondo, N. (1994) One subspecies of the red junglefowl (*Gallus gallus gallus*) suffices as the matriarchic ancestor of all domestic breeds. *Proceedings of the National Academy of Sciences USA* 91, 12,505–12,509.

Giuffra, E., Kijas, J.M.H., Amarger, V., Carlborg, O., Jeon, J.T. and Andersson, L. (2000) The origin of the domestic pig: independent domestication and subsequent introgression. *Genetics* 154, 1785–1791.

Hagedoorn, A.L. (1934) *Animal Breeding*. Lockwoord, London.

Haiger, A. (1991) Ecological animal breeding: dairy cattle as example. In: Boehncke, E. and Molkenthin, V. (eds) *Proceedings of the International*

Conference on Alternatives in Animal Husbandry. University of Kassel, Witzenhausen, Germany, pp. 61–70.
Haiger, A., Storhaus, R. and Bartussek, H. (1988) *Naturgemässe Viehwirtschaft*. Verlag Eugen Ulmer, Stuttgart, Germany.
Hale, E.B. (1962) Domestication and the evolution of behaviour. In: Hafez, E.S.E. (ed.) *The Behaviour of Domesticated Animals*. Ballièrè Tindall, London, pp. 22–53.
Hamm, U., Gronefeld, F. and Halpin, D. (2002) *Analysis of the European Market for Organic Food*. Organic Marketing Initiatives and Rural Development, Vol. 1. School of Business and Management, University of Wales, Aberystwyth, UK.
Hastorf, C.A. (1998) The cultural life of early domesticated plant use. *Antiquity* 72, 773–782.
Hengeveld, G.J. (1865) *Het rundvee: zijne verschillende soorten, rassen en veredeling*. Erven Loosjes, Haarlem.
Idel, A. (1993) Zur Entwicklung der Beziehung zwischen Mensch und Tier. In: Neuerburg, W. and Padel, S. (eds) *Organisch-biologischer Landbau in der Praxis*. BLV Verlagsgesellschaft, Munich.
Jarman, M.R., Baily, G.M. and Jarman, H.N. (1982) *Early European Agriculture*. Cambridge University Press, Cambridge, pp. 58–59.
Jones, G.F. (1998) Genetic aspects of domestication, common breeds and their origin. In: Ruvinsky, A. and Rothschild, M.F. (eds) *The Genetics of the Pig*. CAB International, Wallingford, UK, pp. 17–50.
Ketelaars, E.H. (1992) *Historie van de Nederlandse Pluimveehouderij, van kippenboer tot specialist*. BDU Publishers, Barneveld, The Netherlands.
Klett, M. (1985) Der Mensch als Gestalter der Erde. In: Endlich, B. (ed.) *Der Organismus Erde – Grundlagen einer neuen Ökologie*. Verlag Freies Geistesleben, Stuttgart, pp. 223–239.
Kruijt, J.P. (1964) Ontogeny of social behaviour in Burmese red junglefowl (*Gallus gallus spadiceus*) Bonaterre. *Behaviour Supplement XII*. Leiden, The Netherlands.
Mannion, A.M. (1999) Domestication and the origins of agriculture: an appraisal. *Progress in Physical Geography* 23, 37–56.
Meadow, R.H. (1989) Osteological evidence for the process of animal domestication. In: Clutton-Brock J. (ed.) *The Walking Larder: Patterns of Domestication, Pastoralism and Predation*. Unwin Hyman, London, pp. 81–90.
Naaktgeboren, C. (1984) *Mens en Huisdier*. Thieme, Zutphen, The Netherlands.
Nauta, W.J, Groen, A.F. and Baars T. (2002) Breeding strategies for organic dairy cattle; genotype by environment interaction. In: Thomson, R. and Gibson, K. (eds) *Cultivating Communities. Proceedings of the 14th IFOAM Organic World Conference, Victoria, Canada, 21–24 August 2002*, p. 95.
Rindos, D. (1980) Symbiosis, instability and the origin and spread of agriculture: a new model. *Current Anthropology* 21, 751–772.
Rollin, B.E. (1995) *Farm Animal Welfare*. Iowa State University Press, Ames, Iowa.
Spranger, J. and Walkenhorst, M. (2001) Leitbild Tier; Vergangenheit, Gegenwart und Zukunft der ökologischen Tierhaltung. In: Reents, H.J. (ed.) *Von Leit-Bildern zu Leit-Linien. Beitrage zur 6. Wissenschaftstagung zum Ökologischen Landbau, 6–8 March 2001, Fresing-Weihenstephan.*
Ten Cate, C.L. (1972) *Wan god mast gift ...Bilder aus der Geschichte der Schweinezucht im Walde*. Pudoc, Wageningen, The Netherlands.

Termohlen, W.D. (1968) Past history and future developments. *Poultry Science* 47, 6–26.

Van Putten, G. (2000) An ethological definition of animal welfare with special emphasis on pig behaviour. In: Hovi, M. and Trujillo, R.G. (eds) *Diversity of Livestock Systems and Definition of Animal Welfare. Proceedings of the Second NAHWOA Workshop, Córdoba, 8–11 January 2000*. University of Reading, Reading, pp. 120–134.

Van Putten, G. (2002) Exoten, Exzesse und Ethologie. In: Richter, T. and Herzog, A. (eds) *Tierschutz und Agrarwende; Tierschutz und Heimtier*. Tagung der Fachgruppen 'Tierschutzrecht' und 'Tierzucht, Erbpathologie und Haustiergenetik', pp. 159–178.

Verhoog, H. (2000) Animal integrity: aesthetic or moral value? In: Robinson, P. (ed.) *Eursafe 2000, 2nd Congress of the European Society for Agricultural and Food Ethics*. Centre for Bio-ethics and Risk Assessment, Copenhagen, pp. 269–272.

Visser, T. and Verhoog, H. (1999) *De aard van het beestje. Onderzoek naar de morele relevantie van 'natuurlijkheid' in discussies over biotechnologie bij dieren*. NWO Ethiek en beleid, The Hague.

Vogt, G. (2000) *Entstehung und Entwicklung des ökologischen Landbaus. Ökologische Konzepte 99*. Stiftung Ökologie & Landbau, Bad Dürkheim, Germany.

West, B. and Zhou, B.X. (1989) Did chickens go north? New evidence from domestication, *World's Poultry Science Journal* 45, 205–218.

Wiersma, J. (1989) *Holstein historie*. Misset, Doetinchem, The Netherlands.

Winkler Prins Encyclopaedia (1885) Part 7. Elsevier, Rotterdam, The Netherlands.

Wood-Gush, D.G.M. (1959) A history of the domestic chicken from antiquity to the 19th century. *Poultry Science* 38, 321–326.

Zeuner, F.E. (1963) *A History of Domesticated Animals*. Hutchinson, London.

3

The Diversity of Organic Livestock Systems in Europe

Stephen Roderick,[1] Britt Henriksen,[2] Roberto Garcia Trujillo,[3] Monique Bestman[4] and Michael Walkenhorst[5]

[1]Organic Studies Centre, Duchy College, Rosewarne, Camborne, Cornwall TR14 0AB, UK; [2]Norwegian Centre for Ecological Agriculture (NORSØK), N-6630 Tingvoll, Norway; [3]Instituto de Sociología y Estudios Campesinos, Universidad de Córdoba, Avc. Menéndez Pidal s/n, 14004 Córdoba, Spain; [4]Louis Bolk Instituut, Hoofdstraat 24, NL-3972 LA Driebergen, The Netherlands; [5]Forschungsinstitut für biologischen Landbau, FiBL (Research Institute of Organic Agriculture), Ackerstrasse, CH-5070 Frick, Switzerland

---

Editors' comments

Even at the first NAHWOA meeting, it became clear that organic farming varies very widely across Europe. In part, this is because organic farming as a movement and a concept has developed differently in various European countries, but also because of varying climate, landscape and other farming conditions, as well as differences in traditions and the demand for various products. If one does not realize this and does not understand the need for organic animal husbandry to be locally oriented, it is difficult to discuss meaningfully the practical development of organic farming in Europe. Indeed, adaptation to local conditions is a basic principle of organic farming. This provides the motivation behind this chapter. The authors, who come from different European countries with different farming traditions, show how a variety of organic animal husbandry systems have developed in accordance with local conditions. They use examples of actual farms to illustrate this fascinating diversity and to highlight local solutions to many problems. Some of the same farming systems and individual farms will be used again as examples in later chapters.

© CAB International 2004. *Animal Health and Welfare in Organic Agriculture* (eds M. Vaarst, S. Roderick, V. Lund and W. Lockeretz)

There are significant differences between and within the various countries of Europe in the methods and scale of the systems of production. Climatic, topographic, political and historical factors have given rise to a range of agricultural systems, often producing similar raw products. Although the European organic sector has developed within a common broad philosophy, until recently there have been no unifying standards to regulate and guide production methods. Whilst Scandinavia, for example, has had national production standards since the 1980s (Lund, 2000), in others (e.g. Spain), only regional standards have been in operation. Critically, during 1999, European-wide standards for organic animal husbandry were finally decreed by the European Union, as discussed in Chapter 4 (European Communities, 1999). The range of standards that have existed across the continent, and the long time required to reach agreement on the common EU regulation, indicates the diversity in production conditions and systems.

The previous chapter of this book traced the historical development of livestock farming in Europe. Here we present a picture of the diversity of current organic farming methods by describing in more detail the principle characteristics of selected systems. Critical and unique components that may influence the health and welfare of livestock will be described with the aid of case studies selected not only to illustrate diversity, but also to show how organic farms can be adapted to meet local conditions as well as fulfil broad philosophical aims and legislative requirements. Some examples are discussed within the text to illustrate specific characteristics or constraints. Others are included with little general discussion in the chapter but are presented as supplementary illustrations of how organic farming practices have evolved within a specific region. Where appropriate, the cases also are referred to in other chapters that deal with particular aspects of production.

Setting the Scene – the Livestock Production Zones of Europe

Climatic characteristics have been used to describe variation in typical grassland production systems across Europe (Allen *et al.*, 1982; European Commission, 2001). In northern mountain zones of much of Norway, Sweden and Finland, and including the mountain and moorland areas of the UK and Ireland, predominant conditions include high rainfall, high winds, long winters, thin or peat soils, and forage potential that declines with altitude. Sheep farming is common, and to a lesser extent beef, occasionally complementing forestry. Arable and dairy production tend to be limited to coastal and sheltered areas.

In the more maritime climates of the northern lowland zone (northwestern Spain, western and northern France, lowland UK and Ireland, to the countries bordering the Baltic coast), intensive milk production is most important, mainly from Holstein–Friesian herds. Arable cropping dominates in the more productive central zone and the Po Valley, although opportunities to grow forage maize have resulted in specialized intensive milk and beef production.

The mountains of the Alpine zone, covering the Pyrenees, Alps and Dinaric Alps, are heavily forested, and livestock production generally is limited to small-scale production in fertile valleys with seasonal movement to higher altitudes. Sheep and goats are the most common livestock species in the Mediterranean zone, particularly in the harsher, non-irrigated areas; the area generally is not well suited for cattle.

The diversity across European livestock production zones is illustrated by the varying proportion of land used for different organic systems. In the Mediterranean countries, production of organic livestock is a minor concern, with the emphasis on olives and fruits. For example, in 1999, 79% of organic land in Greece was producing olives, citrus trees, vines and currants (Graf and Willer, 2000). (An example of a sheep enterprise integrated into an organic olive farm in Andalusia, Spain, is provided in Case Study 1.) At the opposite end of the scale, 79% of organically registered land in the UK was grassland during 1999 (Graf and Willer, 2000), and by 2002 this had risen to 92% (Soil Association, 2002). In The Netherlands, there is a tendency for more mixed organic farming, and in the same period 45% of organic land was used for animal production, 24% for horticulture and 23% for arable production (Graf and Willer, 2000).

Institutional and Political Diversity

A farmer's choice of farming system, including organic farming, is influenced by complex personal, social and economic factors (Østergaard, 1998). Personal preferences and perceptions are balanced with a desire for economic returns. These, in turn, are influenced by a range of historical and institutional factors and policies, including the provision of technical information, links with the farming community, financial support through subsidies, and market controls. These forces can profoundly affect the management of organic livestock, and therefore their health and welfare.

Historically, public policies and institutional conditions for organic farming have differed considerably among countries. For example, whilst the organization of organic farming in the UK began in the 1940s, in Greece it only began during the late 1990s. A similar trend, but with a less dramatic time-scale, can be observed in public policy. For example, Denmark has had national support for conversion to organic farming since 1987, when the Danish parliament passed the first law

Case Study 1. Santa Casilda Farm: an integrated organic olive and sheep farm.

The farm is located in Cerro de los Abejuelos, Los Pedroches District, and belongs to D. Jesús Fernández y D. Transito Habas.

The Los Pedroches valley is located in the Sierra Morena Mountains in northern Córdoba, at an altitude of 600–700 m, and covers 3295 km^2. The region receives 500–900 mm of rainfall, mainly in autumn and spring. Average temperature is 26–27°C in summer and 7–8°C in winter. Traditionally, the valley produces olives on the steeper slopes. Livestock are reared in the flatter parts, mainly in the traditional agroforestry Dehesa system.

This organic farm is typical of integrated olive oil and sheep production, operating on two sites. The main farm is a 58 ha holding of olive trees that also is grazed by sheep between January and March. There are 7500 olive trees (129 trees per ha), producing approximately 100,000 kg of olives, yielding 20,000 kg of olive oil at the family-run processor on the farm.

A second site of 37 ha is situated close by, in the Dehesa area. The sheep flock spends most of the time here, grazing on natural pasture and foraging on acorns and prunings from the Encina tree (*Quercus ilex*). Approximately 30 t of wheat or oat/veza hay is cut on one-third of the Dehesa area each year to provide summer forage for sheep.

The sheep flock consists of 150 Merino ewes. The olive farm also supports seven horses and some laying hens for home consumption. The main function of the sheep within the olive system is to provide manure to fertilize 25% of the trees each year in rotation, at a rate of 40 kg per tree. The farm has also started to produce compost from a mixture of manure and a residue from the oil production (*alperujo*). In the same year, during the following October, the fertilized area is sown with veza as a cover crop. Also in rotation, a further 25% of the olive trees are pruned. Sheep are not permitted to graze this area for 2 years to avoid damaging the regrowth.

The ewes are mated naturally. Fifteen days before mating they are 'flushed' on 0.5 kg of concentrate per day each, and a 'teaser' male is introduced to synchronize breeding. The lambing rate is approximately 1.5 per year. Lambs are allowed to suckle uninterrupted for the first 2 months, and are fed *ad libitum* with an organic concentrate. They are weaned at approximately 22–23 kg, each having consumed 50 kg of concentrate over the rearing period. Total annual lamb production is about 5175 kg annually, or 54.4 kg/ha.

The flock is not vaccinated and receives no anti-parasitic treatments. The organic lamb is sold through an organic cooperative and to an organic restaurant in the city of Córdoba.

Since conversion to organic production it is the farmer's opinion that both the productivity and the biodiversity of the farm have improved.

promoting organic production (Holmegård, 1997). In contrast, in some Mediterranean countries this began in the late 1990s, largely as a response to EU legislation.

There are dramatic differences in the growth of organic farming across Europe, illustrated by Foster and Lampkin (1999). Michelsen *et al.* (2001) identified very different pathways and scales of development among six European countries. An important factor was the implementation of EU regulations at the national level: the effective functioning of the regulations depended partly on specific national decision-making. Michelsen *et al.* (2001) also give examples of regional (e.g. Sicily and Marche regions in Italy) and national (e.g. Denmark and Austria) differences in the response to organic farming subsidies. An important driving force behind the development of organic farming in Europe has been consumer demand, but individual countries' responses to market demands have varied considerably, with some, such as the UK, responding only when the general agricultural economy was in decline.

Climate, Resources and Marginal Areas

Climate often overrides other factors in determining the extent, nature and variation in farming systems across Europe, including organic systems. Across the continent organic farmers are faced with specific problems associated with harsh climates and are adapting accordingly. In most EU countries, organic farming tends to be concentrated in areas considered to be of marginal agricultural potential, where prevailing practices tend to be less intensive. Despite the common view that these systems are low input and therefore 'nearly organic', they are under considerable constraints affecting the development of organic farming, and some face obstacles to providing high standards of animal health and welfare. The alpine regions of Switzerland and Austria, the Dehesa of Spain, the Scottish highlands and northern Scandinavia provide examples of areas where the climate is harsh and organic livestock farming is common.

Many of these marginal and extensive farms are located far from markets and abattoirs, for example in areas of Scotland and northern Norway. This can harm health and welfare because of the stress involved in transport over long distances. Furthermore, they tend to be far from sources of cheap and readily available purchased feeds. As a consequence, additional production costs are often incurred, and on occasions poor feed supply increases the risk of sub-optimal nutritional status. Although these problems are not uniquely organic, they may be aggravated by limitations imposed by organic standards, especially the requirement that most of the diet be organically produced.

In the more remote coastal areas of northwestern Europe, the cool, rainy climate is excellent for grass production but poor for grains. In parts of Scandinavia and in the alpine region, the cold climate and short growing season severely limit the kinds of crops that can be grown.

Organic livestock farming in northern Scandinavia is diverse because of variations in natural conditions, cultural traditions, economic conditions and government policy. Typically, the winters are cold, with temperatures as low as –20°C, and lasting for 6–7 months, often with snow. The crop production and grazing seasons are short (2–5 months). In many regions, the growing season is not long enough for cereals to reach maturity. The short season and low summer temperatures also restrict the range of protein fodder that can be grown, especially in the north, where many farmers have to import concentrates or protein supplements.

Unique problems exist in some areas characterized by high rainfall, remoteness, and poor soils and vegetation, such as much of Scotland and the Norwegian west coast. In the highlands of Scotland, the systems of livestock production that have developed rely heavily on sheep and some cattle, often on farms of more than 1000 ha, supplying lowland farms with lambs and calves.

It is very common in the extensive systems of Norway to keep sheep on partly unfenced mountains during the summer (e.g. Bjørg Farm, Case Study 2). The indigenous breeds of sheep have an innate ability to

Case Study 2. A mixed sheep and beef farm in Norway.

Nanna Ebbing and Erik Stenvik, Bjørg Farm

Bjørg Farm is a sheep and beef farm in central Norway, close to Steinkjær (11°E, 64°N), at an average altitude of 50 m. Annual rainfall is 1300 mm and average yearly temperature is about 4°C (12.9°C in June and –5.5°C in January). The growing season lasts from mid-May to mid-September. The soil is mainly silty clay, with some sandy soil and some peat land.

The farmers bought Bjørg Farm in 1978 and started sheep production. The 220 ha farm has about 200 ha of woodland, 15 ha of fully cultivated land and 6 ha of pasture. It was converted to organic production in 1994, and now has 60 wintered sheep, eight beef cows, a bull and a horse. The farm income is supplemented with income from forestry. The farm buildings were originally designed for 180 sheep, and the reduction in stocking rate has been the key to achieve high standards of health care and welfare. During the winter, the cattle are tethered and the sheep housed in groups of ten. All livestock are kept on slatted floors and there is minimal mechanization. There is very little cereal production in this area and hence there is a shortage of bedding material. Winter fodder is mainly grass silage, which results in wet manure. Keeping the sheep lying area clean and dry will present a problem in the future when organic legislation concerning the use of slatted floors is introduced.

The beef cows are cross-bred, including the Norwegian Dairy Cattle, Aberdeen Angus and Hereford. The sheep breed is the Spælsau (Fig. 3.1), a small and hardy breed with good vitality and mothering qualities, but generally poor carcass qualities. The breed was selected for its superior grazing behaviour, in that the animals often stay in large flocks, making it easy to round them up from the mountains in autumn.

Fig. 3.1. Erik Stenvik with some of his Spælsau sheep in the natural environment of Bjørg farm.

Using the complementary grazing behaviour of sheep and cattle enables efficient resource use. The farmers also focus on animal welfare, trying to fulfil the behavioural needs of the animals. They view stocking rate as critical for animal welfare, particularly with regard to stress and disease pressures. Health was improved considerably by reducing the size of the sheep herd and introducing beef cows. Rotational grazing is used, and there are no parasite problems (a strategy discussed in Chapter 14). In contrast to the general practice in the area, no anthelmintics are used. The disease known as *alveld,* a hepatogenous photosensitivity disease connected to consumption of the bog asphodel, *Narthecium ossifragum* (Skurdal, 1997), causes some deaths. Predation, mainly by lynx, can cause annual lamb losses approaching 10%; predation by eagles also is a potential threat. This presents a significant management challenge, met by not releasing weaker lambs on to outlying pasture and by increased surveillance.

In summer, some sheep are grazed in forests, but most are kept in unfenced mountain areas. These outlying areas are an important fodder source, and the livestock serve as a landscape management tool. At the end of May, cattle and sheep are turned out from winter grazing on to pasture areas close to the forests. In June, they move further up in the forest.

Although there is sufficient woodland pasture for the whole sheep flock, conflicts with neighbours restrict the use of this resource. Therefore, about half the sheep are transported to a mountain area at an altitude of 200–500 m. The annual autumn round-up of sheep is a joint activity with other farmers in the area.

Cattle are grazed in spring on cultivated pasture following the sheep, and remain there until August or September, when they are moved to fields close to home. All animals are housed in October. Forage supplementation is necessary in autumn, as grazing quality declines.

recognize and be in the same area of the mountain year after year. The increasing number of predators in these areas presents a potential conflict between the welfare and economic consequences of this free-ranging, natural system as opposed to confinement. Although shepherding would reduce the impact of this problem, high wages, low profitability and a demanding landscape often make this unfeasible.

Of course, extensive grazing is not unique to northern Europe. For example, Spain has a large range of both natural and agricultural ecosystems, most of which are managed extensively with livestock either as the primary product or integrated within cropping systems. These systems maintain a great diversity of plant and animals. Some cover very large areas. The forested Dehesa occupies some 2–2.5 million ha, and the total Dehesa extends to 9 million ha if the deforested areas are considered (Daza, 1998; Ceresuela, 1998; Olea and Viguera, 1998). The Monte Gallego agroecosystem covers 1.35 million ha, and the natural grasses in the mountains occupy more than 3 million ha. According to Garzón (1996), using animals with extensive management is the only efficient form of exploitation without destroying these ancient agroecosystems. Today, there still are systems supporting an adapted form of the ancestral transhumance system of livestock keeping (Garcia Trujillo, 2001). In Chapter 2, pig farming in the Dehesa region was mentioned as an example of a tree-based farming system that has remained largely unchanged since Roman times. Interestingly, the same system is used again in Chapter 5 as an example of the potential conflict between achieving environmental and animal welfare objectives. Case Study 3 provides a more detailed description of an organic farm in this region.

Besides the physical characteristics of the landscape, attitudes and perceptions of the environment are an important influence on a region's agriculture. The alpine areas of Switzerland and Austria are highly valued by the whole population and agriculture is considered essential to the maintenance of the landscape. The biodiversity of the landscape, which evolved from the traditional land use system, clearly depends on the management methods used within this agroecosystem. The specific

Case Study 3. The Dehesa of Spain: integrating organic pig production into an ecologically sensitive area.

The farm is owned by D. Jesús M. Torralba Ocaña and is located 6 km from the Ovejo-Villanueva de Córdoba road in Andalusia, Spain

This farm is situated in a typical Dehesa area. It covers 86 ha divided into seven paddocks used by cattle and pigs. The paddocks contain natural grasses and encina trees (*Q. ilex*), which grow at an average density of 55 per ha. These trees are characteristic of the Dehesa area. The acorns produced between November and February are foraged by fattening pigs. Every 7–10 years one of the paddocks is segregated for hay production. The hay is from natural grass or barley, which is then used during the summer, when grass is scarce. Although normally fed entirely on forage, cattle may be fed concentrates during exceptionally dry summer periods. The total annual hay production is approximately 60,000 kg, fed at a yearly rate of 1800 kg per beef cow. Every 10 years the trees of one paddock are cleared and pruned. The leaves are fed to cows and the branches are used to produce charcoal or directly as fuel.

The pig herd consists of 12 Iberian × Duroc sows plus two Iberian breeding boars. The cattle herd is composed of 31 Brown Swiss cows plus a Limousin bull. The stocking rate is 0.59 cattle equivalent units per ha. At the start of the acorn producing season, in November, three or four paddocks are separated for fattening pigs. The period when pigs are in these paddocks is known as the Montanera. The pigs are moved around these paddocks in rotation once most of the acorns have been consumed. Cattle follow pigs in this rotation, grazing the natural grasses and consuming the remaining acorns. By the end of the Montanera in February, the cattle will have free access to all paddocks. The fattening pigs will have reached slaughter weight at this time. Nose ringing prevents the pigs from rooting.

Sows and the piglets are maintained permanently in a 2-ha paddock, and are fed on commercial concentrates and barley flour. The piglets are weaned at 50–60 days. At this stage they are moved to a second property until the next Montanera. The growing pigs reach a slaughter weight of 105 kg at 10–12 months. The system provides high-quality pork that is used to produce various products such as ham, hard pork sausage and stuffed loin. The annual piglet production is approximately 150 piglets from 12 sows. About 105 of these are sold at weaning, at a weight of 22 kg.

The cattle enterprise achieves a 90% weaning rate. Calves are sold at a weaning age of 4 months and at an approximate weight of 120 kg. Annual live weight production is approximately 104 kg ha^{-1}. Because of the weak domestic market for organic meat products, many animals are sold to the conventional market.

The major health problem associated with the system is bloat in cattle at the start of the rainy season. The principal factors that contribute to the animals' health, well-being and productivity are the use of local breeds, an appropriate stocking rate, the free-ranging activity that allows the animals to express their natural behaviour, and the relationship between animals and stockman.

climatic conditions, high altitude and local environmental factors demand special consideration. In these regions, where arable land plays only a minor role, the basis of agriculture is the use of grassland for the production of milk and meat by ruminants, especially cattle. In these countries, political decisions have been taken to support organic farming as a way to maintain traditional ways of farming.

Climate and Housing

Regions that receive high rainfall are well suited to herbage production but not to cereal production, and livestock frequently requires housing. However, bedding straw can be a limiting factor, because of both unavailability and high cost. An alternative to loose housing on straw beds is either cubicle housing or slatted floors (e.g. Case Study 2: Bjørg Farm in Norway).

Slatted floors were originally developed as a solution to shortages of straw. Today, beef fattening systems in countries such as Germany, Ireland, Spain, Finland and Sweden generally rely heavily on slatted floor housing, whilst in the UK, Belgium and France, deep litter systems are most prevalent (European Commission, 2001).

In many areas of Europe cattle traditionally have been tethered, and this is still practised in several countries. The EU organic livestock regulations on tethering and the proportion of floor area that may be slatted present organic producers with considerable practical restrictions, as well as economic penalties, in order to achieve the higher standards of welfare associated with loose housing. The negative aspects of stalls may be mitigated if animals are allowed access to pasture or an exercise area for a part of each day. However, this is not always possible in high rainfall areas or where land is limited.

In Norway and Sweden, legislation requires all dairy cows to be out on pasture for at least 2 months during the summer. Lund (2000) highlights differences among Nordic countries in response to the legislative requirement for tethered organic livestock to receive regular exercise.

Within-country Diversity

There can be as much variation within a single country as between neighbouring countries. In Italy, for example, we see two very different organic systems. Around the Po Valley and in the coastal plains of central and southern Italy, intensive organic dairy production predominates, characterized by high milk yields and stocking rates, and relying on purchased inputs. In contrast, the higher and less favoured areas in the alpine regions in the north and the Apennine mountains in central and southern Italy traditionally have had extensive systems, with little reliance on purchased inputs (Ansaloni and De Roest, 1997). Each has drawbacks for animal welfare. In the lowlands, the problems are related to the high intensity of production, whilst in the highlands, the problems come from the unavailability of land and the widespread use of tethering.

Within France, Bouilhol (1997) characterizes three distinct organic sheep systems. Between 200 and 300 m in the Massif Central, farms largely have winter lambing and outdoor production. In the central zone, at altitudes between 500 and 1200 m, ewes lamb in two seasons: December to January and August to October, with all lambs reared indoors. In the south, at altitudes between 800 and 1000 m, winter lambing Laucaune ewes are raised for milk production. Parasite risks on these farms are lower in the largely housed systems than in the lower altitude, grass-based outdoor systems.

Indigenous Breeds

The use of well-adapted breeds is encouraged in organic farming, and underpins much of the positive health approach, as discussed in Chapters 12 and 16. In several countries, these are more frequent in organic than in conventional systems. Roderick and Hovi (1999) reported a wide range of sheep and cattle breeds on organic farms in the UK, and in many of the case studies presented in this chapter, the livestock breeds are central to the success of the system. In particular, the Scottish example (Case Study 4) describes a farm that combines the selection of sheep and beef breeds with a breeding and management strategy aimed at reducing health problems whilst enabling animals to live under the natural prevailing conditions.

Spain, the location for two of the case studies presented here, provides an example of a region where indigenous breeds proliferate; in many cases, their adaptability to the prevailing conditions make them well suited to organic production. However, there are constraints to the use of some of them. For example, the milking sheep sector in Spain has an official programme of genetic improvement of local breeds using heat-synchronized artificial insemination. The prohibition of hormone treatments in the organic sector severely constrains the participation of organic farms in this programme (J.M. Ameztoy, Pamplona, Spain, 2001, personal communication).

Nationally, organic production may be dominated by breeds selected for more intensive non-organic systems. The dairy sector is a good example, with Holsteins dominating. However, dissatisfaction with Holsteins, particularly because of their unsuitability for organic production, has led some farmers in The Netherlands to choose more traditional breeds (Nauta, 2001; see also Chapter 16). In Switzerland, there have been successful endeavours to develop a breeding programme for organic producers that focuses upon local breeds, such as the Brown Swiss and the Swiss Red and White (Bapst, 2001).

Case Study 4. An upland beef and sheep farm in Scotland.

Duncan Shell, Godscroft, Abbey St Bathans, Berwickshire, Scotland

Godscroft is a 415-ha mixed upland farm, at an average altitude of 380 m (range 300–470 m). It raises spring cereals, swedes for human consumption and temporary grassland. The farm carries 1100 breeding ewes and 86 suckler cows. The crop rotation involves grass ley for 5 years, followed by swedes, followed by 1 year of cereals before returning to grass. Occasionally cereals are grown in the rotation before swedes. There are 20 ha of permanent pasture and 80 ha of woodland. The farm is entirely enclosed, with no common grazing areas.

The beef herd is pure-bred Aberdeen Angus, managed as a closed herd. This breed is well suited to an all-grass system, being able to maintain itself on grass alone. Other desirable characteristics include ease of calving, good milk production, a good temperament and the production of a high-quality carcass. The herd is currently being enlarged to reach a target of 200 cows, as part of a planned further expansion of the farm. The herd calves from March to May, for several reasons:

- *Health advantages:*
  - Problems of dystokia are reduced, as cows calving at the end of the winter are not too fat.
  - The risk of mastitis is reduced since fly populations are low at this time of the year. Also, milk production is not excessive at this time.
  - Compared with winter calving, there is a reduced risk of calf deaths associated with poor weather.
- *Feeding advantages:*
  - Rapid spring growth of good quality grass means that the maintenance and milk production needs are easily met and fertility rates are maximized, as the cows are in a rising plane of nutrition.
  - Access to good-quality summer grazing means that cows are in good condition at the start of winter, thus reducing winter-feed requirements.
  - Weaned calves achieve good growth on spring and summer grass in their second season, allowing them to be sold for meat between September and November, at approximately 18–20 months.
- *Advantages for calf welfare:*
  - Calves are at least 6 months old when winter starts. This means that they are old enough to withstand harsh outdoor conditions during winter and are sufficiently developed at weaning (December to January) to live on a forage-based diet.

The herd rarely suffers health problems, in part because the system is extensive and the cattle are not housed, thus avoiding respiratory or external parasite problems (e.g. lice) normally associated with housing. Calves typically are treated with an anthelmintic (benzemidazole) before they are 12 months old, as they tend to accumulate a helminth burden during their first year on grass. No animals were treated with anthelmintics in 2001. The farm has chronically low levels of copper, selenium and cobalt. This is typical of farms in this area and is dealt with through conventional supplementation.

The sheep flock mainly consists of Lleyn breeding ewes. The lambing rate was 1.70 and 1.59 per ewe in 2000 and 2001 respectively, and in 2000 the weaned lamb rate was 1.55.

The sheep health strategy has changed on the farm since conversion. Vaccination against clostridial diseases, which is common on many UK sheep farms, has been stopped, and sheep are not dipped against ectoparasites. Occasionally, a pour-on treatment of cypermethrin is required to treat blowfly attacks. Breeding for helminth resistance is an important element of the control strategy. For several years, only 'clean' animals (i.e. no sign of scours) have been selected for breeding. In addition, clean grazing practices have been adopted, a strategy discussed in Chapter 14. Antibiotics are used in emergencies to treat mastitis and acute lambing infections. The incidence of infection-related deaths of young lambs has declined, and breed selection has virtually eliminated lameness problems.

Compared with the previous conventional system of production, the main constraints to organic production are the reduced flexibility in feeding options outside the main grazing season. The provision of high-quality rations can be a problem, and intensive fattening of lambs is no longer an option. Root crops such as swedes and turnips provide a good alternative, but feeding them competes with selling them for human consumption. The farmer views good grassland management as being especially important for organic sheep, on account of their selective grazing habits. If this is not maintained, the sward is likely to deteriorate rapidly, which could result in a long-term decline in productivity as well as a short-term lowering of digestibility and the loss or suppression of the critical clover content of the sward. Organic restrictions on the use of trace element supplementation can lead to frustrating under-performance and welfare problems. (Trace elements in grassland systems are discussed in depth in Chapter 15.)

Chapter 16 also discusses the lack of diversity in poultry breeds, with the sector being dominated by birds bred for intensive production by international breeding companies. However, there are exceptions. In France, where a large proportion of broilers comes from free-range or Label Rouge production (Sauveur, 1997), breeds have been selected for their ranging behaviour as well as meat quality. In other countries, such as the UK, only modern broiler breeds have been available until recently (Gordon and Charles, 2002). In Denmark, the prohibition of battery caged systems before 1980 encouraged investigations on the potential of the local Skalberg laying hen under free-range and barn conditions (Sørensen, 2001). The poultry farm described in Case Study 5 illustrates the importance of selecting poultry breeds that are adapted to the production system.

In some situations, particularly where there are climatic extremes, the

Case Study 5. Organic poultry production in The Netherlands.

Wim Vredevoogd and Bep Jansen, De Grote Kamp, Drempt, Gelderland province, The Netherlands

De Grote Kamp is a bio-dynamic farm of 25 ha of arable land, of which 4 ha is used as runs for a flock of 4000 layers. The arable production is used as feed for the flock, with the poultry manure used as fertilizer for the arable crops. The farm started biodynamic production in 1985. In 1994 five existing bird units were replaced by a single new layer house, at 5 birds per m^2. The flock is divided into two sub-flocks of 2000 birds each, managed as an all-in–all-out system. The flock is stocked at a ratio of one cock for 30 hens. The production cycle is 1 year, and the average laying rate is 75–80%. Birds have access to the outdoors for most of the day. The average daily feed intake is 130 g per bird. Approximately half the feed is provided through automated feeding pans, with the other half offered as whole grain, distributed three times a day inside the poultry house.

The runs are covered with maize and divided into several sections, surrounded by trees (mostly hawthorn, *Crategus oxyacantha*). Attractiveness, shelter against bad weather and birds of prey, and a regular schedule are important factors in achieving a successful hen run. The pop-holes are opened every day and hens stay inside only when the weather is very bad. The farmer considers that training the hens to go outside is more important than the size of the run. The hens should use the whole run area and not only the area closest to the house. It is not easy to keep the runs covered with vegetation. Although rotational grazing is practised, this is not considered to give enough protection against parasites (worms and coccidiosis).

Cocks play an important role in encouraging ranging, as they watch over the flock and lead hens in the search for feed. About 80–90% of the hens actively use the runs, which are planted with maize that contributes to the hens' diet. From early autumn onwards, when the maize is ripening, some cobs are cut daily for consumption by the birds. The maize is left over winter and harvested the following spring, in time to prepare the fields for cropping. The grain yield is up to 4 t/ha. The maize crop offers protection against predation, provides feed, and encourages ranging during autumn and winter (see Fig. 3.2). Crop residues are collected in the spring and left in a heap near the house as an additional attraction for the hens.

The 25 ha of arable land produces 60–65% of the feed required for the laying flock. Home-grown cereals and maize are processed on the farm and mixed with purchased wheat, soybean, lime and grit. Small proportions of home-grown cereals and grass are processed at a feed factory into a concentrate that balances the ration. Outdoor access does not result in a lower feed intake, although intake can drop by 20–25% when birds are introduced to fresh pasture. The roughage consumed is considered to play an important role in the mineral and vitamin supply. During winter, beet or potatoes are occasionally added to the ration. Future plans include growing more high protein crops, such as peas, as these are expensive to purchase.

Fig. 3.2. A maize field used as a run on an organic poultry farm encourages ranging, protects against predation and provides feed.

Several poultry breeds have been tested at the farm. The Bovans Nera breed is the one most recently used. This is a heavy breed performing well under the 'scavenging' conditions prevailing at the farm. Some general features considered important for organic laying hens are hardiness and an appearance that does not encourage feather pecking. Concerning the latter the Nera is at a disadvantage, since it has black feathers and pink skin.

use of locally adapted breeds is frequently the only option. In Norway, the most common sheep breeds are the Dala and Spælsau (Case Study 2), and in Sweden it is the Swedish Landrace. These are local breeds, well adapted to the production conditions in those countries. Breeds such as the Texel and Suffolk are often used as rams to improve carcass quality. The most common dairy breed in Norway is the Norwegian Dairy Cattle (NRF). In Sweden, Swedish Red and White Cattle and Swedish Holstein (SRB and SLB) are the most common breeds.

Securing a Feed Supply

Throughout Europe there are systems where feed supply is limiting the development of organic livestock production. In particular, it presents a major obstacle to using predominantly home-grown organic feeds.

As already noted, the short growing season and cool summers in northern Scandinavia do not allow cereals to reach full maturity, and along the west coast the climate may be too wet for reliable harvesting.

This restricts the fodders that can be grown, making it necessary to import concentrates and protein supplements at a considerable cost. The short vegetation period causes similar problems in alpine regions. The protein supply for monogastric animals poses a particular challenge (Case Studies 5 and 6). The debate over whether to allow the use of synthetic amino acids in organic feeds in part arose from these difficulties.

Case Study 6. An innovative mixed farm in Sweden.

Kjell and Ylwa Sjelin, Hånsta Östergärde Farm, Uppland, Sweden

Hånsta Östergärde Farm, approximately 100 km north of Stockholm, provides an example of a farm that has evolved into an integrated system of production, yielding benefits to the farmers, animals and the natural environment. The farm has 30 ha of cultivated land and 57 ha of forest, and keeps poultry and pigs. An additional 40 ha of arable land is rented from neighbouring farms, making it an average size farm for the area. The farm has been organic since Kjell and Ylwa Sjelin inherited it in 1987. Soils are of average quality, either fertile sandy or silty clays. It is situated at 60°N on the northern rim of a fertile plain. Spring sowing of wheat usually takes place by late April or early May. The grazing period lasts from May to the end of September and the farm is snow-covered from December to March.

A 7-year crop rotation has been developed that involves 2 years of legume–grass ley (a mix of about ten species), winter or spring wheat, oats, peas or horse beans, rye and barley undersown with the legume–grass seed mixture. Some linseed and sunflower also are grown.

The most important enterprise is egg production. Approximately 1000 laying hens are kept in four mobile houses of 9 × 13 m, each housing just over 200 hens, and four smaller houses of 40 hens each. All houses have outdoor runs and the larger houses have verandas. The walls and roof of the outdoor run are covered with plastic netting to protect against predators and to avoid disease transmission from wild birds. The houses rest on runners, and are easily pulled by tractor. The building material is such that a heater can be used to disinfect the house from mites, lice and other parasites that can be a problem in free-range systems. The buildings have automatic feeders, nests, perches, nipple drinkers, dust baths and electric lights and heating. The veranda floor is covered with straw litter and leafy hay. From the end of April until the end of November, the houses are moved weekly across the grass ley, and since only half of it is used, the remainder is harvested as hay for winter feed in June. The ley is ploughed the following year, to be followed by cereals. During the winter, at the end of the grazing season, the houses are not moved and the birds have access only to the veranda but not the outdoor run.

The farm has eight sows that farrow once a year, during April and May. They are also kept in mobile homes, where they farrow whilst using a grass–legume ley. After farrowing, the sows and piglets are kept together for the remainder of the year. The pigs graze in the forest during the summer, so that the arable areas can be cultivated. A new forest area is used each year. The rooting behaviour of the pigs in preparing forest ground for reseeding is extremely important, as the forest soils are of a moraine type with large rocks that are not easily cleared and reforested by conventional methods. Approximately 60% of the Hånsta Östergärde forest land is renewed through natural seeding. The pigs are taken back to the farm to strip-graze the first parasite-free grain field postharvest, where they assist both in seedbed preparation through rooting and in controlling vegetatively propagated weeds. This is followed by drilling winter wheat. Fattening pigs are gradually slaughtered during late autumn.

The Sjelins also keep cattle and sheep and are developing mobile housing for these as well. Besides providing accessible outdoor conditions, mobile housing will help to reduce soil damage, reduce parasitic and other infections, prolong the grazing season and make animal handling easier. The cattle and sheep are kept on arable land all year. During the summer, mobile houses function mainly as sun protection and as a handling pen. Along the one long side of the house there is a feeding area with yokes. During the winter the house is moved when necessary to avoid excessive poaching.

The livestock are generally healthy. Veterinary calls are limited to infrequent antibiotic treatments. Pigs are vaccinated against erysipelas and parvovirus. Salmonella tests of the laying flock, conducted three times per year, have always been negative. There have been incidents of feather pecking in some poultry flocks, which may occur because the layers (of commercial breeds) are bought from conventional breeders at 16 weeks of age. The aim is to rear pullets on the farm in the future.

A major problem is growing suitable high-protein feed crops under the prevailing climatic constraints; the protein supply from home-grown feeds is normally too low for pigs and poultry in particular. Feed intakes could be lowered if the amino acid balance was improved, as diets generally particularly lack methionine.

Periodic shortages of high-quality forage may also occur because poor weather prevents farmers from harvesting at the optimal time. Together with the lack of grain, this increases the chance of nutritional imbalances in cattle, particularly for dairy cows. The situation may be accentuated in organic systems because of the additional cost of purchased grain. Also, in some countries, the market for organic feed is poorly developed.

Some problems can be caused by the agricultural structure. In Norway, for example, climatic extremes, combined with national policies regarding the location of grain or livestock farms, mean that farms with livestock are isolated from farms with crops. Therefore it often is impossible to recycle manure as a fertilizer for organic crops, which limits the effectiveness of organic arable production.

Organic farming standards require ruminant diets to be largely forage-based. But conventional beef production in Spain, for example, often involves high levels of concentrate feeding (Ameztoy, 2000). Adapting such a system to organic production obviously is not straightforward, and organic farmers are trying to adapt to a new system focused on feeding concentrates within the dry matter limits stressed in organic standards. Wright *et al.* (2002) highlight similar restrictions on the development of organic beef production in Greece. In Scotland, both lambs and beef cattle traditionally are reared in the more marginal, extensive, high altitude areas before being sold to more productive lowland farms for fattening. Recently the organic hill farming sector has grown much more rapidly than the arable areas, and in the UK there are insufficient organically produced feeds available (Soil Association, 2002).

The supply of cheap imported animal feeds historically has been important in industrialized production systems, for example in The Netherlands. Therefore, farmers have not needed large areas of land for producing feeds. Farmers who convert to organic production tend to do so on small farms, and cannot provide home-grown feeds in sufficient quantity. This problem is exacerbated by the very high value of land. In contrast to the examples of extensive systems described throughout, an example of the land requirements for outdoor poultry production in The Netherlands is described in Case Study 5.

Restrictions on Animal Health Control Practices

There are significant differences across Europe in the incidence of various diseases, depending upon climate, production systems and many other factors. For example, in Norway, many of the viral diseases of sheep common in countries such as the UK (see Chapter 11) are not prevalent (e.g. footrot and enzootic abortion), but there are problems with others, such as scrapie and some cases of maedi visna. A local problem, particularly among free-ranging sheep on the west coast, is *alveld* (see Case Study 2).

Across the continent, there are disease problems associated directly with the type of production system, and in some cases limiting the development of the organic sector. Although organic farming is often viewed as being well suited to extensive practices, these can also be where it is most difficult to make minimal prophylactic use of conventional veterinary medicines. In Spain, there is only limited opportunity for organic production in the large areas of communally managed rangeland and mountain pastures, because this would result in considerable mixing of organic and conventional flock, thereby creating biosecurity risks, problems in controlling internal parasites, and concerns over certification. There also is a risk that without positive control of animal health and adequate inspection and monitoring, even abandoned animals could be classified as organic (J.M. Ameztoy, Pamplona, Spain, 2001, personal communication).

This feature of open (unenclosed) grazing areas can be found in several countries. It also is occasionally associated with seasonal common grazing areas, where livestock from neighbouring farms can easily meet. In the Scottish highlands, as with the Spanish example, biosecurity then becomes a potential problem, exacerbated by organic regulations limiting chemical control. Ectoparasitic infections, such as sheep scab (*Psoroptes ovis*) and, if there is tall, scrubby vegetation, ticks (*Ixodes ricinus*), become a risk that cannot be solved without targeted prophylactic veterinary medicines. Alternatives such as fencing large areas often are not economically feasible. Other control measures, including the burning of natural vegetation such as heather (*Calluna vulgaris*), may conflict with the environmental objectives of organic farming (Younie, 2000).

In parts of Scotland, farms typically have thousands of hectares, with only small areas of improved grazing that are suitable for cattle and fodder production. In these situations, poor quality vegetation, high altitudes and climatic extremes mean that these farms can only support sheep, which limits opportunities to use alternate mixed grazing to control helminths (see Chapter 14). Without the right balance of sheep to cattle, it can be difficult to operate clean grazing systems to control helminths in young lambs. The risk of infection by liver fluke (*Fasciola hepatica*) in these wet areas may further limit the land available for grazing. Although not entirely a mountain farm, the Scottish livestock farm in Case Study 4 demonstrates the successful integration of cattle and sheep to control helminths.

Across Europe there are variations in the extent of prophylactic antibiotic use in conventional livestock systems that have significantly affected the situation for some diseases in organic systems. Mastitis in the dairy sector provides a good example. In some countries, such as the UK, the use of prophylactic antibiotics to control mastitis, or 'dry cow therapy' (DCT), has formed an important element of the 'Five Point Plan' to control the disease on most farms (see Chapter 12). Whilst the plan has been successful in many respects, the reliance on the use of prophylactic antibiotics can result in potential problems when this practice is stopped after conversion to organic production. In contrast, in Denmark, where the method is not widespread, there is evidence that organic producers are more successful in controlling the disease than are their non-organic counterparts. Similarly, organic dairy herds in Norway and Sweden have lower incidences of mastitis, ketosis and milk fever (Hammarström, 1992; Jonsson, 1997; Strøm and Olesen, 1997; Byström, 2001; Hardeng and Edge, 2001; Hamilton *et al.*, 2002). (See Chapter 11 for a fuller discussion of the comparative incidence of diseases on organic and conventional farms.)

In the alpine regions, as well as in Norway, a traditional form of transhumance is in operation, involving the movement of milking cows to high altitude pastures during the summer. These pastures are normally owned by a community of farmers and managed by employees. The stress associated with this system can substantially harm milk quality and animal health (Walkenhorst *et al.*, 2002). Although these problems are not unique to organic systems, the reduced reliance on antibiotics means that alternative preventive techniques must be adopted, giving the stockperson a key role. In Chapter 12, this situation is discussed in relation to a health plan to reduce these risks.

National legislative structures may also influence the effectiveness of animal health control strategies under organic conditions. Both Norway and Sweden have a National Animal Health Service, working to prevent and control diseases in all livestock systems, not just organic. This service is in widespread use and requires the recording of every disease and treatment, by both veterinarian and farmer. In Sweden, a special programme for organic animals is now being introduced. Effective use of the data provided by such schemes can be very useful for organic farmers developing health plans (again, see Chapter 12).

This type of national programme does not exist in most countries. Although the UK now requires organic farmers to establish health plans, the absence of a culture of health recording in the conventional sector may significantly impede the successful adoption of these plans.

An interesting example of the role of legislation concerns homoeopathy, which is used extensively by many organic farmers and is advocated by the European regulation. In Sweden, however, veterinarians are not permitted to prescribe homoeopathic treatments, as there are doubts regarding their scientific basis (Keatinge *et al.*, 2000; Lund, 2000), and in Denmark, only veterinarians are allowed to treat animals with homoeopathy.

Integrating Livestock Systems

The integration of more than one livestock species and the integration of livestock with cropping can be considered the basis of a balanced and sustainable farming system, allowing nutrient recycling and effective resource use. However, the degree of integration can vary considerably, with many organic farms operating with a single species and little dependence on home-grown feeds beyond forage. This is perhaps most evident with organic pigs and poultry, as these species have a low forage requirement.

On some soil types in the UK, pigs have been successfully integrated into cropping systems, and many pigs are now partially produced outdoors. However, in many conventional systems, prophylactic anthelmintics are important in controlling internal parasites, allowing pigs to be kept confined to the same paddock or field for much longer than would be possible for organic pigs. Furthermore, confining pigs to a specific area can also conflict with environmental goals, including soil conservation and eliminating pollution from leaching of nutrients. As a consequence of these restrictions, organic pig production generally is feasible only on free-draining soils, and when there is sufficient land available to rotate the paddocks frequently. The system described in Case Study 7 is a very good example of how pigs can both benefit from and contribute to other farm enterprises. In Spain, the traditional extensive grazing systems have been converted to organic production by minimal changes, e.g. the Dehesa system (Trujillo and Mata, 2000). Case Study 3 showed how pigs are used within a livestock rotation and enhance land use in an area with important ecological characteristics.

Case Study 7. Integrating pigs into the farming system in the UK.

Helen Browning, Eastbrook Farm, Wiltshire, UK

This large tenanted farm of 540 ha has been organic since the 1980s. The farm runs in a long strip to a maximum altitude of approximately 250 m. The soil type ranges from a silty clay loam over chalk to heavier clay soils at a lower altitude. Besides a pig herd, the farm has two dairy herds, a beef herd, a sheep flock and 263 ha of arable production.

The pig herd consists of 140 sows raised entirely outdoors. There are plans to enlarge the herd. The breeding herd is predominantly of the British Saddleback breed. One of the most important advantages of this breed is its hardiness and suitability to outdoor systems. Saddlebacks are good mothers, produce excellent quality meat and, importantly, are relatively resistant to sunburn. The breed does have a tendency to become over-fat at finishing, so Duroc boars are used to reduce the backfat levels of finished pigs. The farm currently has 14 Duroc and one Saddleback breeding boar. Recently, the farm has been involved in a research project that has been examining other breeds' suitability for organic production, including the PIC Camborough 15 breed line.

The system employs two people, whereas a conventional system would employ only one person to care for the same number of pigs. The main additional labour input is for moving pigs and paddocks, and the outdoor finishing system. Water pipes are above ground, and in freezing weather water has to be carried to the paddocks, which requires additional labour.

At any one time, there are up to 80 separate paddocks of varying sizes, including approximately 30 used as farrowing paddocks, fitting into a 5-year crop rotation. The pigs are followed by a crop that can use the additional fertility, such as potatoes, maize or wheat. The rotation involves three seasons of arable crops, the last of which is undersown with grass. Grass leys are used by the sheep and cattle, and for silage before pigs are introduced. This system is designed to break disease and parasite cycles and to improve soil fertility.

The system is based on a simple series of small, electrified paddocks, with pigs never going back on to the same land for at least 5 years. The farm is large enough to absorb this frequent movement of pigs. Dry sows are moved three times each year, and are kept in paddocks of approximately 0.2 ha, with four to eight sows per paddock. The timing of each move depends on the weather and the state of the pasture. In theory, the pigs are moved as soon as all the green cover has gone from the pasture. The pens holding batches of growing pigs are used the longest, about 3–4 months.

Each farrowing paddock of approximately 15–25 × 20 m accommodates one sow with her litter. Paddocks are arranged in blocks of eight, with piglets allowed free movement among them. Mixing among litters is encouraged, with just a single-strand electric fence separating paddocks. This allows easier grouping, and less fighting and competition at weaning. Previously, sows were accommodated at two per paddock, but this created too much competition at feeding, with the weaker sow frequently suffering. The current practice has resulted in better overall condition of the sows. The single wire separating the sow is removed 5 weeks after farrowing to allow the sows to mix, and thus encourage oestrus. Sows are then served at the second oestrus of the cycle, which may help litter size.

At 8 weeks of age, the piglets are weaned, sorted by sex, and moved into outdoor weaning 'chalets' for about 2 weeks in groups of 30–50. They remain in these groups until slaughter. As the piglets are all socialized before weaning, the social stress is minimized because they are never mixed with truly 'unfamiliar' animals. At 10–12 weeks, weaning groups are moved to finishing paddocks until slaughter at 5–6 months. Most are slaughtered as bacon pigs at approximately 100 kg. Slower growers are removed as pork pigs at approximately 70 kg. Pigs ready for slaughter are housed for 1 or 2 weeks in preparation for slaughter. This allows feeding them high forage diets, such as silage, to improve carcass grading. Concentrate feeding is restricted during this period. This arrangement also makes sorting and loading easier.

Gilts are first put to the boar at 8–10 months of age, when weighing 90–100 kg, and farrow for the first time at 12–14 months. The aim is to get gilts and sows in good condition before farrowing, since it is very difficult to improve sows' condition during lactation. Breeding management is geared to avoiding the loss of condition of sows and gilts, as it has been noted that a severe loss in condition can result in excessive feeding inefficiency for the rest of the animal's productive life.

Lameness is the main health problem, although it is infrequent. The land is flinty and pieces of flint occasionally cut the feet of sows. Most cases of lameness are treated with homoeopathy. Very bad cases are treated with antibiotics, but are extremely rare. There were problems with meningitis (*Streptococcus suis*) and *Escherichia coli* when the pig unit was started; they were solved through homoeopathic nosodes. Farrowing on clean sites also has helped rid the farm of an *E. coli* problem. Predation by foxes is an occasional problem.

The pigs are always moved on to clean pasture. Regular faecal egg sampling allows the strategic use of anthelmintics, thus preventing clinical disease and the build-up of a worm burden. Deworming was formerly routine, but under the present system of frequent movement it is only necessary at 1–3-year intervals, when there is an indication of a problem (e.g. raised faecal egg counts or reports of milkspot from the abattoir).

Pigs on this farm play a very important part in the crop rotation, and their fertility-building role enables three cereal crops to be grown before grass leys are sown.

In other countries, it is not as feasible to integrate outdoor pig production into an organic farm. In The Netherlands, which has a large conventional indoor pig industry, outdoor systems are not economically feasible because of the high price of land. In such situations, to be profitable the outdoor access provided for organic pigs may be restricted to meet the minimum requirements of organic standards, and may frequently be in the form of concrete exercise areas. Intensive production units in The Netherlands have been converted to organic systems by allowing access to outdoor pens (Vermeer *et al.*, 2000).

In Austria, small herds of pigs are frequent on organic farms, mainly for home consumption and local sales (Baumgartner *et al.*, 2001); in contrast to these small farms, more than half of national production occurs on only 7% of organic farms (Eder *et al.*, 2000). The herds are rarely effectively integrated into the farming system, relying on being housed in existing buildings with concrete exercise areas. In such situations (mainly small, housed herds), animal health can be a potential problem, particularly with regard to endo- and ectoparasites (Baumgartner *et al.*, 2001). Similarly, Klump and Härring (2002) described the fattening of organic pigs in southwestern Germany as largely indoors, occurring mainly in converted conventional housing. In Denmark, where conventional pig production is important, the few organic herds that exist tend to combine genuine outdoor conditions for breeding sows with concrete exercise areas for fattening pigs. Although disease levels tend to be good, lameness associated with rain and frost-affected paddocks can occur, as can mortality associated with poorly designed exercise areas (Vaarst and Thamsborg, 2001). Because of the problems associated with achieving frequent rotation, nose ringing is permitted within national standards as a precaution against environmental damage. (The nose ringing controversy is discussed further in Chapter 8.)

Commercial poultry production systems that are easily and effectively incorporated in a mutually beneficial manner with the rest of the farm are even more difficult to achieve, for various reasons. Climate, availability of resources, risk of predation and the existing farming systems and enterprises are likely to be limiting factors. For poultry meat, consumers' taste preference may be important, as in France, where the high demand for free-range chicken meat has resulted in a proliferation of both organic and non-organic systems. The system for layers described in Case Study 6 is an example of how egg production can be successfully integrated with a cropping programme designed partly to provide feed for the flock. It potentially could be implemented in some other countries.

In Conclusion

Across Europe, we see diversity in climate, topography, legislation, farm structure, markets, animal disease pressures and many other factors. These have resulted in the evolution of very different organic production systems, frequently producing the same product, and certainly producing to the same basic standards, albeit with some variation in interpretation across national boundaries. Although there are growing political and economic pressures to develop common legislative structures, the differences portrayed here also call for a degree of flexibility in the adoption of organic standards. Or perhaps the opposite is true. If local conditions are such that a uniform set of standards cannot apply, or the spirit of the law is in jeopardy, then perhaps the region should forgo certain organic enterprises and focus on those that are better suited to the prevailing conditions.

Climate and the availability of resources have been major factors determining the kinds of farming systems across Europe. Historically, farmers were forced to adapt their farming techniques to the limits set by the landscape and climate. Modern farming has at least partly overcome these obstacles, using cheap fossil energy to move resources around the world and to overcome climatic obstacles. Organic farming principles support the concept of land-based production adapted to local conditions. The inherent difficulties in adhering to these principles have been highlighted in the various case studies and discussions.

We also see commonalities across regions, particularly with regard to production in marginal areas, where much organic farming tends to be concentrated, although conditions are far from ideal. Within the various systems with their inherent pressures, we also see a great deal of adaptation and innovation, sometimes based on traditional practices, sometimes novel. Some of these have been highlighted as case studies within the text. Whilst many of the specific methods may not be appropriate in other regions, there is certainly considerable scope for sharing ideas across the continent.

Acknowledgements

We thank the farmers who contributed to this chapter: Kjell and Ylwa Sjelin, Helen Browning, D. Jesús Fernández, Duncan Shell, Wim Vredevoogd and Bep Jansen, D. Jesús M. Torralba Ocaña, Nanna Ebbing and Erik Stenvik. Thanks also to José Miguel Ameztoy, Hilary Kelly, Vonne Lund, David Younie, Simon Jonsson, Ton Baars, Werner Zollitsch and Lisa Grøva for providing details of production in individual countries.

References

Allen, D.M., Bougler, J., Christensen, L.G., Jongeling, C., Petersen, P.H. and Serventi, P. (1982) Cattle. *Livestock Production Science* 9, 89–126.

Ameztoy, J.M. (2000) Organic animal husbandry conversion in Navarra, Spain. In: Hovi, M. and Trujillo, R.G. (eds) *Diversity of Livestock Systems and Definition of Animal Welfare. Proceedings of the Second NAHWOA Workshop, Córdoba, 8–11 January 2000*. University of Reading, Reading, pp. 35–39.

Ansaloni, F. and De Roest, K. (1997) Use of resources and development of organic livestock farming in Italy: first results of an ongoing study. In: Isart, J. and Llerena, J.J. (eds) *Resource Use in Organic Farming. Proceedings of the Third ENOF Workshop, Ancona, 5–6 June 1997*, LEAAM-Agroecologica (CID-CSIC), Barcelona, pp. 299–310.

Bapst, B. (2001) Swiss experiences on practical cattle breeding strategies for organic dairy herds. In: Hovi, M. and Baars, T. (eds) *Breeding and Feeding for Animal Health and Welfare in Organic Livestock Systems. Proceedings of the Fourth NAHWOA Workshop, Wageningen, 24–27 March 2001*. University of Reading, Reading, UK, pp. 44–50.

Baumgartner, J., Leeb, T., Guber, T. and Tiefenbacher, R. (2001) Pig health and health planning in organic herds in Austria. In: Hovi, M. and Vaarst, M. (eds) *Positive Health: Preventive Measures and Alternative Strategies. Proceedings of the Fifth NAHWOA Workshop, Rødding, Denmark, 11–13 November 2001*. University of Reading, Reading, UK, pp. 126–131.

Bouilhol, M. (1997) Livestock farming systems and parasite risk for sheep in organic farming. In: Isart, J. and Llerena, J.J. (eds) *Resource Use in Organic Farming. Proceedings of the Third ENOF Workshop, Ancona, 5–6 June 1997*, pp. 149–157.

Byström, S. (2001) *Djurhälsan i Öjebyn-projektet 1996–2000*. Nytt från institutionen för norrlänsdsk jordbruksvetenskap ekologisk odling, no 1. 2001. SLU, Umeå, Sweden.

Ceresuela, J.L. (1998) De la Dehesa al bosque mediterráneo. In: Hernández, C.G. (ed.) *La Dehesa. Aprovechamiento Sostenible de los Recursos Naturales*. Editorial Agrícola Española, S.A., Madrid, pp. 45–51.

Daza, A. (1998) El ganado ovino en el ecosistema de la Dehesa. In: Hernández, C.G. (ed.) *La Dehesa. Aprovechamiento Sostenible de los Recursos Naturales*. Editorial Agrícola Española, S.A., Madrid, pp. 175–193.

Eder, M., Kirner, L. and Zollitsch, W. (2000) Animal husbandry in alpine organic farming – regional diversity and critical obstacles in Austria. In: Hovi, M. and Trujillo, R.G. (eds) *Diversity of Livestock Systems and Definition of Animal Welfare. Proceedings of the Second NAHWOA Workshop, Córdoba, 8–11 January 2000*. University of Reading, Reading, UK, pp. 26–34.

European Communities (1999) Council Regulation (EC) No. 1804/1999 of 19 July 1999 supplementing Regulation (EEC) No. 2092/91 on organic production of agricultural products and indications referring thereto on agricultural products and foodstuffs to include livestock production. *Official Journal of the European Communities 24.98.1999*. Brussels. L 222 1–28.

European Commission (2001) *The Welfare of Cattle Kept for Beef Production*. A report by the Scientific Committee on Animal Health and Animal Welfare.

Foster, C. and Lampkin, N. (1999) *European Organic Production Statistics 1993–1996. Organic Farming in Europe: Economics and Policy*, Vol. 3. Universität Hohenheim, Stuttgart, Germany.

García Trujillo, R. (2001) Bases ecológicas de la ganadería extensiva de España: In: Labrador, J. and Altieri, M.A. (eds) *Agroecología y desarrollo: Aproximación a los fundamentos agroecológicos para la gestión sustentable de los sistemas mediterráneos*. Universidad de Extremadura – Mundi Prensa, Madrid.

Garzón, J. (1996) Proyecto 2001: Conservando la naturaleza mediante el desarrollo rural. In: *Las Cañadas Viejas, Caminos para el Futuro de la Naturaleza. Congreso 2001, Madrid 20 al 21 de noviembre*. Fundación 2001.-Fondo Patrimonio Natural Europeo.

Gordon, S.H. and Charles, D.R. (2002) *Niche and Organic Chicken Products: Their Technology and Scientific Principles*. Nottingham University Press, Nottingham, UK.

Graf, S. and Willer, H. (2000) *Organic Agriculture in Europe – State and Perspectives of Organic Farming in 25 European Countries*. Stiftung Ökologie und Landbau, Bad Dürkheim.

Hamilton, C., Hansson, I., Ekman, T., Emanuelson, U. and Forslund, K. (2002) Health of cows, calves and young stock on 26 organic dairy herds in Sweden. *Veterinary Record* 150, 503–508.

Hammarström, M. (1992) Djurhälsostudier i mjölkbesättningar anslutna till KRAV. Länsstyrelsen i Värmlands län. Stencil 3. 15 s.

Hardeng, F. and Edge, V.L. (2001) Mastitis, ketosis, and milk fever in 31 organic and 93 conventional Norwegian dairy herds. *Journal of Dairy Science* 84, 2673–2679.

Holmegård, J. (1997) *Økologiens pionertid*. Erhvervsskolernes forlag, Odense.

Jonsson, S. (1997) Ekologiska fodermedels inverkan på fruktsamhet och djurhälsa i en mjölkkobesättning inom Öjebynprojektet. Jämförelse med en konventionell besättning. Slutredogörelse till Jordbruksverket. April 1997.

Keatinge, R., Gray, D., Thamsborg, S.M, Martini, A. and Plate, P. (2000) EU Regulation 1804/1999 – the implications of limiting allopathic treatment. In: Hovi, M. and Trujillo, R.G. (eds) *Diversity of Livestock Systems and Definition of Animal Welfare. Proceedings of the Second NAHWOA Workshop, Córdoba, 8–11 January 2000*. University of Reading, Reading, UK, pp. 92–98.

Klump, C. and Härring. A.M. (2002) Finishing pigs: conversion is more than respecting the standards. In: *Proceedings of the 14th IFOAM Organic World Congress, Victoria, Canada, 21–24 August 2002*. Canadian Organic Growers, Ottawa, p. 82.

Lund, V. (2000) What is ecological animal husbandry? In: Hermansen, J.E., Lund, V. and Thuen, E. (eds) *Ecological Animal Husbandry in the Nordic Countries. Proceedings from NJF-seminar No. 303, Horsens, Denmark, 16–17 September 1999*, pp. 9–12.

Michelsen, J., Lynggaard, K., Padel, S. and Foster, C. (2001) *Organic Farming Development and Agricultural Institutions in Europe: A Case Study of Six Countries. Organic Farming in Europe: Economics and Policy*, Vol. 9. Universität Hohenheim, Stuttgart, Germany.

Nauta, W. (2001) Breeding strategies for organic animal production, an international discussion. In: Hovi, M. and Baars, T. (eds) *Breeding and Feeding for Animal Health and Welfare in Organic Livestock Systems. Proceedings of the Fourth NAHWOA Workshop, Wageningen, 24–27 March 2001*. University of Reading, Reading, UK, pp. 7–13.

Olea, L. and Viguera, F. J. (1998) Pastizales y cultivos. In: Hernández, C.G. (ed.) *La Dehesa. Aprovechamiento Sostenible de los Recursos Naturales*. Editorial Agrícola Española, S.A., Madrid, pp. 95–114.

Østergaard, E. (1998) Ett skritt tilbake og to frem. En fenomenologisk studie av bønder i omstilling til økologisk landbruk. Doctor Scientiarum thesis, Department of Horticultural and Crop Sciences, Agricultural University of Norway, Ås.

Roderick, S. and Hovi, M. (1999) *Animal Health and Welfare in Organic Livestock Systems: Identification of Constraints and Priorities*. A report to the Ministry of Agriculture, Fisheries and Food. The University of Reading, Reading, UK.

Sauveur, B. (1997) Criteria and factors of the quality of French Label Rouge chickens. *Productions Animales (Paris)* 10, 219–226.

Skurdal, E. (1997) *Beiting i utmark – i praksis og i plansamanheng*. Norsk sau- og geitanslag, Landbruksforlaget, Oslo.

Soil Association (2002) *Organic Food and Farming Report*. Soil Association, Bristol, UK.

Sørensen, P. (2001) Breeding strategies in poultry for genetic adaptation to the organic environment. In: Hovi, M. and Baars, T. (eds) *Breeding and Feeding for Animal Health and Welfare in Organic Livestock Systems. Proceedings of the Fourth NAHWOA Workshop, Wageningen, 24–27 March 2001*. University of Reading, Reading, UK, pp. 51–61.

Strøm, T. and Olesen, I. (1997) *Mjølkekvalitet, helse og holdbarhet på kyrne ved omlegging til økologisk mjølkeproduksjon*. Norwegian Centre for Ecological Agriculture, Tingvoll.

Trujillo, R.G. and Mata, C. (2000). The Dehesa: an extensive livestock system in the Iberian Peninsula. In: Hovi, M. and Trujillo, R.G. (eds) *Diversity of Livestock Systems and Definition of Animal Welfare. Proceedings of the Second NAHWOA Workshop, Córdoba, 8–11 January 2000*. University of Reading, Reading, UK, pp. 50–61.

Vaarst, M. and Thamsborg, S.M. (2001) Danish organic dairy and pig herds – brief introduction to the farms and the background for farm visits. In: Hovi, M. and Vaarst, M. (eds) *Positive Health: Preventive Measures and Alternative Strategies. Proceedings of the Fifth NAHWOA Workshop, Rødding, 11–13 November 2001*. University of Reading, Reading, UK, pp. 159–167.

Vermeer, H.M., Altena, H., Bestman, M., Ellinger, L., Cranen, I., Spoolder, H.A.M. and Baars, T. (2000) Monitoring organic pig farms in The Netherlands. In:

Proceedings of the 51st Annual Meeting of the European Association of Animal Production, The Hague, 21–24 August 2000. Wageningen Pers, Wageningen, The Netherlands.

Walkenhorst, M., Spranger, J., Klocke, P., Jörger, K. and Schären, W. (2002) Evaluation of risk factors contributing to udder health depression during alpine summer pasturing in Swiss dairy herds. In: *Proceedings of the 14th IFOAM International World Congress, Victoria, Canada, 21–24 August 2002.* Canadian Organic Growers, Ottawa, p. 98.

Wright, I.A., Zervas, G. and Louloudis, L. (2002) The development of sustainable farming systems and the challenges that face producers in the EU. In: Kyriazakis, I. and Zervas, G. (eds) *Organic Meat and Milk Production from Ruminants. Proceedings of a Joint International Conference Organised by the Hellenic Society of Animal Production and British Society of Animal Science.* EAAP Publication No 106, Athens, 4–6 October 2001, pp. 27–37.

Younie, D. (2000) Integration of livestock into organic farming systems: health and welfare problems. In: Hovi, M. and Trujillo, R.G. (eds) *Diversity of Livestock Systems and Definition of Animal Welfare. Proceedings of the Second NAHWOA Workshop, Córdoba, 8–11 January 2000.* University of Reading, Reading, UK, pp. 13–21.

Organic Livestock Standards

4

Susanne Padel,[1] Otto Schmid[2] and Vonne Lund[3]

[1]Institute of Rural Studies, University of Wales, Aberystwyth SY23 3AL, UK; [2]Forschungsinstitut für biologischen Landbau, FiBL (Research Institute of Organic Agriculture), Ackerstrasse, CH-5070 Frick, Switzerland; [3]Department of Animal Welfare and Health, Swedish University of Agricultural Sciences, PO Box 234, SE-532 23 Skara, Sweden

Editors' comments

Organic farming is subject to international standards. For most European farmers, this includes the EU regulations regarding organic plant production and animal husbandry. Legally binding on the member countries, the regulations must be implemented and followed. Understanding how the standards and regulations work is crucial for everyone with a potential interest in organic animal husbandry. The first task of the authors of this chapter was to give an introduction to the EU standards. In light of the rich diversity of organic animal husbandry in different EU member states and candidates, as described in the previous chapter, the standards need to be examined critically: what difficulties do various regions and countries face when following these rules, do they reflect basic values in organic farming, and do they support the development of good animal welfare? The authors also present the historical development of the EU standards in the context of national rules for organic animal husbandry in Europe over the past several decades. An understanding of how the standards work and how they were developed in turn can point to future adjustments and development of the regulatory structure for organic animal husbandry in Europe.

Historical Development of Organic Livestock Standards

The ideas on which organic agriculture is based can be traced back along different lines. During the first half of the 20th century, several pioneers from a range of backgrounds (farmers, researchers and medical doctors) began to develop 'biological' cropping methods, based on the idea that healthy soils would be the key to producing healthy crops and thus help

© CAB International 2004. *Animal Health and Welfare in Organic Agriculture* (eds M. Vaarst, S. Roderick, V. Lund and W. Lockeretz)

to improve human health. In general, animal husbandry during this period was looked upon primarily as a means to improve plant production – the animals were manure producers necessary for the important composting process. In addition, the ruminants had the role of digesting forage legumes in order to permit a balanced crop rotation that built soil fertility. The animals themselves and the milk and meat produced were of secondary importance, particularly since specialized market outlets for crops developed well before those for organic milk and meat. The limited amount of research that was carried out was mainly concerned with issues of soils, crop production and fertilization.

A notable exception was Rudolf Steiner's agricultural course of 1924, which led to the development of biodynamic farming, an alternative agricultural concept in which livestock were important, especially cows. Steiner viewed farm animals as spiritual beings and argued that humans owe them great respect and care (Steiner, 1929). It is interesting that in the 'agricultural course', Steiner mentioned that feeding ruminants with meat of their own species would cause symptoms similar to those that are now known as BSE (bovine spongiform encephalopathy or mad cow disease).

Towards the end of the 1960s, a new interest in 'biological' agriculture emerged from increasing environmental awareness and the desire to create a livelihood in harmony with nature. Available alternatives in agriculture were taken as departure points for what was to become organic agriculture as we know it today (Christensen, 1998). The protests against industrialized and polluting agriculture also included protests against 'factory farming', i.e. intensive livestock production based on purchased feeds. Ideas of more 'natural' and animal friendly ways to raise livestock were embraced by organic farmers in developing alternatives to this.

IFOAM

As a result of this interest, the International Federation of Organic Agriculture Movements (IFOAM) was founded in 1972. IFOAM is a worldwide umbrella organization coordinating the international network of organic agriculture organizations. It has developed internationally accepted principles and basic standards of organic farming, which are then implemented locally by national or regional certification and inspection organizations. The Basic Standards were first published in 1980 and have been revised every 2 years (most recently in 2002, henceforth every 3 years). They do not provide the direct basis for certification. Rather, they set out the criteria that accredited certifiers must fulfil, and give guidance for national standard-setting organizations.

The overall aims of organic animal husbandry, reflected in the IFOAM standards, can be summarized as falling into the following broad areas:

- *Ecological/Environmental:* respecting the principles governing the relationships between living organisms and their surroundings and seeking to preserve and enhance environmental quality, e.g. by recycling of nutrients and biomass; maintaining or increasing soil fertility, biodiversity and biological activity; avoiding the use of toxic inputs and minimizing pollution; and creating a harmonious balance between crop production and animal husbandry.
- *Ethological:* respecting the physiological and ethological needs of animals, such as providing living conditions that allow animals to express the basic aspects of their innate behaviour (as discussed in Chapter 6).
- *Ethical:* respecting nature, the intrinsic value of farm animals, animal interests and human rights; production of high-quality food; aiming for an aesthetically attractive cultural landscape; providing everyone involved in organic farming and processing a quality of life that satisfies their needs (see Chapter 5).
- *Ergonometric:* providing a safe, secure and healthy working environment for the farmer and farm staff;
- *Economic:* using renewable and farm-derived resources as far as possible, and achieving an appropriate income for the farmer at a fair price for the consumer.

An important principle of organic farming is to rely mainly on the management of internal farm resources rather than external inputs. This applies equally to organic livestock and crop farming. As far as possible, livestock enterprises should be land-based and supported from the farm's own resources; this excludes very intensive systems that depend heavily on brought-in feeds (Lampkin *et al.*, 2002). Furthermore, the manures produced should be capable of being absorbed by the agricultural ecosystem without leading to problems of waste disposal or pollution. Stocking rates should therefore reflect the inherent carrying capacity of the farm and not be increased by reliance on 'borrowed' land. In addition, organic animal husbandry should respect the physiological needs of animals and allow species to express their natural behaviour. In short, the animals should be part of an agricultural production system that is environmentally sound and considers the needs of the animals as well as the humans.

Later development

Since the first IFOAM Basic Standards were published in 1980, the standards and legislation for organic livestock production have developed considerably. Organic farming developed through a close alliance between organic farmers and consumers, and the content and emphasis of the standards was influenced by a growing awareness of problems in conventional agricultural production. On several occasions, the public

Table 4.1. Analysis of the historical development of organic livestock standards.

Time period	Public debate/problem area	Views of critical consumers and producers	Influence on the content of organic livestock standards
Before 1970	Deficiencies with micro-nutrients due to the intensive use of chemical fertilizers in food and feed	Healthy soil – healthy plants – healthy animals – healthy food; animal manure is an important source for fertilizing soil/plants	Animals should be an important part of an organic farm; there should be a balance between plant and animal production (biodynamic concept of 'farm organism')
1970–1998	Problems with chemical residues in food (accumulation of organochlorides in the food chain, residues in mother's milk and animal products)	The risk of contamination by using conventional feed must be minimized	No prophylactic addition of antibiotics and hormones in feeds; maximum allowances of feed from conventional sources (10–20% on dry matter basis)
1980–	Problems of industrial animal production systems, animals suffering in intensive systems (battery chickens, etc.)	Conditions for 'happy' animals (animal welfare) have to be established, particularly on organic farms	Minimum requirements for outdoor access for all animals; sufficient space in housing; no slatted floors; straw as bedding material, natural light
1994–	Animals suffering during transport and slaughter	Animal welfare must include transport and slaughter	More detailed standards for transport and slaughter
1994–1997	BSE crisis, debates about hormone use, etc.	Risks from feeding animals with animal compounds must be excluded	Stricter regulations regarding the use of animal feed compounds
1998–	Problems with antibiotic resistance	Risks for human beings from the therapeutic use of antibiotics to combat animal diseases or to promote growth must be reduced	The use of antibiotics must be much more strongly restricted (minimum of 2× official withholding period, maximum of three courses of treatments with allopathic medicines)
1998–	GMO debate	Risk of contamination of organic products through GMOs	Prohibition of GM-derived components in the diet of organic livestock
1999	BSE cases in continental Europe, and resulting second wave of the debate	Risk of human infection through livestock	Ruminants for meat production have to be born on organic farms rather than undergo a conversion period; further restrictions on feed ingredients

Source: Schmid (2002), amended.

debate affected the awareness of organic farmers, advisers and certification bodies in searching for solutions to the problems, which in turn influenced the development of the organic livestock standards (Table 4.1). For example, concern about the widespread presence of residues of DDT and other organochlorines in the 1970s led to the introduction of limits on the proportion of conventional feeds in organic livestock diets. A debate about the industrialization of animal production in the 1980s led to the prohibition of battery cages for organic chickens and additional welfare requirements for other farm animals, such as prohibition of slatted floors in the lying area, a requirement for the use of bedding material such as straw, minimum space requirements in some organic standards and a requirement for access to the outdoors for all animals.

This search for closer links between organic farmers and consumers may be one reason for a critical debate about farming practices and may explain why organic livestock standards often were adapted to emerging issues very quickly. The practical implementation of alternative livestock husbandry on a broad scale, however, depends on economic and technical feasibility, which in turn depends on the willingness of consumers to pay for animal welfare with a higher price, or the willingness of states to support organic farmers through subsidies. National standard-setting is influenced by public attitudes towards farm animals and livestock production, the development of markets for organic milk and meat, animal welfare legislation, animal welfare certification schemes (such as Freedom Food in the UK, Neuland in Germany, Freiland in Austria) and current practices of organic producers. Considerable variation among countries was one reason behind lengthy negotiations until a common European legal framework for organic livestock production could be agreed upon in EC Regulation 1804/1999, which extended EU Regulation 2092/91 on organic plant production (European Communities, 1999).

Functions of Organic Farming Standards

Before going further into the content of the standards for livestock production, it is important to clarify the different functions that organic farming standards have. These apply equally to all areas of organic production, including both crops and livestock. (Standards also cover other operations such as the processing of organic products, but these are not considered here.) The overall aim of the standards is to maintain and enhance the quality of organic food production and to provide guarantees to the consumer.

Standards therefore represent a commitment from the producer to the consumer and give producers guidance on how the organic principles are to be applied to the different enterprises on the farm. The certification bodies act as mediators between these two main interest groups, whereby the standards are the basis for a legal contract that is binding on both the producer and the certification body.

As mentioned in the previous section, the standards are not static, and different areas have been revised frequently in light of new developments on farms, research on organic farming systems, general trends in the agricultural industry and concerns of the consumers. Decision-making about standards varies among different organizations. For example, changes in the IFOAM Basic Standards must be approved by the biennial (henceforth triennial) General Assembly.

Conflicts and Trade-offs

At any one time, international and national standards and those of certifying organizations represent a compromise between what is desirable from an idealistic point of view and what is technically and economically feasible at the farm level, with due consideration to the different interests involved. Also, in many countries, grant aid is paid only to those producers who adhere to the standards. It therefore is necessary to maintain an element of stability for the producers. This means balancing the need for refinement and development in light of new trends, on one hand, against producers' need for some degree of continuity on the other.

Summarizing the different functions: the standards have to serve the interests of both consumers and producers and have to balance different principles of organic production, such as maximizing the protection of the environment and enhancing animal welfare as well as the other basic principles outlined earlier. They also must give the organic farmer reasonable working conditions and allow an income adequate to survive economically. These different interests and principles do not always coincide, and it is a delicate task for the certification bodies to decide on necessary preferences. A few examples will illustrate conflicts that may occur between animal welfare and other objectives (a point discussed further in Chapter 8):

- The ban in the EU regulation on routine use of antibiotics and anthelmintics grants the consumer 'clean' food and minimizes the impact on the environment, but may put animal welfare at risk (*consumer and environmental considerations vs. animal welfare objectives*).
- Nose-ringing of pigs, allowed by some certification bodies, will keep a protective plant layer on the soil and reduce the loss of nutrients, but will prevent pigs from performing a basic behaviour (*environmental vs. animal welfare objectives*).
- Artificial insemination (AI) is used in organic farming to achieve fast progress in breeding, to enhance the animals' health, to increase pro-

duction capacity, and to provide a safe working environment. However, AI conflicts with the basic principle of allowing animals to express their natural behaviour (*producer and consumer interests vs. animal welfare objectives*).

Not surprisingly, different standards vary in their emphasis on general principles. The IFOAM Basic Standards emphasize that all management techniques, especially where production levels and speed of growth are concerned, must aim for good animal health and welfare and must be governed by the physiological and basic ethological needs of the animal in question (IFOAM, 2002). This contrasts somewhat with the main principles of organic livestock production laid down in EU Regulation 1804/1999, which put more emphasis on the environmental impact of organic livestock production (European Communities, 1999), and with the organic guidelines in the FAO/WHO Codex Alimentarius framework from 2001 (described below in more detail), which states that animals contribute to closing the nutrient cycle, improve the fertility of the soil through their manure, control weeds through grazing and diversify the farm (FAO, 2001).

In developing these standards further, certification bodies need to consider whether the rules laid down can realistically be enforced. They need to allow existing organic farmers to learn about and implement new requirements, particularly if costly investments are necessary. On the other hand, consumer protection is an important guiding principle behind any development of the standards and regulations, and high priority is given to ensuring that consumers are confident in the integrity of the organic production system. However, how to decide between conflicting interests is an ethical question whose answers depend on what basic values are embraced. To maintain consumer confidence it is important for the organic movement to have an open and ongoing debate regarding the guiding principles behind the standards, especially when conflicting interests are involved.

Summary of the Main Areas Currently Regulated in the EU

EC Regulation 1804/1999 was passed in August 1999, extending the legal definition of an organic, ecological or biological food product (EC Regulation 2092/91 and amendments) to cover livestock products, with the aim of protecting consumers and harmonizing trade. (A consolidated version of the Regulation and all amendments is available at the website; European Communities, 1999.) Governments and private sector certification bodies in all EU countries were given 1 year to adapt their livestock standards to the requirements of the new regulation. This was an important step in setting up a common legal framework for organic livestock production in Europe. However, the regulation allows member states to maintain higher national standards in certain areas, and several issues appear to require further interpretation and clarification, leaving room for considerable differences in certification practices among the member states. The following section summarizes the main areas that are currently regulated under the organic livestock production standards in the EC Regulation (European Communities, 1999).

Land management and crop production

Livestock farms must follow organic farming principles in the management of both their land and their stock. The general principle is that the number of animals on the farm should correspond to the amount of feed that a unit can produce and to the amount of manure that can be spread on the land without causing environmental damage. For forage and grassland production, the most important principles are reliance on legumes for nitrogen fixation and, where appropriate, the development of a crop rotation. Fertilization of the land should rely on farm-derived resources, but where this is not possible, additional farmyard manure from extensive husbandry systems (excluding factory farming) may be brought in to the extent that the total amount of N applied annually does not exceed 170 kg/ha. Most other organic and non-organic fertilizers need the permission of the certification body and should be applied only if a need has been proven. Under certain circumstances the collaboration of farms in a particular region is a good mechanism to meet the requirements for feed and manure usage.

Conversion of the holding

A holding will need to undergo a conversion period during which necessary changes in crop and livestock production are implemented but no products are marketed as fully organic. However, some countries allow the use of a special label for products raised during conversion. Livestock may be marketed as organic provided the land has completed its conversion and the stock have undergone minimum conversion periods (possibly simultaneously). The main aim in specifying minimum time periods for conversion is consumer protection, so that the likelihood of residues from the previous conventional management is reduced. It is important to recognize that this requires careful planning of the timing of conversion of the livestock enterprises. The conversion period also is an important period for the farmers to learn about and adjust to organic farming methods.

Stock origin and conversion periods

Ideally, organic farms should aim for closed flocks, and most stock should be born and raised on organic holdings, because this has clear advantages regarding traceability and prevention of a wide range of communicable diseases. Suitable breeds for an organic regime should be used, but further research is needed to clarify which breeds fulfil this (a point discussed in detail in Chapter 16). Particularly in pig and poultry production, there appears to be a need to develop such breeds, as stressed in the NAHWOA Recommendations (see also Padel and Keatinge, 2000). The standards have allowed a range of exemptions from the principle that livestock should be born and raised under organic management, with conversion periods specified during which brought-in animals must be managed fully in accordance with the standards. The risk of BSE infections in several European countries in early 2001 led to changes in the requirements for ruminants intended for meat production, which no longer can be converted, but must be born under organic management. The producer must keep records of stock movement.

Husbandry and housing

The standards contain several guidelines and rules regarding husbandry and housing of livestock, with the principle aim of establishing livestock systems that are appropriate to each species' physiological characteristics and natural behaviour. This is specified in several general ways, such as access to free range and pasturage, clean bedding, natural light and clean air, the prohibition of unnecessary mutilations to avoid unnecessary suffering (see Chapter 8), and preference for natural methods of reproduction. Rules specify maximum stocking density inside buildings, minimum space requirements per animal and some specifications as to the design of buildings, such as restrictions on the floor area that can be covered by slats and restrictions or prohibition of tethering.

However, organic livestock standards need further development in this area, for example clearer guidelines regarding the housing for specific species, specification of the restrictions on mutilations (e.g. age limits for castration of pigs) and more detailed rules for transport of animals. For some species, there also appears to be a lack of scientific information on building design, group sizes and ethological needs of the animal, as well as practical experience with housing systems that fulfil all requirements laid down in the standards.

Feeding

Two basic principles govern the feeding of livestock: the animal should be fed a diet appropriate to its species, and ideally all the animals on an organic holding should be fed a fully organic diet from the holding itself. Apart from reflecting closed production cycles, the later principle also reflects the belief that the improved 'health' of the soil provides the basis for a healthy crop, which in turn is the basis for healthful nutrition of the animals. Furthermore, the principles reflect the desire to minimize the risk of residues entering the food chain, but also the land-based character of animal husbandry systems. However, because of practical limitations, such as the unavailability of certain feedstuffs on the farm and the need to achieve a balanced diet for the stock, the standards allow exemptions to these principles, which are reviewed and revised regularly. EU Regulation 2092/91 (in line with the IFOAM 2002 Basic Standards and the FAO/Codex Guidelines) specify that not more than 10% of the diet of herbivores may come from conventional sources; for pigs and poultry the figure is 20%. They further specify a proportion of the diet that can come from in-conversion sources (30%), the reminder having to come from fully certified land. The conversion proportion can increase to 60% if it comes from the holding itself, and the regulation recommends giving preference to feed from the same holding, but the proportions have not yet been specified; the NAHWOA Recommendations call for clearer guidelines governing exemptions from the requirements for home-grown feeds.

The goal of species-appropriate diets includes that diets should be balanced, avoiding high levels of concentrated protein or energy feed sources associated with intensive production (see Chapter 15). Some specific rules for particular species are laid down, such as the need to include at least 60% of roughage in the diet of most ruminants, to make sure that the rumen can function effectively and that the ruminants' nature as a grass-fed animal is respected. Offspring from mammals should be fed for a certain period on milk, preferably their mothers'. To protect consumers, feed should not contain solvent-extracted feeds or those originating from other processes involving chemical agents. Trace elements and concentrated vitamins are permitted only where they are necessary to maintain nutrient balances; preference should be given to natural sources. Pure amino acids of synthetic origin are not allowed, a matter of considerable debate (Chapter 15). A positive list specifies what components may be used, but currently there is variation among countries and sector bodies, and as yet there are no sufficiently clear criteria in the EU regulation for deciding on which feeds may be used on organic farms; the NAHWOA Recommendations specifically call for this shortcoming to be corrected.

Health management

In principle, emphasis should be given to prevention rather than treatment; that is, effective management should be employed so that animals can be kept in good health (as described in Chapter 12). This is achieved in two main ways:

- Improving the immunity and resistance of the stock through good husbandry, such as appropriate nutrition, selection of robust breeds, selection of resistant strains within breeds, and reduction of stress levels in the herd by allowing the animals to express their natural behaviour.
- Reducing the disease challenge by limiting the mixing of animals from different herds, reduced indoor and outdoor stocking densities, good hygiene and grazing management to prevent parasitic infections (see Chapter 14).

The goal of underpinning health and welfare with husbandry and good management practice is well established (Boehncke, 1997). However, for some diseases, there are gaps in our knowledge of suitable management approaches. Furthermore, because of other factors, situations may arise where effective control through management alone might not be possible, so that veterinary intervention is needed. In this case a range of control measures may be used.

The standards prohibit the routine preventive use of chemically synthesized allopathic treatments. Preventive treatment should be understood as any treatment that is not based on the diagnosis of an individual animal (or group of animals) or an assessment of a particular risk to animal health and welfare according to local conditions and epidemiology of the particular disease (Keatinge *et al.*, 2000). For example, the routine use of dry cow therapy is prohibited, but if individual animals are diagnosed with a condition for which this is the most appropriate treatment, it may be used. The rule should not be interpreted as prohibiting conventional veterinary intervention until there is clinical evidence of disease, as this would conflict with the interests of animal welfare and consumer protection in organic farming. The EU regulation states that in the case of disease, preference should be given to traditional and complementary treatments, although this conflicts with national legislation in Sweden, which does not allow veterinarians to use homeopathy. The EU regulation also limits the number of applications of most synthetic veterinary drugs that can be used during a year (for breeding stock) or the lifetime of an animal. The NAHWOA Recommendations suggest reconsideration of this requirement, possibly replacing it by a limit on the average number of treatments for the entire herd or flock. The regulation further specifies that the producer must maintain accurate records of all veterinary treatments and make them available to the certifying body at the time of inspection.

Some national standards (e.g. UKROFS in the UK) require a specific animal health plan to be agreed upon by the certifying authority at the start of conversion. This represents good practice and should be included in all standards, as discussed in Chapter 12.

Comparison of Different International Livestock Standards

Organic standards are further developed at the international level. IFOAM has been very important in facilitating communication concerning the principles and concepts of organic agriculture worldwide and in harmonizing organic standards internationally (Rundgren and Lockeretz, 2002). In this task, IFOAM has been supported by many people, including several researchers and organizations involved with organic farming.

With the aim of promoting international harmonization and development of the standards for organic livestock production, the Swiss Research Institute of Organic Agriculture (FiBL) in Frick, Switzerland, compared different standards for organic livestock production at different times. A comparison of 12 sets of organic livestock standards (including the IFOAM Basic Standards) in 1998 (Schmid *et al.*, 1998; Schmid, 2000) showed that many aims and items were similar, especially the limit on feed from non-organic farms and the prohibitions on the use of growth promoters and of veterinary drugs in the absence of illness. Areas where the rules differed were the requirements for housing, the grazing/free-range area, the withholding period after the use of veterinary drugs, proportion of brought-in animals, conversion times, the proportion of feed from non-organic sources, nose-ringing of pigs, and minimum times before weaning of piglets and ruminants.

The EU Regulation also has had an impact on international harmonization of organic livestock standards, because all countries intending to export livestock products to the EU need to meet these standards. A further step in the international harmonization of organic livestock standards was the development of organic guidelines within the framework of Codex Alimentarius (FAO, 2001). This is a worldwide guideline for countries and others developing national standards and regulations, but, as with the IFOAM Basic Standards and the other international standards, these guidelines should not be used directly for the certification of producers. However, in case of a trade dispute between countries, the Codex Alimentarius Guidelines and the IFOAM Basic Standards can be taken as international references under WTO rules.

The following section briefly compares the requirements of EC Regulation 1804/1999, the IFOAM Basic Standards for animal production and the Codex Alimentarius Guidelines (FAO, 2001), in order to establish areas with different approaches and where further development and international harmonization might be necessary (Schmid, 2002).

Because of the different purposes of these various standards, those of the Codex and IFOAM are more general, outlining principles and criteria that have to be fulfilled. They are less detailed than the EC regulation, which deals specifically with the European context.

The conditions for *conversion of a holding* with livestock are very similar in all three sets of standards. In the IFOAM Basic Standards, the feed can be considered as 100% organic after 1 year of organic management, whilst the EU Regulation has a 2-year conversion period, which can be reduced to 1 year in special circumstances, depending on the previous land use (for example, land managed under an agrienvironmental scheme with almost identical restrictions on inputs).

The EU Regulation has set a limit for *manure application* at the equivalent of an annual N supply of 170 kg/ha (with a list of the maximum allowed stocking densities for each species) and specifies that the outdoor stocking densities on pasturage must be low enough to avoid overgrazing. In contrast, IFOAM and Codex only state the general principle of having appropriate stocking rates.

Differences occur regarding the *conversion periods* for dairy cows and layers. The IFOAM Basic Standards require 30 days for dairy cows and layers, whereas the EU Regulation and Codex specify longer conversion periods (3–6 months for dairy cows, 6 weeks for layers).

All three sets of international standards specify an allowance for *conventional feedstuff* of 10% for herbivores (ruminants and horses), but differ in the allowance for other non-ruminants (15% in the IFOAM and Codex standards and 20% in the EU regulation). The EU regulation currently sets a time limit (2005) as to when this derogation should be reviewed. In setting guidelines for organic food production worldwide, Codex and IFOAM have not settled on any time limits, as organic production in many countries is at an early stage of development.

IFOAM sets a minimum *fodder self-sufficiency* (50% from the farm or produced in cooperation with other farms in the region, at least for ruminants), whereas Codex and the EU Regulation, although aiming for the same principle, do not quantify such restrictions. Codex and IFOAM set out more clearly the criteria regarding the *feed ingredients* that may be used, such as feed additives and processing aids. The EU has a detailed list of feedstuffs and feed additives.

All three standards double the withholding period after administration of conventional *veterinary drugs*. In addition, the EU Regulation and Codex require stock or products to be excluded from organic labelling after two or three courses of treatments per year or (if shorter) the lifetime of the animal.

The IFOAM basic standards on *mutilations* of animals are stricter than the EU Regulation and Codex, allowing fewer derogations.

All the standards prohibit permanent *tethering* of animals with no access to pasturage or to outdoor runs for regular exercise, but the

IFOAM regulation allows tethering for a certain period, whereas the EU Regulation permits tethering only on smallholdings. In Codex the tethering of livestock is not permitted without the approval of the competent authority. During the fattening period, animals should have access to an outdoor run, but the standards differ in the exemptions that may be granted to individual producers. The EU Regulation provides more detailed rules regarding the *housing* requirements for poultry than do the other two standards, but none of the three provides detailed rules for the keeping of pigs. The EU Regulation further sets out detailed minimum requirements for *indoor area* and *outdoor runs*. IFOAM and Codex only call for sufficient size, recognizing differences among local breeds and the varying risk of environmental problems in different climatic regions.

IFOAM contains more detailed rules on *transport and slaughter* than the EU Regulation and Codex.

Only the EU regulation contains a positive list of *cleaning agents*.

Future Development

There still are several areas where appropriate practice is not well defined and others where further harmonization is needed. The new EU regulation represents an improvement in setting ethological standards compared with previous organic livestock standards (Schmid, 2000). Organic livestock systems as defined by the organic standards need further development in the following areas, which should be informed by research that reflects the practice and experience of organic farmers and by consultations with experts on ethnology and animal welfare (this list closely parallels the NAHWOA Recommendations).

- It is important to debate the philosophical basis and ethical principals for animal welfare in organic farming (as discussed in Chapter 5) and to identify areas of conflict between animal health, welfare and other aims (Chapter 8).
- Clearer guidelines for species-appropriate husbandry, housing, transport and slaughter of all farm animals should be included in the standards, and objective methods need to be developed to assess whether the guidelines have been met. However, organic standards should be sufficiently flexible to allow continuous improvement of animal welfare and to meet the different needs in the different regions (see Chapter 3).
- The development and implementation of animal health plans should be included in the standards (Chapter 12). These should emphasize the need to move towards management strategies for disease prevention. More research is needed to support implementation of the standards' recommendation of greater use of alternative therapies when an animal must be treated.

- The requirement to select suitable breeds and strains needs further clarification as to which breeds, species and breeding strategies should be considered suitable in organic systems in general and under specific circumstances (see Chapter 16). In particular, there is a need to develop both standards and breeding practices for organic poultry production to avoid the welfare problems of more intensive systems. The standards could give more emphasis to excluding external sources of disease, largely within a closed flock.
- The recently introduced ban on protein sources from animal origin and on synthetic amino acids and vitamins needs to be implemented by identifying natural sources and alternative feeding regimes, particularly for monogastric animals. This may lead to a need to review the list of permitted feeds. Criteria need to be developed as to which feeds may be used in organic livestock production.

As in the past, any further development of standards must aim for a balance between consumers' desire for products of high quality and considerations of ethical and ecological integrity on the one hand, versus the practical and financial needs of producers on the other. Researchers can make a very important contribution to this development, and should aim to communicate their results, particularly in the areas listed above, to bodies that develop and interpret standards at the national and international level.

References

Boehncke, E. (1997) Preventive strategies as a health resource for organic farming. In: Isart, J. and Llerena, J.J. (eds) *Proceedings of the 3rd ENOF workshop, Ancona, 5–6 June 1997*. European Network for Scientific Research Coordination in Organic Farming, Barcelona, pp. 25–35.

Christensen, J. (1998) *Alternativer – natur – landbrug*. Akademisk forlag, Viborg.

European Communities (1999) Council Regulation (EC) No. 1804/1999 of 19 July 1999 supplementing Regulation (EEC) No. 2092/91 on organic production of agricultural products and indications referring thereto on agricultural products and foodstuffs to include livestock production. *Official Journal of the European Communities* 24.98.1999. Brussels. L 222 1–28. See http://europa.eu.int/comm/agriculture/qual/organic/reg/index_en.htm

FAO (2001) *Guidelines for the Production, Processing, Labelling and Marketing of Organically Produced Foods,* GL 32 – 1999, Rev. 1 – 2001. Rome. http://www.fao.org/organicag/doc/glorganicfinal.pdf

IFOAM (2002) *IFOAM Basic Standards*. International Federation of Organic Agriculture Movements, Tholey-Theley, Germany. http://www.ifoam.org/standard/

Keatinge, R., Gray, D., Thamsborg, S.M., Martini, A. and Plate, P. (2000) EU Regulation 1804/1999 – the implications of limiting allopathic treatment. In: Hovi, M. and Trujillo, R.G. (eds) *Diversity of Livestock Systems and Definition of Animal Welfare. Proceedings of the Second NAHWOA Workshop, Córdoba, 8–11 January 2000*. University of Reading, Reading, UK, pp. 92–98.

Lampkin, N., Measures, M. and Padel, S. (2002) *2002/03 Organic Farm Management Handbook*. University of Wales, Aberystwyth, and Organic Advisory Service, Newbury, UK.

Padel, S. and Keatinge, R. (2000) Discussion of the EU Reg. 1804/1999 at the 2nd NAHWOA workshop. In: Hovi, M. and Trujillo, R.G. (eds) *Diversity of Livestock Systems and Definition of Animal Welfare. Proceedings of the Second NAHWOA Workshop, Córdoba, 8–11 January 2000*. University of Reading, Reading, UK, pp. 99–107.

Rundgren, G. and Lockeretz, W. (eds) (2002) *Reader. IFOAM Conference on Organic Guarantee Systems: International Harmonisation and Equivalence in Organic Agriculture. Nuremberg, 17–19 February 2002*. IFOAM, Tholey-Theley, Germany.

Schmid, O. (2000) Comparison of European Organic Livestock Standards with national and international standards – problems of common standards development and future areas of interest. In: Hovi, M. and Trujillo, R.G. (eds) *Diversity of Livestock Systems and Definition of Animal Welfare. Proceedings of the Second NAHWOA Workshop, Córdoba, 8–11 January 2000*. University of Reading, Reading, UK, pp. 63–75.

Schmid, O. (2002) Comparison of EU Regulation 2092/91, Codex Alimentarius Guidelines for Organically Produced Food 1999/2001, and IFOAM Basic Standards 2000. In Rundgren, G. and Lockeretz, W. (eds) *Reader. IFOAM Conference on Organic Guarantee Systems: International Harmonisation and Equivalence in Organic Agriculture. Nuremberg, 17–19 February 2002*. IFOAM, Tholey-Theley, Germany. pp. 12–18.

Schmid, O., Schüpbach, K. and Beltrami, R. (1998) Nationale und internationale Richtlinien der Tierhaltung im ökologischen Landbau. Unpublished report. Forschungsinstitut für biologischen Landbau, Frick, Switzerland.

Steiner, R. (1929) *Landwirtschaftlicher Kursus*. Section for Natural Science of the Anthroposophic Society, Dornach, Switzerland.

5

Animal Welfare, Ethics and Organic Farming

Henk Verhoog,[1] Vonne Lund[2] and Hugo Fjelsted Alrøe[3]

[1]Louis Bolk Institute, Department of Organic Animal Husbandry, Driebergen, The Netherlands; [2]Department of Animal Welfare and Health, Swedish University of Agricultural Sciences, PO Box 234, SE-532 23 Skara, Sweden; [3]Danish Research Centre for Organic Farming, Danish Institute of Agricultural Sciences, Research Centre Foulum, PO Box 50, DK-8830 Tjele, Denmark

Editors' comments

Organic farming is built on values that form a coherent ethical framework. Animal welfare is an important part of that framework, and is a concept that has been discussed in many different contexts. How can animal welfare be conceived within organic farming? Can the animals have a good quality of life, and what ethical framework is relevant when dealing with this question? The emphasis on naturalness in organic farming is well known, and of course will greatly influence how animals' lives and welfare are perceived. The authors explore the connection between organic values and the perception of the quality of the lives of the animals in the organic herd. They present different ethical frameworks for thinking about animals in organic farming systems, then go on to discuss the dilemmas that can arise when the particular ethical approach chosen in organic farming must be applied to very down-to-earth choices in practical husbandry. Such dilemmas are challenging conflicts that may never be resolved to everyone's satisfaction, but force us to think about the choices we face.

Introduction

A striking aspect of man's relation to animals is its ambiguity. Not only may animals be perceived differently in different cultures (think of the sacred cows in India, for instance, compared with how the cow is seen in

© CAB International 2004. *Animal Health and Welfare in Organic Agriculture* (eds M. Vaarst, S. Roderick, V. Lund and W. Lockeretz)

Western culture) – we sometimes see striking differences even within the same culture. Pigs in our Western culture offer a good example. Pigs have a long history of domestication (see Chapter 2). Today wild pigs still roam the forests. There they sometimes are hunted, but most of the time they can live their own lives – which also include less pleasant things, such as encounters with predators or freezing cold winters.

We further see, at one extreme, how pigs have been almost totally and massively instrumentalized into short-lived production machines, with natural science helping to increase the efficiency of this process. (However, it is also argued that compared with their wild cousins, farm pigs are provided a more protected environment, and given daily and scientifically composed rations of nutritious feed; as a result they display good physical fitness, which is said to be indicated by their high growth rates.) As production animals they are anonymous and treated as things, as objects. All that counts is their body, the meat. They can hardly be called 'Haustiere' any longer. Millions are killed every year, and when infectious diseases such as foot-and-mouth disease or swine pest occur, they are massively destroyed for purely economic reasons.

At the other extreme, there is a recent trend to keep pigs as pets. As such, pigs are very clean, friendly, intelligent animals, even loving music. They respond to individual names, and humans communicate with them as subjects. For those who keep pigs as pets, there is no doubt that these animals have consciousness, that pigs are individuals. It seems that the individual character of each pig gets a chance to manifest itself under the conditions of pet keeping. In this situation, it is not the body only, but also the pig's mind that is important. However, it is not uncommon that pet pigs are caused distress because of their owners' ignorance, such as regarding pig nutrition.

This example raises several questions: what is the essence of a 'good pig life', and what obligations do humans have to make such a life possible? One can also ask if human obligations to organic pigs should be any different from our obligations towards conventional pigs. This chapter is meant to be a non-technical introduction to these kinds of philosophical issues, relating to animal welfare and organic farming. Philosophy deals with issues usually left out of empirical science, which aims to be objective, value-free and rational. The philosopher reflects on the meanings of those terms. The reflections may concern a particular view of or attitude towards nature or animals, which underlies a certain definition of animal welfare – or the question whether the animal welfare concept really is value-free. Are not science and ethics necessarily connected when we discuss animal welfare?

Other important elements of a philosophical reflection regarding animal welfare in relation to organic farming are the role of the human in nature, the differences and similarities between humans and animals, and their moral relevance. Are farm animals looked upon differently in organic and conventional farming, and if so, why? Are different values involved, or different views of man's place in nature?

We first give an overview of different interpretations of animal welfare and their relation to different lay and expert approaches to it. The importance of values to animal welfare is discussed and confronted with the role of science and the scientific ideals of being value-free and objective. The interpretations of animal welfare are then compared with different ethical frameworks, which can help clarify the moral aspects of animal welfare. We then discuss the values of organic farming as a departure point for a discussion of what a relevant view of animal welfare might be in the context of organic farming. This reveals obvious conflicts between different values within organic farming, with some of these dilemmas illustrated by examples from organic practice. Finally, we indicate the relevance of such philosophical questions on values and the meaning of animal welfare for the future course of the organic movement.

Different Understanding of Animal Welfare

The three situations of the pig described above exemplify three main elements in the concept of animal welfare. Appleby (1999) calls them *nature* (manifested in the wild pig), *bodies* (the pig as production animal) and *minds* (coming to the fore with the pig kept as a pet).

Accordingly, three types of animal welfare definitions are often distinguished, depending on what is considered important for the well-being of the animal (Fraser *et al.*, 1997):

- *The natural living approach*: the welfare of an animal depends on its being allowed to perform its natural behaviour and live a life as natural as possible.
- *The biological functioning approach*: animal welfare is related to the normal functioning of physiological and behavioural processes (often expressed as the animal's ability to cope with its environment).
- *The subjective experience approach*: the feelings of the animal (suffering, pain and pleasure) determine the welfare of the animal.

This distinction raises the questions of how these three aspects relate to each other, how animal welfare can be evaluated, and whether or not this evaluation can leave out moral issues, as some natural scientists believe. Another question is how animal welfare should be defined in organic agriculture. The answer to this last question has direct consequences for what the organic standards should prescribe.

Returning to the question of the relation among the above definitions, we find that complicated philosophical issues form the background. Depending on our basic outlook on life (what we think is important) and the different philosophical presuppositions that we carry, such as about the nature of science and the relations among behaviour, mind (consciousness) and body, we can arrive at different conclusions regarding what is important for animal welfare. Occupation may also influence how we perceive it: thus, veterinarians tend to emphasize physical functioning, whereas ethologists tend to look at animal behaviour in the first place and physiological functioning second. When ethologists talk about disturbed behaviour in domestic animals, they usually refer to the natural behaviour of their wild relatives as the standard. This links to the animal welfare concept in which natural living is the most important criterion.

Do Animals have Minds or Feelings?

The existence and nature of subjective experience in animals is a central question in relation to animal welfare: how can we know that animals actually suffer? It has been an issue in Western intellectual debate since antiquity. Aristotle caused a philosophical crisis in the 4th century BC by stating that animals do not have reason, intellect or thought, which was a common perception of that time. The Stoics and the Epicureans also argued that only humans have reason and belief, and because of this they excluded animals from moral concerns. Even so, there were other ancient schools of thought that spoke against the denial of reason to animals. (An initiated review of the ancient debate is given by Sorabji, 1993.)

The dominating, traditional Western Christian view of animals and their moral status is substantially based on Aristotle and the Stoic tradition, whose thinking was refined in a Christian context by two influential Church Fathers: Augustine in the 4th and Thomas Aquinas in the 13th century (Sorabji, 1993). With a few exceptions,[1] European philosophers until our time have adhered to the dogma that animals cannot reason, and as a consequence have no rights in relation to humans (Taylor, 1999).

When modern science emerged, Descartes (1637 [1985]) set the direction with his notorious statement that animals are to be compared to machines, without minds or feelings, although this statement did not remain undisputed. Taking a quick leap through history, one can contrast this with Charles Darwin (1890), who saw the differences in mental capacity between humans and other highly developed mammals not as fundamental but gradual, both among species in the course of evolution and among existing species.

[1]The most notable is perhaps Jeremy Bentham, the founder of modern utilitarianism. He argued that the issue of animal mind and reasoning was not relevant to how humans should treat animals. In a famous passage he stated that 'the question is not Can they reason? nor Can they talk? but, Can they suffer?' (Bentham, 1789 [1970]).

When ethology emerged as a scientific discipline in the late 1800s, early ethologists generally recognized animal subjectivity (Rollin, 1989; Burkhardt, 1997). By the mid-1920s, two distinct approaches could be discerned: behaviourism and ethology. Behaviourists sought to apply the methods of 'hard science' such as chemistry and physics, and thus confined their research to the animal's physical behaviour and measurable biological functions associated with it. Some took a step further and adhered to the logical positivist standpoint that drew a hard distinction between the 'meaningful' statements of empirical science and the unverifiable and therefore 'nonsensical' statements of disciplines such as religion and ethics. They also regarded questions about consciousness as unscientific – thus leaving the question of animal feelings to the philosophers. Darwin's notion of mental continuity was strongly criticized by this group.

Classical ethology emerged from zoology, mainly through the work of Konrad Lorenz and Niko Tinbergen in the mid-20th century (Eibl-Eibesfeldt, 1975). They also took issue with the vitalism and subjectivism of the early ethologists. Tinbergen (1942), especially, claimed that ethology has to be uncompromisingly 'objectivistic'. Still, ethologists use a 'mentally loaded' language when describing this behaviour (aggression, fear, stress, etc.). This is often defended with the so-called *analogy argument*: when certain behaviours or physical functions appear in animals and similar features can be observed in humans, where we know from our own experience that they correspond to certain feelings, these feelings are assumed to be present in animals also. Thus, although the animal's 'state of mind' cannot be measured directly, it can be measured indirectly through such parameters. However, not all scientists agree this is a valid argument. For example, Bermond (1997) states that the brain's prefrontal cortex is crucial for the human experience of suffering. Since phylogenetically it is the most recent structure and absent in most animals except primates, other animals would be unable to experience suffering. (However, others argue that other parts of the brain can have the same function in those species.)

The emergence in the 1970s of cognitive ethology (Griffin, 1976), animal psychology, and animal welfare science have put the issue of animal consciousness on the agenda with new force (see, e.g., Allen and Bekoff, 1997). The debate continues among philosophers as well as scientists, with defenders of each doctrine coming from both camps. Part of the current discussion is given in Dol *et al.* (1997). Even so, the standpoint that animals do not have minds has become increasingly outmoded among philosophers (Taylor, 1999), and among ethologists, Tinbergen's strict approach has been softened. Thus, the study of individual reactions and behavioural nuances is no longer excluded as a useful approach for understanding the subjective lives of animals, and theories that include subjective experience are conceivable (Fraser, 1999).

Values and Science

How one views science is crucial to one's understanding of animal welfare. If you state that science has no access to animal consciousness, you already presuppose a kind of dualism between mind and body, as if they were two totally different categories. However, the work of the Dutch ethologist Wemelsfelder (1993, Chapter 2), for example, has shown that this is not a necessary assumption. (Her research is an example of the new ethological approach suggested by Fraser above.) After providing an interesting overview of different views on animal subjectivity she pleads for an 'empirical approach towards animal subjectivity'. She maintains that 'the subjective aspects of animal behaviour are primarily manifest in our first-hand, common sense interaction with animals'. To get direct access to the subjective aspects we therefore have to keep close to our immediate (common sense) experience of animal behaviour, especially the animal's capacity to interact with the environment. She has developed a method of qualitative assessment of animal behaviour, demonstrating that observers from quite diverse backgrounds, including farmers, show significant agreement in their spontaneous assessment of the expressive behaviour of pigs (Wemelsfelder *et al.*, 2000).

Not all biologists agree with this approach, as becomes clear when we read the English biologist Wolpert (1993). He considers the task of science to be to abstract 'objective facts' from our common sense or 'phenomenological' experience of nature. This means that subjective, qualitative human experience (including value judgements) must be removed. Science's concept of 'nature' is not the same as the nature we see with our eyes, touch with our hands, or smell with our nose. Following this definition of science there seems to be an inherent opposition between the scientific and popular understanding of animal welfare (the way consumers view it, for instance).

A dualistic view like Wolpert's, which separates the mind (consciousness) from the body and assigns to natural science the task of studying only the physical aspects of the world, has consequences for the relation between science and ethics, as well as for the definition of animal welfare. It also affects whether moral values are involved in the scientific description of animal welfare. It means that values cannot be part of the material world, and they can only arise in the human mind. They are subjective, not objective. Scientists should aim for value-free (objective) research; the goal of science should be to describe animal welfare objectively as the animals' state of being, without making any value judgements. In line with this, Broom (1991) argues that the assessment of animal welfare can be carried out in an objective way, quite independently of any moral considerations.

In opposition to this view, the American veterinarian Tannenbaum (1991) has shown quite convincingly that this 'pure science model of animal welfare' has unexamined moral values implied in the definition of animal welfare, or already regarding the question of what objects and processes should be studied. He argues that the belief in the pure science model allows scientists to separate scientific and ethical questions, and to believe that their work is unaffected by political and ethical debates. Since the appearance of Tannenbaum's paper several studies have confirmed and reinforced his analysis (Fraser, 1995, 1999; Sandøe and Simonsen, 1992; Alrøe *et al.*, 2001). All these authors agree that it is important to make the underlying values explicit (later we will attempt to do this for the definition of animal welfare in organic farming). This means, for example, that each time it should be clearly stated which definition of animal welfare is being used, and why.

Different definitions of the animal welfare concept are thoroughly discussed in an article by Stafleu *et al.* (1996), who describe the historical development of attempts to make animal welfare objectively assessable. In the search for objective quantities, science removed itself from public concerns about animal welfare, which were based on the subjective experience approach. The scientific operationalization of the animal welfare concept also implied 'an erosion of the moral element of the welfare concept'. This elimination of feelings and moral aspects from scientific definitions of animal welfare leads to a gap between 'common sense concepts', having a political and social frame of reference, and scientific measurements. To be helpful for policy, scientists should make the values and moral choices behind scientific facts explicit, and also the relation between various scientific variables and overall welfare as a subjective experience.

Animal Welfare and Different Ethical Frameworks

Dimensions of values

One way of assisting the process of restoring values and moral issues to animal welfare science is to relate the issues of animal welfare to different ethical frameworks. If we agree that ethical values are intrinsically involved when we talk about animal welfare, the next step is to ask what we mean by 'values' or 'valuing'. In the literature about 'value clarification', three dimensions of valuing usually are distinguished (Simon *et al.*, 1978):

- *The cognitive dimension*: the idea of a morally desirable end-state, and the arguments why this state is preferable compared to other end-states (alternatives). A free moral choice is involved. An example is the idea that human beings should not treat farm animals as mere instruments, because they are sentient beings who can suffer.

Rational arguments may serve to provide us with a justification of what ought to be done. They are socially important for communicating our values to others.

- *The emotive dimension*: the affective component of values, related to one's inner experiences and feelings. For example, one may feel negative emotions when farm animals are treated as instruments, since such treatment does not comply with one's basic attitude towards animals or life. Emotions are very important in the sense that they draw our attention to the fact that certain values are involved. Feelings often serve as a necessary link between our cognitive knowledge and our action.
- *The normative dimension*: values telling us what we ought to do or not do (values as standards). For example, they may forbid actions such as torturing animals or promote actions that enhance the animal's welfare. A person who believes that animals should not be treated as mere instruments is also expected to act upon this idea, with a certain consistency.

These different dimensions of the valuation process can be recognized more or less explicitly in the different ethical frameworks.

Ethical frameworks

An ethical framework of thinking involves a certain basic attitude towards animals. It decides which animals or animal experiences should have moral status, and how normative concepts such as animal welfare should be defined. They integrate what we think about, feel about and do with animals. Thus, the meaning of normative concepts such as intrinsic value, animal welfare and animal integrity depends on the framework in which they function. There are several possible ethical positions (see Box 5.1). Below, the main positions are described.

Box 5.1. Ethical frameworks.

1.	Anthropocentric:	only human beings have intrinsic value or are morally significant
2.	Zoocentric:[a]	sentient animals also have intrinsic value or are morally significant
3.	Biocentric:	all living entities with a good of their own have intrinsic value or are morally significant
4.	Ecocentric:	species (not individual animals) and ecosystems have intrinsic value or are morally significant

[a]Also called sentientistic.

1. The anthropocentric framework

In a strict anthropocentric framework, only human beings have moral status, which means that we have no direct moral responsibilities towards animals. We could have indirect responsibilities to the owners of animals or, as in traditional anti-cruelty laws, because we believe that people who are cruel to animals may also be cruel to humans. Sometimes this is interpreted as the view that cruelty to animals may hurt other people who see it. Most anti-cruelty laws go one step further, however, prohibiting deliberate mistreatment of an animal without a good reason. Animal experimentation, for instance, is considered to be a good reason by most people.

An anthropocentric view of animal welfare is that as long as the animal is functioning well with respect to human goals (for example, as long as it grows and reproduces), it has good welfare. The emphasis is on biological functioning.

The basic emotive attitude related to this framework is man as a ruler or steward, having the (God-given) right to rule over the rest of nature. Human well-being is the most important thing, and animals may be used solely to serve human ends.

2. The zoocentric framework (also called 'sentientistic')

This theory differs from the anthropocentric one in saying that animal suffering is prima-facie wrong (that is, wrong unless supervened by another normative principle). When human beings deal with sentient animals, they have the responsibility to minimize animal suffering; the use of animals may be forbidden if it imposes too much suffering. The argument most often used to defend this view is the analogy argument mentioned earlier. Thus, only 'higher animals' are given moral status in this view.

Good examples of zoocentric thinkers are the philosophers Singer (1975) and Rollin (1995). Rollin introduces the concept of 'telos' to refer to the species-specific nature of animals: 'animals like humans have natures, and respect for the basic interests that flow from those natures should be encoded in our social morality' (p. 159). It is important to realize that it is not the nature itself that should be respected, but the interests determined by it. Genetic modification of animals, for instance, is not wrong in itself, according to Rollin: 'I never argued that the telos itself could not be changed' (p. 171). To change the telos of chicken through genetic engineering, so that they no longer have a nesting urge, means to remove a source of suffering for animals held in battery cages. They are better off than before. Rollin agrees that it may be better to change the rearing conditions, but as long as this is not expected to occur in our present societies, it is better to decrease the suffering. The presence of consciousness is a necessary and sufficient condition for moral relevance. Species cannot suffer. Thus, crossing species barriers is not a morally relevant issue.

Preventing animal suffering is not the only important element in Rollin's concept of animal welfare. It also is important to make animals happy in a positive sense (augmenting animal happiness involves satisfaction of the telos): 'Well-being involves both control of pain and suffering and allowing the animals to live their lives in a way that suits their biological nature.' Here we see a combination of the subjective experience approach and the natural living approach to animal welfare. Both Singer and Rollin represent a utilitarian approach to the zoocentric framework, weighing human versus animal benefits according to a principle of proportionality. The decision is then based on what gives the best results (cost–benefit analysis), for example the least suffering or the greatest fulfilment of individual interest. This means that the greater the animals' suffering, the greater the human benefits must be in order to justify the action.

3. The biocentric framework

In this framework not only sentient beings should be morally considered, but all living beings. In the most common interpretations of this framework 'integrity' is an important concept. This goes beyond the subjective experience approach to animal welfare. Mutilations of animals and the genetic modification of animals are seen as a violation of the animal's integrity. The concept of integrity refers to the 'wholeness' of the animal, which as a member of a particular species has a species-specific nature. The natural living approach to animal welfare fits into this framework.

A typical example of this way of theorizing we find in the work of Taylor (1984, 1985). Taylor introduced the concept of 'inherent worth' (comparable to the concept 'Würde der Kreatur' as used in Germany and Switzerland), which he defined as follows:

> the value something has simply in virtue of the fact that it has a good of its own. To say that an entity has inherent worth is to say that its good [welfare, well-being] is deserving of the concern and consideration of all moral agents and that the realization of its good is something to be promoted or protected as an end in itself and for the sake of the being whose good it is
> (1984, p. 151)

The domain of morally relevant natural entities is widened to all animals and also plants, thus all living things (teleological centres of life). Plants, Taylor says, do not have interests such as sentient animals, but we can say that something contributing to their good is of interest to them. The basic emotive attitude involved here is man as a partner to all living beings. As far as the normative dimension is concerned, we see a shift from the utilitarian to a deontological approach. In a deontological approach the act itself is judged by whether or not it conflicts with a normative principle such as 'respect for life'. 'We ought to respect life' then becomes a prima-facie ethical norm.

4. The ecocentric framework

The biocentric framework just described applies to individual organisms; species or ecosystems are not considered to have moral status. Only individual organisms can be seen as partners. In an ecocentric theory, in contrast, populations, species and ecosystems deserve moral consideration, while individual organisms are subordinate. In this framework, the killing of individual animals is not a problem as long as the survival of the species is not endangered. This is different from a biocentric ethic, where killing an animal may be seen as the final destruction of the integrity of the organism. The natural living approach to animal welfare fits into this framework when it is defined at the level of populations (or herds). An example is someone who pleads for grazing of cattle for ecological reasons, but who sees no problems with dehorning cows or castrating bulls. From an ecological point of view, it is difficult to defend the biocentric position of no killing. In an ecosystem, organisms are continuously being replaced, and from an evolutionary point of view, dynamics within populations are necessary for evolution.

In some interpretations of the ecocentric view, human beings are excluded. The ecologist Colwell (1989) is one example, stating that the human position in the ecosystem is ecologically and evolutionarily 'unnatural'. Human intervention in nature leads to a degradation (devaluation) of the intrinsic value of species. Natural (wild) species have intrinsic value, because they are essentially irreplaceable and independent of human will. As a consequence, individual organisms belonging to a certain species have no intrinsic value, and domestic animals or plants are excluded because they are not 'natural'. Regarding genetic modification of already domesticated species, Colwell (1989) sees 'no ethical justification for any bar on genetic alteration of domesticates, by whatever technical means'. Verhoog (1992) has criticized this radical dualism between wild and domesticated species. He suggests that domestication can better be described as a gradual process of 'denaturalization'.

In other ecocentric frameworks, it is an explicit purpose to include man in the ecosystem (Zweers, 2000). In these frameworks, man should foster an attitude of being a participant within the whole ecosystem, not a dominant ruler. Other species also have a right to existence.

Values in Organic Farming

Since values and specific ethical frameworks are involved in people's conception of animal welfare, it is reasonable to ask which of these relate to organic farming and what the implications are for the conception of animal welfare in organic farming. These issues have been raised by Alrøe *et al.* (2001), Lund and Röcklinsberg (2001), and Verhoog *et al.* (2003). The principles for organic farming suggested by DARCOF (2000) also relate to this discussion.

DARCOF (2000) suggests three basic normative principles of organic farming as a means to guide decisions, for example on new research, use of new technologies and development of organic rules. In Denmark, there has been a comprehensive discussion of this ethical question in connection with a major new organic research effort. Three related principles have crystallized in the discussion: the cyclical principle, the precautionary principle and the nearness principle. So far, these principles have been used in an analysis of the acceptance of different technologies in organic farming and as a basis for making suggestions on future developments.

- *The cyclical principle* concerns how to interact with nature. It says that organic food cycles should emulate and benefit from nature's systems and cycles, fit into them and help sustain them. This is the oldest and most established organic principle. Related concepts are the ecological principle and the idea of naturalness.
- *The precautionary principle* concerns how to make decisions on changes in technology and practice. It says that action should be taken to prevent harm, even if there is no conclusive scientific evidence that this harm will occur. The principle also calls for the active promotion of cleaner, safer technologies and comprehensive research to detect and reduce risks.
- *The nearness principle* concerns how to learn and communicate. It says that possibilities for personal experience and close contact among consumers, producers, researchers and other organic actors should be created and maintained. All relevant actors should be encouraged to take part in the development of organic agriculture. This participation should be facilitated by promoting transparency and cooperation in the production and communication processes in the organic food cycles.

In interdisciplinary studies in Denmark, the concepts of harmony, naturalness, freedom of choice and care emerged as central to the understanding of animal welfare in organic farming (Alrøe *et al.*, 2001). The systemic conception of agriculture, which emphasizes the interaction between human and nature, is found to be fundamental to an examination of animal welfare in organic farming. The central ecological notion of 'harmony' refers to the interplay between the farm and the environment (harmony with nature), among the different elements of the farm, and among the individual animals in the herd, and to the integrity of an organism. This systemic ('holistic') ecological view is considered to be characteristic for organic farming. For instance, the Danish Association for Organic Farming states that one of the aims for organic farming is 'to do everything possible to ensure that all living organisms ... from micro-organisms to plants and animals, become "allies".' (quoted by Alrøe *et al.*,

2001) 'Naturalness' concerns the conditions in the production system for expressing natural behaviour and for natural reproduction and growth. 'Freedom of choice' is an element in the expression of natural behaviour that concerns the individual and dynamic preferences of the animals. 'Care' is understood as the counterpart of naturalness, which expresses the special responsibility that humans have towards domestic animals (contrary to wild animals).

Lund and Röcklinsberg (2001) identify important values in the IFOAM principles and standards of organic farming. They recognize the difficulties in IFOAM never having explicitly identified the core values behind the principles, and in organic farming being a dynamic movement, with the IFOAM standards having developed over time (since 1980). Nevertheless, when organic farming is compared with conventional farming, several basic differences in the relation between human and nature have been formulated in the literature. Lund and Röcklinsberg distinguish three core values related to the understanding of the animal welfare conception in organic farming systems. The first two are:

- Aim for a holistic view, in contrast to a mechanistic and reductionistic approach to the understanding of nature and the way of living in and with nature.
- Aim for sustainability (social, economic and ecological).

Since these two are considered to motivate welfare only indirectly (inasmuch as welfare is important for sustainability and the functioning of the whole ecosystem) the need arises to extract yet another core value that can be used more directly as an argument for the far-reaching animal welfare aims that always have been present in organic farming. Lund and Röcklinsberg suggest that it can be based on the understanding of interdependence and deep respect for other living entities in nature, as expressed in the bio- and ecocentric frameworks. They formulate this third core value as:

- Respect for nature

Thus, farm animals are valued as fellow members of the biotic community and as having some kind of intrinsic value, independent of their instrumental value in agricultural production. Another consequence is that solutions to problems should be sought that mimic nature as much as possible.

The use of the agroecosystem as a concept for agriculture implies that organic farming is a form of culture in which the intrinsic value of nature is respected and the wisdom of nature is a source of inspiration, made concrete in the use of ecological principles, natural substances, etc. (Verhoog *et al.*, 2003). It implies that technical interferences in and control over nature are restrained in order to develop an ecologically based (thus not high-tech) sustainable relation between man and nature.

They conclude that elements of all the ethical frameworks are included in the 'ethos' (the system of norms and values) of organic farming. They therefore characterize the ethical framework of organic farming as 'holocentric'. Although we cannot really speak of an anthropocentric element, it cannot be denied that the agroecosystem is meant to feed human beings. The zoocentric element is clearly present because subjective feelings of individual animals are taken into account, because they are part of the characteristic nature of vertebrate animals. That the biocentric element is very important is clear from the statement that 'all living organisms should become allies' (Danish Association for Organic Farming, quoted by Alrøe *et al.*, 2001). In the integration of nature and culture, we can find the ecocentric attitude of the human being a participant in the larger ecosystem.

Verhoog *et al.* (2003) came across three different approaches within organic farming:

- The no-chemicals approach.
- The agroecological approach.
- The integrity approach.

The *no-chemicals approach* is prominent in official legal standards. It is a negative approach in the sense that organic agriculture is distinguished from conventional because no chemical pesticides, no artificial fertilizers and no GMOs are permitted. 'Artificial' inputs should be replaced with more 'natural' ones, for example biological pest control, organic manure or mechanical weed control. The approach often is the first step in the conversion process. This way of thinking is often very conventional.

The *agroecological approach* reflects the belief that organic farming should be more than the no-chemicals approach, and that a more fundamental change in the way of thinking of problems is needed. Contrary to conventional farming that tries to become independent of nature, exerting control by technology in an artificial environment, organic farming tends to integrate agricultural activities into nature (defined as an ecological system) and agricultural practice is modelled as a complex, sustainable and balanced agroecosystem. After conversion, farmers often start to think in a more ecological way, looking for the broader context of problems. Terms like *closed system, mineral cycles, self-regulation* and *biodiversity* are important keywords in this approach.

The *integrity approach* is a consequence of the integration of nature into the agrocultural system. Whereas in a dualistic view of the relation between culture and nature, nature is seen as a material object without value in itself, a non-dualistic view includes such a valuational aspect. An interview study by Verhoog *et al.* (2003) found that many organic farmers imply this valuational dimension when they speak about the 'natural'. The term 'natural' here refers to taking into account the characteristic nature of plants, animals, humans and ecosystems. It manifests itself, for example, as respect for the integrity of life, of the agroecosystem, and of human needs (including social and economic integrity). This is the result of an inner process of involvement with natural entities' ways of being. Farmers begin to realize that the way they see problems and solutions is connected with their personal attitude and relationship with either the soil or the cultivated plants or animals. They experience that the farming system is more than merely a complex ecological mechanism and more than the sum of the parts. This feeling is also present in relation to the plants or animals they take care of. They develop a respect for the wholeness, harmony or identity of a living entity based on a personal involvement with the life of plants or animals.

Animal Welfare in an Organic Context

The concept of animal welfare refers to one or several aspects of the animal's quality of life. It thus makes sense to make the choice of animal welfare definition reflect the basic values in the farming system in which it is to be used.

Lund (2000) suggests the natural behaviour approach as the focus of the animal welfare concept in the organic context. She also asks if this approach is enough, or if additional aspects should be included in order to satisfy the demand for a holistic view that is basic to organic farming.

Lund and Röcklinsberg (2001) elaborate this question further. They state that the animal welfare concept should mirror the values in the agricultural system in which it is to be used. They conclude that the animal welfare concept should be broad and include multiple components, and suggest that animal welfare should be considered on several systemic levels, in accordance with the systemic view favoured by proponents of organic agriculture. Natural behaviour (understood as species-specific behaviour, feed, and environment) is seen as a central feature on the individual level and on the level of the agroecosystem. This should be interpreted so that management and environment must offer functional feedback systems, corresponding to those present in nature and to the animals' needs. On the next level, respecting animal integrity becomes an important issue. On higher systemic levels, individual welfare can be related to the herd or farm, where the focus is on breeding strategies and management systems. For a complete assessment, higher levels also need to be considered.

The conception of animal welfare formulated by these authors is not basically different from the view presented by Alrøe *et al.* (2001):

- Animal welfare depends on the greatest possible accord between the innate nature of the animal and the conditions provided. When

problems occur, solutions should primarily be sought at the systemic level (adapting the system to the animals instead of adapting the animals to the system). This is a basic principle underlying organic livestock health plans, as discussed in Chapter 12.
- The animal's species-specific (characteristic) nature and integrity should be respected in breeding and reproduction as well as in operations on individual animals.
- The animal's quality of life increases with a wider opportunity for the expression of natural behaviour. This goes beyond the satisfaction of the animal's physiological and behavioural needs, and beyond negative definitions of animal welfare as the absence of suffering.
- The consequences of the (ecological) natural living approach may conflict with the traditional emphasis on absence of suffering (prevention of aggressive behaviour, reduced risks of illness, etc.) and with other organic goals, such as reduced pollution of the environment by grazing herds. Therefore the emphasis on natural behaviour may put greater demands on management and human care.
- The organic food system can be sustained only if there is a shared perception of animal welfare among consumers, farmers and experts. The popular understanding of animal welfare can be taken into account by including experiential forms of assessment besides scientific assessments. (Various approaches to assessment of animal welfare are compared in Chapter 9.) Such forms of assessment are closer to the immediate perception of animal welfare made by ordinary people.

Verhoog *et al.* (2003) also underline the importance of natural behaviour within the context of the agroecosystem, and the concept of animal integrity related to the animal's species-specific nature. They conclude that the value of naturalness can be an important guiding value for the future of organic farming only when a broad conception of the natural is taken, including the no-chemicals approach, the agroecological approach and the integrity approach. With respect to animal welfare, this means that it is not enough that the animals get organic feed and can fulfil their needs in a minimal fashion: they should also be able to express their natural behaviour in a balanced agroecosystem. Finally, the harmony and integrity of the whole ecosystem, with the plants, animals and human beings living in it, should be taken into account. Together they are the 'ethos' of organic farming, not yet to be interpreted juridically as strict standards, but ethically, as a source of inspiration and self-reflection. It should be an ideal that is continually reflected upon within the organic movement. How it is to be interpreted within a specific situation always has to be a free moral choice for the individual farmer, but the farmer must also be willing to defend it in a dialogue with others in the organic movement.

Dilemmas

The great challenge for organic farming is how to integrate the elements of the various ethical frameworks into a balanced holocentric framework. Conflicts may arise between the different values in organic farming (Alrøe *et al.*, 2001; Lund and Röcklinsberg, 2001); the NAHWOA Recommendations begin with a call 'to formulate a philosophical definition and basis for animal welfare' to help resolve such conflicts.

Some examples may be helpful as an illustration:

The pig in the Dehesa system

The so-called 'Dehesa', an extensive agroforestry (oaks) and pasture livestock system including free-range Iberian pigs, is a sustainable agroecosystem existing since Roman times on the Iberian peninsula (Trujillo and Mata, 2000). However, to protect the soil and the trees, the pigs' noses are ringed. (For further details regarding the Dehesa system, see Chapter 3.) This is a case of conflicting sets of values: on the one hand the welfare (in the sense that the pig is being prevented from performing a basic behaviour) and the integrity of the pig, and on the other hand the conservation of a sustainable agroecosystem and preservation of the cultural values inherent in an ancient agricultural system.

For an ethical evaluation, it is important to consider alternative solutions (for a discussion of the effects of nose ringing, see Chapter 8). The actual decision can then be made, either by evaluating the consequences of each alternative, or by establishing a priority (ranking) of ethical principles and applying them to the alternatives. In this particular case, one could consider nose ringing to be the best solution, given the special ecological and historical situation. The argument could be that ecological and cultural values should be ranked higher than the interests of the pig, especially since the effect of nose ringing on animal welfare is debated.

An animal rights activist would probably decide that respect for the integrity of the individual animal must have the highest priority. This person would then rank the ethical principles differently (i.e. the principle of respecting 'animal integrity' would rank higher than those of respecting 'culture' or 'ecology'). The conclusion would then be that a system such as the Dehesa must be changed in favour of the pigs, or that pigs should not be kept there at all.

Castration of fattening bulls

The organic standards require that animals should be put on pasture in summer, but it is extremely difficult to keep sexually mature bulls fenced if there are cows in oestrus nearby. Many organic farmers therefore prefer to keep such bulls in small, well-fenced outdoor enclosures rather than letting them graze large areas. Castration would thus allow the bulls a more natural behaviour and in this sense a better quality of life, and would also increase the safety of the stockperson. This is an example of a conflict between the principles of allowing the animal a more natural life (well, at least certain aspects of it) and protecting the safety of the stockperson versus animal integrity.

Free-range poultry

Organic poultry are required to have access to outdoor runs. This is a result of the demand for naturalness, aiming to allow the animals a more natural behaviour. It also involves other values related to the ecological and cultural role of farm animals in the landscape and on the farm. However, the demand for natural behaviour may conflict with health and welfare aims. For example, the effects of an outbreak of cannibalism will be much more severe in free housing systems, where many more birds will be exposed, than in a cage system. If the most important principle is to minimize suffering (a common utilitarian view that is expressed in the subjective experience definition of animal welfare) or maximize health (as expressed in the biological functioning definition), keeping the hens outside probably would not be required. But the leading principle for organic farming, expressed in the standards, is 'natural behaviour'. Free range poultry will also be exposed to parasites, and in some cases to predators, although the risk of cannibalism would be reduced (see Chapter 12 for a discussion of this point, and Chapter 8 for a discussion of the related dilemma of beak trimming as a way of dealing with feather pecking and cannibalism). Outdoor keeping may also conflict with environmental concerns, since indoor keeping causes less damage to the soil and less leakage of nutrients. This is an example of the inherent conflict between the ecocentric and biocentric (or zoocentric) approaches, where the former give priority to system needs (i.e. conservation of the environment) while the latter two focus on the needs of individuals. The conflict increases when 'natural' is interpreted as 'wild' (as little human interference as possible, or even none), rather than as having respect for the species-specific nature of the animals. Domestic animals also have a species-specific nature. Organic agriculture wants to integrate nature into the agricultural system, not to separate the two.

It is clear that here again we need a delicate balancing of different values and ethical principles. The farmer has a big responsibility (care and management), but so has the consumer of the products. An acceptable solution may imply smaller herd size both to give the farmer more time to care for the animals and for environmental reasons; the resulting economic costs must be shared.

Future Challenges

The animal welfare conception frames how organic animal husbandry should be organized, but it may also challenge other central ideas and values in the organic movement, as illustrated by the dilemmas presented in this chapter. This is where we need ethics: normative ethics can help us rank values and give guidelines for our actions. In order to make progress in any area of organic farming, but perhaps particularly in animal husbandry, it is important to continue to deepen the discussion regarding what values organic farming should be based on and should promote. The early organic movement especially was characterized by such discussions (Christensen, 1998), but on the professional academic level this has begun only recently. It is valuable for philosophers to engage in the debate. Philosophers can make important contributions by clarifying and systematizing the arguments. They also can question supposedly self-evident principles and beliefs, pointing out the presuppositions underlying them. However, it is as crucial for the philosophers to interact with the organic farmers as it is for the organic farmers to involve philosophers in these discussions.

One aim of this chapter has thus been to make the organic movement aware of values involved in organic farming and animal welfare, allowing choices to be made more consciously. The value-laden and moral aspects of animal welfare need to be considered, and they cannot be treated independently from the general values of organic farming. One may conclude that there is a basic difference in value orientation between the organic movement and conventional agriculture (however, this alternative orientation may not be shared by all organic farmers!). Thus, the organic answer to the question raised in the introduction – 'what is the essence of a good pig life?' – focuses on natural behaviour. In turn, this is mirrored in the organic understanding of the animal welfare concept. Similarly, human obligations to farmed animals become different if they are based on a fundamental respect for the wholeness, harmony or identity of the animal as a living creature rather than on purely economic interests.

Another aim has been to spotlight the fact that animal welfare is an integral part of organic farming. It can be deduced from organic values. The organic movement is constantly developing, which poses challenges. The growth in consumer demand and the establishment in several countries of subsidies for organic production have created economic incentives for conversion. New categories of farmers and traders have been influenced to take an interest in organic production, although they might not be as idealistic or accepting of the same values as farmers in the early organic movement; this development has come in for considerable criticism (Lund and Röcklinsberg, 2001). Organic farming is facing new challenges, with many different competing interests, each demanding

Askham Bryan College
LIBRARY BOOK

attention and priority when setting standards. If organic animal husbandry is to grow further, conflicts with other areas of interest as well as within livestock production must be solved in a way that is transparent and makes sense to farmers as well as consumers. Here the ethical discussions have an important task. Hopefully, the clarification of values can contribute to intelligent and practical solutions to emerging problems, such as by shifting the focus to reconciling conflicting interests on a systems level rather than only on the individual level. A broader approach to problems will also make a wider range of solutions available. Perhaps the organic approach to animal welfare can also shed new light on the animal welfare concept as such. The values of animal welfare pose a challenge to organic farming and the values of organic farming pose a challenge to the discussion of animal welfare.

References

Allen, C. and Bekoff, M. (1997) *Species of Mind. The Philosophy and Biology of Cognitive Ethology*. The MIT Press, Cambridge, Massachusetts.

Alrøe, H.F., Vaarst, M. and Kristensen, E.S. (2001) Does organic farming face distinctive livestock welfare issues? A conceptual analysis. *Journal of Agricultural and Environmental Ethics* 14, 275–299.

Appleby, M.C. (1999) *What Should We Do about Animal Welfare?* Blackwell, Oxford.

Bentham, J. (1789 [1970]) *An Introduction to the Principles of Morals and Legislation.* The Athlone Press, London.

Bermond, B. (1997) The myth of animal suffering. In: Dol, M., Kasanmoentalib, S., Lijmbach, S., Rivas, E. and Van den Bos, R. (eds) *Animal Consciousness and Animal Ethics*. Van Gorcum, Assen, The Netherlands, pp. 125–143.

Broom, D.M. (1991) Animal welfare: concepts and measurement. *Journal of Animal Science* 69, 4167–4175.

Burkhardt, R.W. Jr. (1997) The founders of ethology and the problem of animal subjective experience. In: Dol, M., Kasanmoentalib, S., Lijmbach, S., Rivas, E. and van den Bos, R. (eds) *Animal Consciousness and Animal Ethics.* Van Gorcum, Assen, The Netherlands, pp. 1–13.

Christensen, J. (1998) *Alternativer – natur – landbrug*. Akademisk forlag, Viborg.

Colwell, R.K. (1989) Natural and unnatural history: biological diversity and genetic engineering. In: Shea, W.R. and Sitter, B. (eds) *Scientists and Their Responsibilities*. Watson Publishing International, Canton, Massachusetts, pp. 1–40.

DARCOF (2000) *Principles of Organic Farming*. Discussion Document prepared for the DARCOF Users Committee. Danish Research Centre for Organic Farming. [Available on-line at http://www.darcof.dk]

Darwin, C. (1890) *The Descent of Man, and Selection in Relation to Sex*, 2nd edn. John Murray, London.

Descartes, R. (1637 [1985]) Discourse on the method of rightly conducting one's reason and seeking the truth in the sciences. In: *The Philosophical Writings of Descartes*, Vol. I. Cambridge University Press, Cambridge.

Dol, M., Kasanmoentalib, S., Lijmbach, S., Rivas, E. and van den Bos, R. (eds) (1997) *Animal Consciousness and Animal Ethics*. Van Gorcum, Assen, The Netherlands.

Eibl-Eibesfeldt, I. (1975) *Ethology: The Biology of Behavior*, 2nd edn. Holt, Rinehard, and Winston, New York.

Fraser, D. (1995) Science, values and animal welfare: exploring the 'inextricable connection'. *Animal Welfare* 4, 103–117.

Fraser, D. (1999) Animal ethics and animal welfare science: bridging the two cultures. *Applied Animal Behaviour Science* 65, 171–189.

Fraser, D., Weary, D.M., Pajor, E.A. and Milligan, B.N. (1997) A scientific conception of animal welfare that reflects ethical concerns. *Animal Welfare* 6, 187–205.

Griffin, D. (1976) *The Question of Animal Awareness: Evolutionary Continuity of Mental Experience*. Rockefeller University Press, New York.

Lund, V. (2000) Is there such a thing as 'organic' animal welfare? In: Hovi, M. and Trujillo, R.G. (eds) *Diversity of Livestock Systems and Definition of Animal Welfare. Proceedings of the Second NAHWOA Workshop, Córdoba, 8–11 January 2000*. University of Reading, Reading, UK, pp. 151–160.

Lund, V. and Röcklinsberg, H. (2001) Outlining a conception of animal welfare for organic farming systems. *Journal of Agricultural and Environmental Ethics* 14, 391–424.

Rollin, B.E. (1989) *The Unheeded Cry: Animal Consciousness, Animal Pain and Science*. Oxford University Press, Oxford, UK.

Rollin, B.E. (1995) *The Frankenstein Syndrome: Ethical and Social Issues in the Genetic Engineering of Animals*. Cambridge University Press, Cambridge.

Sandøe, P. and Simonsen, H.B. (1992) Assessing animal welfare. Where does science end and philosophy begin? *Animal Welfare* 1, 257–267.

Simon, S.B., Howe, C.W. and Kirschenbaum, H. (1978) *Values Clarification: A Handbook of Practical Strategies for Teachers and Students*. Hart Publishing Company, New York.

Singer, P. (1975) *Animal Liberation*. The New York Review, New York.

Sorabji, R. (1993) *Animal Minds and Human Morals: the Origin of the Western Debate*. Cornell University Press, Ithaca, New York.

Stafleu, F.R., Grommers, F.J. and Vorstenbosch, J. (1996) Animal welfare, evolution and erosion of a concept. *Animal Welfare* 5, 225–234.

Tannenbaum, J. (1991) Ethics and animal welfare: the inextricable connection. *Journal of the American Veterinary Medical Association* 198, 1360–1376.

Taylor, A.M. (1999) *Magpies, Monkeys, and Morals: What Philosophers Say about Animal Liberation*. Broadview, Peterborough, Ontario.

Taylor, P.W. (1984) Are humans superior to animals? *Environmental Ethics* 6, 149–160.

Taylor, P.W. (1985) *Respect for Nature: a Theory of Environmental Ethics*. Princeton University Press, Princeton, New Jersey.

Tinbergen, N. (1942) An objectivistic study of the innate behaviour of animals. *Bibliotheca Biotheoretica* 1, part 2, 39–98.

Trujillo, R.G. and Mata, C. (2000) The Dehesa. An extensive livestock system at the Iberian peninsula. In: Hovi, M. and Trujillo, R.G. (eds) *Diversity of Livestock Systems and Definition of Animal Welfare. Proceedings of the Second NAHWOA Workshop, Córdoba, 8–11 January 2000*. University of Reading, Reading, UK, pp. 50–61.

Verhoog, H. (1992) The concept of intrinsic value and transgenic animals. *Journal of Agricultural and Environmental Ethics* 5, 147–160.

Verhoog, H., Matze, M., Lammerts van Bueren, E. and Baars, T. (2003) The role of the concept of the natural (naturalness) in organic farming. *Journal of Agricultural and Environmental Ethics* 16, 29–49.

Wemelsfelder, F. (1993) Animal boredom. Towards an empirical approach to animal subjectivity. MSc Thesis, The University of Leiden.

Wemelsfelder, F., Hunter, E.A., Mendl, M.T. and Lawrence, A.B. (2000) The spontaneous qualitative assessment of behavioural expressions in pigs: first explorations of a novel methodology for integrative animal welfare measurement. *Applied Animal Behaviour Science* 67, 193–215.

Wolpert, L. (1993) *The Unnatural Nature of Science*. Faber and Faber, London.

Zweers, W. (2000) *Participating with Nature*. International Books, Utrecht, The Netherlands.

6

Understanding Animal Behaviour and its Use in Organic Animal Husbandry

Vonne Lund[1] and Dan Weary[2]

[1]Department of Animal Welfare and Health, Swedish University of Agricultural Sciences, PO Box 234, SE-532 23 Skara, Sweden; [2]Animal Welfare Program, Faculty of Agricultural Sciences, The University of British Columbia, 2357 Main Mall, Vancouver, BC, V6T 1Z4, Canada

Editors' comments

In the organic understanding of animal welfare, natural behaviour is a prerequisite for a good animal life. The organic standards emphasize the requirement to respect the animals' behavioural needs. This makes ethology, the science of animal behaviour, an important tool for implementing animal welfare in practical farming. A basic understanding of ethology is indispensable for the development of husbandry systems in line with organic ideals. The challenge taken on by the authors of this chapter is to introduce the science of ethology and discuss its role as a tool for improving animal welfare in organic animal husbandry. The focus is how we study animal behaviour and the general concepts that have emerged from such work.

Introduction

In many ways, good farmers have always required a good understanding of the behaviour of the animals with which they work. For example, knowledge of the animals' natural behaviour helps in herding and moving them, in identifying those that are ready for breeding and in identifying others that may require treatment for an illness or injury. The study of behaviour (ethology) may be particularly important in organic farming. This is because organic practices tend to rely less on technology and more on the farmer's direct contact with the animal for the purposes outlined above. Indeed, organic animal production aims to make all aspects of

© CAB International 2004. *Animal Health and Welfare in Organic Agriculture* (eds M. Vaarst, S. Roderick, V. Lund and W. Lockeretz)

livestock production more natural, and expressly encourages that animals be kept in a way that lets them express their natural behaviour. Thus one of IFOAM's 17 principal aims for organic farming is 'to give all livestock conditions of life with due consideration for the basic aspects of their innate behaviour' (IFOAM, 2000).

Naturalness in general, and natural behaviour in particular, are therefore of inherent value to those in organic animal production. However, the ability to express natural behaviour is also considered to be of instrumental value in assuring a high quality of life for animals; maintaining a good quality of life for farm animals is another general principle of organic animal husbandry. For example, the IFOAM principles of animal husbandry state that:

> management techniques in animal husbandry should be governed by the physiological and ethological needs of the farm animals in question. This includes: That animals should be allowed to conduct their basic behavioural needs. That all management techniques, including those where production levels and speed of growth are concerned, should be directed to the good health and welfare of the animals.
>
> (IFOAM, 2000)

Consumers also have high expectations regarding the welfare of animals in organic systems. For example, Swedish consumers believe organic livestock experience better animal welfare than animals in conventional farming (Holmberg, 2000), and a study of consumers from several European countries showed the same tendency (Harper and Henson, 2001). The Danish Ministry of Food, Agriculture and Fisheries (1999) states that:

> as one of the important tenets of organic farming is to provide all livestock with good living conditions in keeping with their natural behaviour and needs, it must be accepted that consumers naturally assume that organic livestock enjoy high standards of welfare.

Thus an improved knowledge of animal behaviour is important in organic husbandry for three reasons: (i) to solve immediate challenges in animal care and production; (ii) to understand what conditions on the farm are needed to allow animals to express their natural behaviour; and (iii) to improve the welfare of farm animals. In this chapter, we provide examples of how ethological research can contribute to each of these aspects.

Improved Husbandry through Improved Knowledge of Animal Behaviour

In conventional farming, production methods have often been developed in response to technological and economic constraints, and the animals have more or less been expected to cope with the living conditions created in these systems. However, problems can occur when rearing animals under conditions for which they lack appropriate behavioural responses, or when animals have reponses that no longer function in the modern rearing systems (Fig. 6.1; Fraser *et al.*, 1997). A better understanding of behaviour can help create management and housing conditions that minimize these problems.

For example, one of the most serious welfare problems with pigs is the high rate of deaths and injuries of piglets from crushing by the sow. About 10% of piglets die during the first few weeks of life, many from crushing. Crushing is largely due to large litters: it is rare in the small litters (about five piglets) typical of the wild boar, but much more common in the litters of ten or more that are typical of modern domestic breeds. Thus with selection for larger litter sizes came the need for producers to develop ways of preventing deaths and injuries by crushing. Crushing is the result of both the piglet's behaviour in staying close to the sow, and the sow's in lying down on top of the piglet. The approach developed during the 1950s and 1960s was to use farrowing crates to constrain the sow's movements, but these crates have disadvantages for the sow.

A more sophisticated appreciation of the behaviour of sows and piglets can help to create management and housing systems that minimize crushing while allowing sows much more freedom of movement. For example, detailed behavioural studies have been used to identify body movements by sows that crush piglets when sows are kept in open pens, and to learn how changes in penning can help to prevent the most dangerous movements. In one study, time-lapse video recordings showed that when sows were housed in open pens they were particularly likely to crush piglets when they rolled from lying on their udder to lying on their side (Weary *et al.*, 1996a). Any physical structure that slows this rolling movement, allowing piglets time to escape, should reduce crushing. This

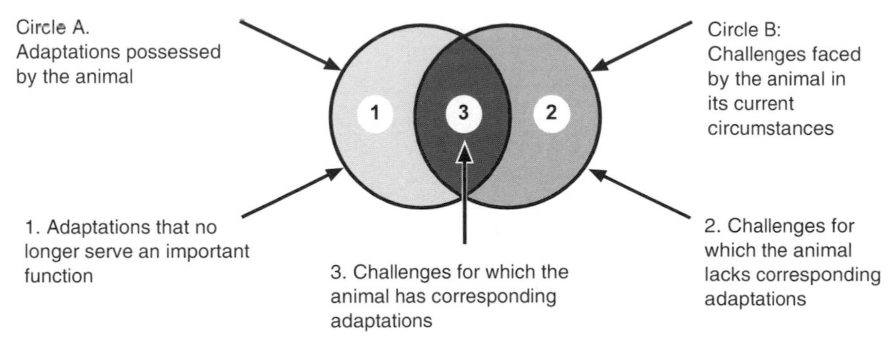

Fig. 6.1. Many of the welfare challenges in contemporary farming occur either because the animal has an adaptation that no longer can find a function in modern rearing systems or because the animal lacks adaptations to such systems (after Fraser *et al.*, 1997).

is achieved by the high level of confinement used in farrowing crates, but these results indicate that a simple modification to the design of pens or outdoor farrowing huts, such as adding bars or other impediments along the floor of the pen, should prevent or slow these dangerous rolling movements by the sow.

Other problems can be avoided by identifying behavioural adaptations that have no appropriate outlet in current rearing conditions, one example being sucking in milk-fed calves. The young calf is highly motivated to suck, as under natural conditions calves rely on this behaviour to obtain almost all their required nutrients (de Passillé, 2001). Also, on organic farms calves are often separated from the cow soon after birth, and are often observed sucking other calves in the same or adjacent pens. One might imagine that this motivation to suck is related to hunger, but detailed behavioural studies have shown that calves actually are more motivated to suck after getting milk than before, and have little motivation to suck outside of meal times. This work indicates that efforts to reduce cross-sucking in calves should be focused on times when calves are being fed milk. Nipple feeders allow calves to express this natural sucking behaviour while consuming their milk meals, and their use reduces cross-sucking. Cross-sucking can still be a problem when calves must compete for access to nipples, but this can be reduced by providing more nipples or using enclosed feeding stations for calves that allow them to use the nipple unmolested by group mates. Further benefits in calf growth and behaviour can be achieved by allowing calves to remain with their own cow (Flower and Weary, 2001) or another 'suckler' cow that nurses several calves, as discussed in Chapter 7 (Loberg and Lidfors, 2001). Again, a better understanding of the behaviour can help create management and housing conditions that minimize the problem.

Early rearing conditions and learning may also be important in this context. Chapter 7 gives a detailed example: the problems such as feather pecking and cannibalism that arise when laying hens in houses with perches did not learn how to use perches when they were young.

Livestock producers commonly use behaviour to identify sick animals. In extreme cases, such diagnoses can be straightforward, such as using obvious gait abnormalities to identify hoof and leg injuries in cattle. However, much more subtle behavioural differences can also be useful, especially in early diagnoses and in identifying pre-clinical conditions. For example, several sensitive gait assessment scales have been developed that successfully identified animals for early treatment (Table 6.1; Sprecher *et al.*, 1997).

Similarly, any good herdsman will know that a drop in food intake indicates illness, but with group-fed animals such changes can be difficult to monitor. Now, simple electronic transponders attached to an ear tag allow much more sophisticated monitoring of feeding behaviour, by telling researchers and producers which animals are eating and which are

Table 6.1. Use of behaviour to evaluate lameness in dairy cattle.

Gait score	Clinical description	Assessment criteria
1.0	Sound	*Smooth and fluid movement.* Flat back. No swinging out. Legs bear weight equally and joints flex freely. Hind claws land on or in front of fore-claw prints (tracking up). Head remains steady.
2.0	Imperfect gait	*Ability to move freely not diminished.* Back is flat or mildly arched. Minimal swinging out. Legs bear weight equally but joints show slight stiffness. Hind claws do not track up perfectly but shortened strides are uniform. Head remains steady.
3.0	Lame	*Capable of walking but ability to move freely is compromised.* Arched back. Swinging out. Legs bear weight equally but a slight limp in one limb can be seen. Joints show signs of stiffness. Hind claws do not track up and strides are shortened. Head remains steady.
4.0	Moderately lame	*Ability to move freely is diminished.* Arched back. Swinging out. Reluctant to bear weight on at least one leg but still uses that leg to walk. Strides are short, hesitant and deliberate. Stiff joints. Head bobs as animal walks.
5.0	Severely lame	*Ability to move is severely restricted.* Animal must be strongly encouraged to stand and/or move. Extreme arched back. Swinging out. Inability to bear weight on one or more legs. Strides are short, hesitant and deliberate. Stiff joints. Head jerks as animal walks.

Adapted from Sprecher *et al.* (1997).

not. Research using such equipment has shown that reduced time at the feed trough can be used as an early indicator of morbidity in beef cattle. Such techniques identify sick steers 4 days before they can be identified as such by experienced herdsmen (Quimby *et al.*, 2001). Even when such equipment is not practical for producers to use, research with these tools allows ethologists to craft management techniques that minimize the incidence of disease.

As the above examples illustrate, good animal producers must to some extent be amateur ethologists – they need the patience to watch their animals and the skill to properly interpret what they see. However, their work can also be helped greatly by professional ethologists whose research using more sensitive behaviour methods can identify better ways to house and manage farm animals.

Behaviour of Wild Animals

For the organic movement, the ability of farm animals to express their natural behaviour is considered to be of inherent value. One way to understand the natural behaviours of domestic animals is by studying their wild counterparts (e.g. Collias and Collias, 1996). The wild ancestors of some farmed species still exist, while others have become extinct. The latter species may still be studied as feral populations.

Effects of domestication on behaviour

Considering that most of the modern genera of mammals first appeared on earth some 35 million years ago, the period of domestication can be considered no more than a short moment in history (Stricklin, 2001; Price 2002). Most of our traditional farm animals probably were domesticated some 8000–12,000 years ago (Clutton-Brock, 1981; see also Chapter 2). Moreover, intensive breeding under conditions that are very different from the wild has taken place only during the last 50 years. Not surprisingly then, studies of pigs and fowl have shown a remarkable resemblance between the behaviour of domestic animals and their wild counterparts, with no losses or additions to the species' repertoires (Andersson, 2000). This remarkable stability may be explained by the fact that behavioural traits are often controlled by complex genetic systems. Behavioural changes during domestication are mostly quantitative rather than qualitative and result from changes in the threshold for responding to stimuli (Price, 1984, 1999). For example, compared to their wild ancestors, most domestic animals are less aggressive, reach sexual maturity earlier and show greater behavioural plasticity (Boice, 1973).

Some studies have shown that domestic animals can exhibit many of the same behaviours as their wild counterparts when kept under naturalistic conditions. For example, Stolba and Wood-Gush (1989) observed the behaviour of groups of domestic pigs in large forested enclosures, and catalogued an 'ethogram' of some 103 different behavioural elements. They found few differences between this ethogram and that from studies on the European wild boar, suggesting that domestication and artificial rearing had little effect on the pigs' rich behavioural repertoire.

Indeed, many domestic animals have retained their natural behaviours so well that escaped or released individuals have been able to establish free-living 'feral' populations in the wild (Baker and Manwell, 1981), as has happened for pigs, goats, dogs, buffalo, camels and chickens. Although not genetically identical to their wild ancestors, these feral animals often behave very similarly. For example, the social organization of feral domestic fowl is similar to that of the wild jungle fowl (Duncan *et al.*, 1978).

However, there are several examples of behaviours seriously altered by artificial selection. This is especially true for poultry, as intensive selection is facilitated by short generation intervals, many offspring and cheap rearing of breeding stock. Comparative genetic and behavioural studies of the red jungle fowl, the Swedish bantam (a breed that has not undergone strong selection for production traits) and hybrid white leghorns showed that behaviours that use a lot of energy, such as extensive foraging and social interactions, were less frequent in the leghorns than in the other two breeds, allowing these hens to reallocate resources to egg production (Schütz and Jensen, 2001). Interestingly, some behaviours have become more frequent during domestication, such as aggressiveness following mixing with new birds (Håkansson, 2003). Other studies have indicated that selection for lean growth in pigs has affected their behaviour, resulting in pigs that are more excitable (Grandin, 1994). The lean breeds are less active, more anxious (Shea-Moore, 1998) and more aggressive (Busse *et al.*, 1999), resulting in practical difficulties for handlers (Grandin and Deesing, 1998) and increased fear and fighting among pigs.

Using behavioural studies of wild or semi-wild animals to design alternative rearing systems

The knowledge gained from studies of animals living under wild or semi-wild conditions can be used when developing alternative rearing systems to better allow animals to express their natural behaviour. For example, studies of domesticated sows living in semi-wild conditions have demonstrated the importance of nest-building behaviour (Jensen, 1986). Even sows kept on concrete flooring spend hours engaged in rooting behaviour similar to that of nesting sows (Thodberg *et al.*, 1999). Many alternative production systems provide straw or other building materials so that sows can perform the natural behaviours and construct a nest.

Studies of wild counterparts have also suggested improved rearing practices for poultry. For example, males are typically considered superfluous in conventional poultry production, but studies of jungle fowl under semi-wild conditions have shown that roosters can play important roles. The dominant male defends the flock's territorial boundaries (Collias and Collias, 1996), guards the group against predators, and helps to maintain the social hierarchy (McBride *et al.*, 1969). More recent research has shown that incorporating roosters into flocks of hybrid laying hens significantly reduces the hens' fear and aggression (Odén, 2003). This positive effect disappears if the male–female ratio is too low – a minimum of one rooster per 50–100 hens is recommended. The calming effect is due mainly to social dominance, but roosters also facilitate the formation of sub-groups within the flock, similar to those preferred by the red jungle fowl.

Social Behaviour

Many of the most common welfare problems in domestic animals, such as cross-sucking, feather pecking and tail biting, stem from misdirected social behaviour. Conventional producers' common responses to these problems, such as individual pens for calves, beak trimming for hens and tail docking for pigs, are not allowed in the organic standards or are considered inappropriate by many organic farmers. Thus, improved production practices will often require special attention to aspects of social behaviour.

Given the choice, most farm animals would live in groups for at least parts of their lives, and with some exceptions such as breeding males and the sick, the IFOAM standards require that herd animals be housed in groups (IFOAM Basic Standards 2000, 5.1.5). In the wild, group living can provide many advantages for the individual, such as increased efficiency in detecting and acquiring food and better protection against predators, but there can also be disadvantages, such as increased exposure to parasites and pathogens (Mendel and Held, 2001). Thus, natural populations have optimal group sizes, and the communication and recognition abilities of the animals are probably adapted to this group size. Farming methods that take advantage of these more natural group sizes may therefore avoid some problems in social behaviour. Moreover, group life varies within and between species, and the group structure may vary with time of the year and developmental stage of the individual. For example, some feral cattle live in mixed groups composed of cows, heifers and young bulls, with older males living in separate groups, but during the mating season, breeding groups form that consist of one or more adult bulls and adult females (Daycard, 1990).

The signals used to mediate social interactions often require space or visual barriers, and problems can occur when animals are kept in ways that prevent these signals. Animals normally maintain a certain individual distance or 'social space' from their herd mates, but this varies depending upon circumstances. For example, hens stay approximately 3 m apart when walking but only 1.5 m apart when preening (Keeling, 1994); cattle stay about 4–10 m apart when grazing but only 2–3 m when lying down in pasture (Fraser and Broom, 1990). Space or visual barriers can be particularly important when animals compete for resources such as food. For example, barriers between places at the trough will reduce aggression among pigs (e.g. O'Connell *et al.*, 2002), as discussed in more detail in Chapter 7.

Allowing animals to use natural signalling behaviour can be important for more than just reducing aggressive interactions. One important class of signalling is between mother and young, and one of the most interesting examples comes from the domestic sow. Sows produce a series of grunts during the nursing episode, and the grunting rate increases just seconds before the milk 'ejection' that makes milk available for the piglets. These calls both attract piglets to the udder and help synchronize the piglets' change from nuzzling to sucking that needs to occur for piglets to get the milk (Fraser, 1980). In housing with loud background noise from ventilation fans, the sows' grunts can be masked, resulting in less synchronized nursing behaviour and poorer weight gains by piglets (Algers and Jensen, 1991).

Given the importance of social interactions, separation from social partners can be stressful for many farm animals. Good farmers are aware of this, and will take advantage of herding to move animals more calmly and efficiently. When one animal needs to be treated, it is often easier for handlers and less stressful for the animals if at least a few animals are kept together. Conventional producers repeatedly regroup animals to achieve production or management efficiencies associated with grouping animals by size or production level, but this regrouping can result in high levels of aggression and other problems (e.g. Robert and Martineau, 2001).

Some disruption of social bonds is inevitable, such as in weaning, and organic certification bodies set minimum weaning times based on the natural behaviour of the species (IFOAM Basic Standards 2000, 5.6.9). For example, piglets are naturally weaned at about 4 months (Jensen and Recén, 1989), although weaning really is a gradual process, with piglets beginning to nurse less often, consume solid food, and spend time away from the sow within 2–3 weeks after birth. The natural weaning age may not be ideal for either production or welfare reasons. For example, continued nursing slows the sow's return to oestrus, thus reducing the farm's production, and sows increasingly attempt to separate themselves from piglets or avoid nursings as the piglets age (Pitts *et al.*, 2002). There is some research indicating that the earliest allowed weaning age for piglets (40 days according to the EU standards) may work reasonably well for both sow and piglets (Andersen *et al.*, 2000). In some cases, there may be benefits to very early separation, before mother and young develop a strong social bond. This is one reason that some dairy farmers separate the calf from the cow within hours after birth. Indeed, cows and calves show a reduced response to separation if this is done early (Flower and Weary, 2001), although there can be production advantages associated with delayed separation (Chapter 7).

Natural Behaviours as an Asset in Production

In conventional farming, animal behaviour can cause problems. For example, under natural conditions, pigs and chickens spend most of the day looking for food, but such foraging behaviour is greatly reduced for intensively reared animals fed concentrated diets from a feeder. In the absence of natural outlets for foraging, components of this behaviour can lead to behavioural disturbances such as belly nosing and tail biting for pigs, and feather pecking for chickens. The organic requirement to provide animals with roughage helps to mitigate some of the problems related to foraging behaviour, but a challenge remains for production systems to use animal behaviour as an asset, rather than simply treating problems. The aim of organic farming is to develop viable agroecosystems, where the different parts support each other to create a better and more productive whole (Lund *et al.*, 2003). In an ecosystem, animals interact with other system components through their behaviour, often to their mutual benefit. For example, a squirrel hides acorns in the autumn to create a food hoard for the future, but the behaviour can also benefit the oak and other organisms in the forest ecosystem. To achieve these more holistic aims, organic producers will need to focus more on the capabilities of the animals with which they work, and not simply the animals' requirements (Andresen, 2000).

There are some practical examples of systems in which animal behaviour is utilized as an asset. Some traditional forms of animal agriculture, such as the Dehesa system in Spain (described in Chapter 3), harnessed natural behaviours in a way that was beneficial to both the producers and the pigs. Unfortunately, there has been little work on developing such solutions for more modern production systems, particularly for monogastric farm animals. In one example, researchers at the Swedish University of Agricultural Sciences have developed a system for integrating pigs into the crop rotation (Fig. 6.2; Andresen, 2000). In this case, the pigs' natural rooting behaviour is used for weed control and to till the land after crops have been harvested. From a crop production perspective, rooting tends to be as effective as mechanical tillage, but has the added advantages of allowing pigs to consume the residual forage and reducing environmental impact. As noted in Chapter 8, this approach also addresses the animal welfare dilemma concerning nose-ringing to prevent rooting.

There are also good examples of poultry rearing practices that take advantage of the birds' natural behaviour to the benefit of the system, such as a mixed raspberry and layer hen farm (6500 birds) in British Columbia (Reid, 2002). In this farm, fields are partitioned using cross-fencing and access runs so that the chickens' movements can be regulated according to the needs for crop production. The chickens access raspberry fields freely all year around, except from May to early June to protect the development of the primal canes, and again during the picking season. Whilst in the raspberry fields, the birds eat insects, weeds and foliage from the sides of the canes, thus eliminating the need for hand weeding between raspberry rows, and providing effective primal cane management in the raspberries. The tall raspberry plants also provide excellent cover for the birds, encouraging the animals to forage underneath the plants. Another example of a poultry system that makes good

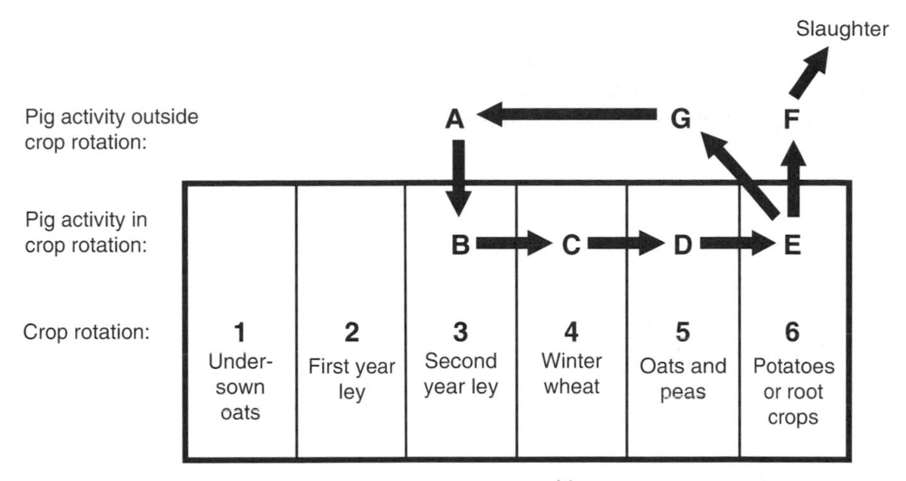

Fig. 6.2. Integration of pigs in the crop rotation at Ekhaga Organic Experimental Farm, Swedish University of Agricultural Sciences (Andresen, 2000; Paul Ciszuk, Swedish University of Agricultural Sciences, Uppsala, Sweden, 2003, personal communication).

In the Ekhaga system, sows farrow once a year in spring. Ideally they are transferred to forest grazing after farrowing. Alternatively, they are put on a ley pasture (A), although they don't perform much work at this time of year. After the harvest of the second year ley the pigs are 'put to work' in the crop rotation (early July). First they strip-graze a field of second year ley (B), which is sown with winter wheat after they are removed. After the wheat harvest they continue to the wheat fields, to eat crop residues and weeds (C). They are then moved to harvested fields of oats and peas (D), and finally they work their way through fields with potatoes or root crops such as turnips or carrots (E). During the last 2–3 weeks before slaughter (at about 7 months of age), fattening pigs are kept in a smaller pasture and fed a grain supplement in addition to roughage (F), while sows are moved to their winter shelter (G).

use of the hens' natural behaviour is the Hånsta Östergärde farm in Sweden, described in Chapter 3.

Using the Study of Animal Behaviour to Assess and Improve Animal Well-being

The animal welfare literature identifies three types of welfare concerns: about natural living (such as the ability to express natural behaviour); about biological functioning (such as maintaining good health); and about affective states (such as the negative states of pain and distress, and positive states of pleasure) (Fraser *et al.*, 1997). The study of animal behaviour can be used to assess and improve all these aspects of animal well-being.

For organic farmers, natural living concerns are of inherent interest

(Lund, 2002; also Chapter 5), but the ability to express natural behaviours can be of instrumental value to the extent that it promotes the animal's well-being as indicated by good biological functioning and positive affect. There is some research that supports such a link. For example, allowing young dairy calves to ingest their milk using a nipple rather than a bucket allows them to express their natural sucking behaviour and thus helps avoid the problem of cross-sucking (Fig. 6.3). In addition, sucking for milk also improves biological functions such as the release of digestive hormones (de Passillé, 2001). Sucking may also contribute to positive affective states (if the behaviour is pleasurable for the calf), and prevention of this highly motivated behaviour may lead to negative states (such as frustration).

However, just allowing animals to express natural behaviour does not guarantee their welfare. In fact, natural behaviour sometimes can diminish welfare. For example, loose housing for gestating sows allows more natural movements and social behaviour, but, depending upon how such systems are managed, loose housing can also increase injuries, pain and distress because of fighting between sows (Arey and Edwards, 1998).

Natural behaviours include activities that are adaptations to difficult conditions (Fraser *et al.*, 1997). The escape behaviour of a hen in response to a fox, and the wallowing by pigs when hot, are both natural responses to adverse conditions. However, it would not be in the best interest of these animals to expose them to these conditions just so they could express their natural responses. From a welfare perspective, a problem

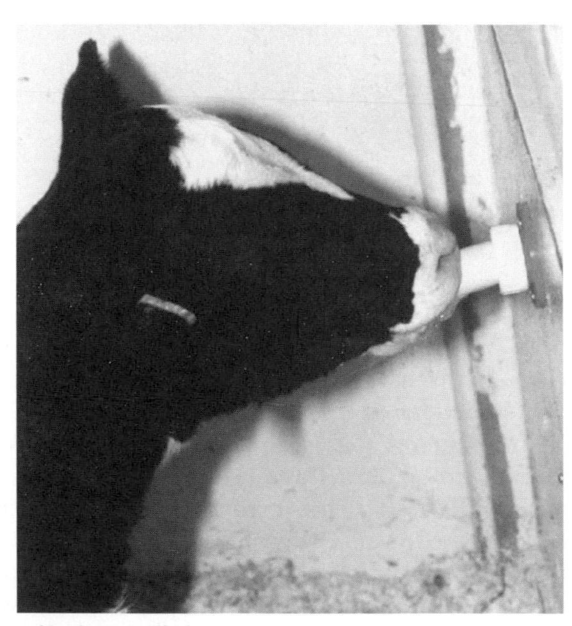

Fig. 6.3. Dairy calf taking milk from a nipple.

emerges if animals are kept in conditions where they cannot show their evolved responses when adverse conditions require them. Thus the absence of a wallow where temperature is well controlled may not be a welfare problem, but it becomes a problem if the temperature is allowed to rise.

Organic producers are therefore unable to rely on naturalness as a guarantee of good welfare. The science of animal welfare assessment can be used to determine which behaviours and conditions are actually important to the animal. Of the three concerns outlined above, two remain to be discussed: biological functioning and affective states. Behavioural indicators can be useful indicators of biological functioning, such as in the early detection of lameness and the other examples described above, but these can also often be assessed directly through measures of disease and injury. However, behavioural approaches can be especially useful in assessing affective states, as illustrated in the examples below.

Behavioural Indicators of Subjective States

The welfare of animals is greatly affected by their subjective experiences, whether these are positive (joy, pleasure, happiness) or negative (pain, boredom, frustration). The difficulty comes in assessing these states. One intuitively appealing ethological approach is the use of the preference test, which allows animals to choose to among two or more alternative environments (see Fraser and Matthews, 1997, for a review). In this type of test, the animal's behaviour allows the experimenter to determine which option the animal likes the most, or, when presenting the animal with aversive options, which it dislikes the most. For example, in several experiments dairy cows have been allowed to choose among lying surfaces. Cows almost always choose the softest surfaces, such as those with copious bedding (e.g. Tucker *et al.*, 2003).

Knowing that animals express a preference for one treatment over another is useful, but there are limitations in interpreting this type of result. Although most animals may choose one option, certain individuals may not share that preference, and indeed may strongly prefer an alternative. Even consistent preferences may provide only limited information, because the animals are restricted to the options provided. Therefore, a preference for one option does not mean that it is actually good, only that it is better than the alternative. Also, tests can detect differences that may not be important to the animal.

One way to strengthen this approach is to examine how animals respond to the various options when these are presented alone. In the above example, cows not only choose to lie down on the softer surfaces, they also spend more time lying on them than on harder surfaces when there is no choice available.

A second way to strengthen a preference test is to impose some cost on the animal's choice. In a well-known experiment on laying hens, Dawkins (1983) found that recently fed hens strongly preferred to enter a cage with litter (in which they could dust bathe), but with no food, over one with food but a plain wire floor that did not allow dust bathing. In a series of trials, she withheld food from the hens for various periods and retested them. After about 3 h of food deprivation, hens were equally likely to enter both cages, but after 12 h of food deprivation were much more likely to enter the cage with food.

In motivational experiments, the idea of imposing a cost on the animal's behaviour is taken one step further (see Dawkins, 1990). For example, animals can be trained to perform a task such as pressing a bar or pecking a key to gain access to some item. Once they have learned to do this, the 'price' of the item can be increased. We therefore can measure how much the animal values the item by how much work it is willing to do to get it. In one experiment, Mason *et al.* (2001) trained farmed mink to push variably weighted doors to gain access to several different environmental features, including a tunnel, a raised platform, an alternative nest box, and a small pool of water where they could swim. They found that the mink were willing to push much heavier doors for access to the pool than for the other options, suggesting that the mink valued this feature. Similar experiments have shown that hens will work to be able to perch at night (Olsson and Keeling, 2002) or to get access to litter material (Gunnarsson *et al.*, 1999).

Animal handling and restraint techniques are the source of many welfare complaints, but little is known about how animals perceive the various options. Researchers can use avoidance learning to examine conditions that animals find unpleasant, such as being restrained. For example, sheep are normally restrained manually for shearing, but during the 1980s, some Australian producers began to use the new technology of electro-immobilization, which was thought to be a more humane method of restraining sheep. This technique involves passing a low electrical current through the sheep's body to immobilize its muscles. To determine how sheep perceived this method, Rushen (1986) moved sheep along a runway leading to a treatment area where they were sheared using manual restraint or electro-immobilization. After a series of repeated exposures, the sheep exposed to the electro-immobilization moved much more slowly along the runway than did those restrained manually, indicating that the electro-immobilization was actually making shearing more unpleasant for these animals.

To the casual observer, among the most compelling evidence of welfare problems are abnormal behaviours such as stereotyped rocking or pacing. There are many examples of these behaviours in intensively reared farm animals, and these abnormal behaviours have often been studied by scientists interested in animal welfare. These studies can be used in two ways. In the first, the average incidence or time spent performing the behaviours is used as a general indicator of welfare. For example, Broom and Johnson (1993) suggest that a low incidence of these behaviours is consistent with high welfare, but that once the incidence rises above 5%, welfare begins to be compromised.

In contrast to this general (and arguably arbitrary) approach, a second method is to study the incidence of the behaviour as a symptom of specific problems in the animals' environment. For example, it has been suggested that the oral stereotypies performed by sows kept in stalls are related to sows being fed much less food than they would consume voluntarily. In a test of this idea, Robert *et al.* (1993) fed sows three diets identical in major nutrients but differing in bulk. Compared to the sows on the low-bulk diets, those on the high-bulk diet spent about 50% less time engaged in the most common stereotyped behaviour observed in the experiment. Thus, the organic requirement to include roughage in pig diets probably helps to decrease this type of abnormal behaviour.

We can also use normal behaviours to assess welfare. One promising class of normal behaviours are signals such as vocalizations and displays. Signals are by definition directed towards other individuals, and in many cases have evolved to provide information about the signaller's state. The public nature of signals allows the enterprising ethologist to use them to learn about the animals' state (Weary and Fraser, 1995). Particularly informative are signals that indicate the signaller's need for certain resources.

This is often the case for signals produced by dependent young, such as the begging behaviour of piglets. Sows typically nurse their piglets about once an hour, but milk is only available for about 10 s during the milk ejection. If piglets are away from the udder or are displaced from their teat during this milk ejection, they will not have access to milk until the next nursing. Piglets that miss one or more such nursings often signal this fact to the sow with characteristic begging behaviour. The piglets signal their need for the sow's milk both by calls and by nuzzling her udder. All piglets nuzzle the udder during a nursing episode, but piglets that have missed previous milk ejections are much more likely to nuzzle between nursings, and this signalling by piglets can be used by both the sow and the farmer as an indication that a piglet needs milk. Unfortunately, the signalling comes at a cost to piglets, as it places them in a position that increases their risk of being crushed by the sow (Weary *et al.*, 1996b).

Some of the principal concerns about welfare are those associated with negative affective states such as pain and distress, and several behavioural techniques have been developed to assess such states. Many practices routinely performed on farm animals, such as branding, castration, tail docking and dehorning are painful, and behavioural methods of pain assessment have been developed to identify less painful alternatives. In North America, many beef cattle are hot-iron branded, and in some jurisdictions, this is the only legal means of permanent identification. During branding, an iron heated to about 500°C is held to the skin for 3–5 seconds. One alternative is freeze branding, which instead uses an element cooled in liquid nitrogen or dry ice. In one study, beef steers were either hot-iron or freeze branded while restrained in a chute and video recorded (Schwartzkopf-Genswein *et al.*, 1998). Steers branded by either method were much more likely to flick their tails, kick, fall and vocalize than were steers that were restrained in the chute but not branded. However, those that were freeze branded showed fewer and less rapid head movements. Thus both methods are painful, but freeze branding is the preferred method if some form of branding is required.

In addition to the immediate pain, behavioural responses can be used to evaluate pain that occurs in the hours or days that follow a procedure. For example, hot-iron dehorning of dairy calves causes a pain response that continues for some 12–24 h. During this time, calves that have been dehorned are more active and perform several characteristic behaviours that are rare under normal conditions, including head shakes, ear flicks and head rubs. These responses have been used to evaluate a simple intervention, putting ketoprofen in the calf's milk before and after dehorning (Faulkner and Weary, 2001). The responses characteristic of post-operative pain were almost eliminated in the calves that received ketoprofen, indicating that this treatment is effective.

Summary and Conclusions

We have illustrated several ways in which improved understanding of farm animal behaviour can benefit both producers and their animals. Good animal producers should also be amateur ethologists, for example using changes in feeding behaviour as an early indicator of disease. Natural behaviours, a special concern in organic production, can be studied using wild counterparts of our domesticated breeds. Such studies have shown many ways in which natural behaviours can be enhanced under farm conditions. Social behaviour is particularly important for most farm animals and deserves special attention. In some cases, such as hens foraging in raspberry plots, allowing animals to express natural behaviours can benefit both the animal and other aspects of the production system.

The behaviour of farm animals also can be used in assessing their welfare. Allowing animals to perform natural behaviours is of inherent value in organic production, but this alone is no guarantee of welfare. Preference testing can be used as a simple method of assessing which environments animals prefer, and more can be learned by observing how they use the different options provided and how hard they are willing to 'work' to gain access to certain options. Some abnormal behaviours, such as stereotyped chewing and bar-biting by sows, can indicate problems in the animal's environment. We can also use normal behaviours, such as the begging of dependent young, to identify animals in need. To date, little ethological research has been applied to organic farming, and much more remains to be learned about how behaviour can be used to improve conditions on organic farms. Such research, combined with farmers' increased awareness of their animals' behaviours, may provide many improvements in welfare and production in the years to come.

References

Algers, B. and Jensen, P. (1991) Teat stimulation and milk production during early lactation in sows: effects of continuous noise. *Canadian Journal of Animal Science* 71, 51–60.

Andersen, L., Kjær, K.J., Jensen, K.H., Dybkjær, L. and Andersen, B.H. (2000) Weaning age in organic pig production. In: Hermansen, J.E., Lund, V. and Thuen, E. (eds) *Ecological Animal Husbandry in the Nordic Countries*. DARCOF Report 2/2000. DARCOF, Tjele, Denmark, pp. 119–123.

Andersson, M. (2000) Domestication effects on behaviour: foraging, parent-offspring interactions and antipredation in pigs and fowl. PhD thesis, Swedish University of Agricultural Sciences, Uppsala.

Andresen, N. (2000) The foraging pig – resource utilisation, interaction, performance and behaviour of pigs in cropping systems. PhD thesis, Swedish University of Agricultural Sciences, Uppsala.

Arey, D.S. and Edwards, S.A. (1998) Factors influencing aggression between sows after mixing and the consequences for welfare and production. *Livestock Production Science* 56, 61–70.

Bäckström, L. (1973) Environment and animal health in piglet production. A field study of incidences and correlations. Thesis, *Acta Veterinaria Scandinavica*, suppl. 41.

Baker, C.M.A. and Manwell, C. (1981) 'Fiercely feral': on the survival of domesticates without care from man. *Zeitschrift für Tierzuchtung und Zuchtungsbiologie* 98, 241–257.

Baxter, M.R. (1991) The design of the feeding environment for pigs. In: Batterham, E.S. (ed.) *Manipulating Pig Production III*. Australian Pig Science Association, Attwood, Victoria, Australia, pp. 150–157.

Boice, R. (1973) Domestication. *Psychological Bulletin* 80, 215–230.

Broom, D.M. and Johnson, K.G. (1993) *Stress and Animal Welfare*. Chapman and Hall, London.

Busse, C.S. and Shea-Moore, M.M. (1999) Behavioral and physiological responses to transportation stress in three genetic lines of pigs. *Journal of Animal Science* 77 (Supplementum 1), 147.

Chen, H. and Leibengut, F. (1995) Restriction patterns of mitochondrial DNA in European wild boar and German Landrace. *Comparative Biochemistry and Physiology* 110, 725–728.

Clutton-Brock, J. (1981) *Domesticated Animals from Early Times*. Heineman and British Museum of Natural History, London.

Collias, N.E. and Collias, E.C. (1996) Social organization of a red junglefowl, *Gallus gallus*, population related to evolution theory. *Animal Behaviour* 51, 1337–1354.

Crawford, R.D. (1990) Origin and history of poultry species. In: Crawford, R.D. (ed.) *Poultry Breeding and Genetics*. Elsevier, Amsterdam.

Danish Ministry of Food, Agriculture and Fisheries. (1999) *Action plan II. Developments in organic farming*, English summary. Danish Directorate for Development, Copenhagen. http://www.dffe.dk/in_english/index.htm

Daycard, L. (1990) Structure sociale de la population de bovines sauvages de l'île d'Amsterdam, sud de l'Océan Indien. *Revue d'Ecologie (La Terre et la Vie)* 45, 35–53.

Dawkins, M.A. (1983) Battery hens name their price: consumer demand theory and the measurement of ethological 'needs'. *Animal Behaviour* 31, 1195–1205.

Dawkins, M.A. (1990) From an animal's point of view: motivation, fitness, and animal welfare. *Behavioral and Brain Sciences* 13, 1–9 and 54–61.

de Passillé, A.M. (2001) Sucking motivation and related problems in calves. *Applied Animal Behaviour Science* 72, 175–187.

Duncan, I.J.H., Savory, C.J. and Wood-Gush, D.G.M. (1978) Observations on the reproductive behaviour of domestic fowl in the wild. *Applied Animal Ethology* 4, 29–42.

Epstein, H. and Bichard, I.L. (1984) Pig. In: Mason, I.L. (ed.) *Evolution of Domesticated Animals*. Longman Inc., New York.

Faulkner, P. and Weary, D.M. (2000) Reducing pain after dehorning in dairy calves. *Journal of Dairy Science* 83, 2037–2041.

Flower, F.C. and Weary, D.M. (2001) Effects of early separation on the dairy cow and calf: 2. Separation at 1 day and 2 weeks after birth. *Applied Animal Behaviour Science* 70, 275–284.

Fraser, A.F. (1968) *Reproductive Behaviour in Ungulates*. Academic Press, London.

Fraser, A.F. and Broom, D.M. (1990) *Farm Animal Behaviour and Welfare*. Ballière Tindall, London.

Fraser, D. (1980) A review of the behavioural mechanisms of milk ejection of the domestic pig. *Applied Animal Ethology* 6, 247–255.

Fraser, D. and Matthews, L.R. (1997) Preference and motivation testing. In: Appleby, M.C. and Hughes, B.O. (eds) *Animal Welfare*. CAB International, Wallingford, UK, pp. 159–173.

Fraser, D., Weary, D.M., Pajor, E.A. and Milligan, B.N. (1997) A scientific conception of animal welfare that reflects ethical concerns. *Animal Welfare* 6, 187–205.

Fumihito, A., Miyake, T., Sumi, S.-I., Tadaka, M., Ohno, S. and Kondo, N. (1994) One subspecies of the red junglefowl (*Gallus gallus gallus*) suffices as the matriarchic ancestor of all domestic breeds. *Proceedings of the National Academy of Sciences USA* 91, 12505–12509.

Grandin, T. (1994) Solving livestock handling problems. *Veterinary Medicine* 89, 989–998.

Grandin, T. and Deesing, M.J. (1998) Genetics and animal welfare. In: Grandin, T. (ed.) *Genetics and the Behavior of Domestic Animals*. Academic Press, San Diego, California.

Grubb, P. (1974) Social organization of Soay sheep and the behaviour of ewes and lambs. In: Jewell, P.A., Milner, C. and Morton-Boyd, J. (eds) *Island Survivors – the Ecology of the Soay Sheep of St Kilda*. Athlone Press, London, pp. 131–159.

Gunnarsson, S., Keeling, L.J. and Svedberg, J. (1999) Effects of rearing factors on the prevalence of floor eggs, cloacal cannibalism and feather pecking in commercial flocks of loose housed laying hens. *British Poultry Science* 40, 12–18.

Hafez, E.S.E. (ed.) (1962) *The Behaviour of Domestic Animals*. Williams and Wilkins, Baltimore, Maryland.

Håkansson, J. (2003) Domestication effects on aggressiveness in fowl. A comparison of agonistic behaviour in red junglefowl (*Gallus gallus*) and white leghorn layers. Specialarbete 19. Department of Animal Environment and Health, Swedish University of Agricultural Sciences, Skara.

Harper, G.C. and Henson, S.J. (2001) Consumer concerns about animal welfare and the impact on food choice – The final report. EU FAIR CT98 3678. The University of Reading, Reading.

Harrison, R. (1964) *Animal Machines*. Vincent Stuart, London.

Hemsworth, P.H., Barnett, J.L. and Hansen, C. (1987) The influence of inconsistent handling by humans on the behaviour, growth and corticosteroids of young pigs. *Applied Animal Behaviour Science* 17, 245–252.

Holmberg, H.-E. (1999) *Konsumentundersökning om ekologisk produktion/KRAV*. LUI ref. number 30-7866, published 1999–12–21. http://www.krav.se/arkiv/rapporter/luiund.pdf

IFOAM (2000) *IFOAM Basic Standards*. International Federation of Organic Movements, Tholey-Theley.

Jensen, P. (1986) Observations on the maternal behaviour of free-ranging domestic pigs. *Applied Animal Behaviour Science* 16, 131–142.

Jensen, P. and Recén, B. (1989) When to wean – observations from free-ranging domestic pigs. *Applied Animal Behaviour Science* 23, 49–60.

Jensen, P. and Toates, F.M. (1997) Stress as a state of motivational systems. *Applied Animal Behaviour Science* 53, 145–156.

Karlsson, L., Andresen, N. and Ciszuk P. (1996) Svinproduktion i odlingssystemet. *Forskningsnytt om økologisk landbruk i Norden* 1, 12–13.

Keeling, L.J. (1994) Inter-bird distances and behavioural priorities in laying hens: the effect of spatial restriction. *Applied Animal Behaviour Science* 39, 131–140.

Kruijt, J.P. (1964) Ontogeny of social behaviour in Burmese red junglefowl (*Gallus gallus spadiceus*) Bonaterre. *Behaviour Supplement XII*. Leiden, The Netherlands.

Loberg, J. and Lidfors, L. (2001) Effect of stage of lactation and breed on how dairy cows accept foster calves. *Applied Animal Behaviour Science* 74, 97–108.

Lund, V. (2002) Ethics and animal welfare in organic animal husbandry – an interdisciplinary approach. PhD thesis, Department of Animal Environment and Health, Swedish University of Agricultural Sciences, Skara.

Lund, V., Anthony, R. and Röcklinsberg, H. (2003) The ethical contract as a tool in organic animal husbandry. *Journal of Agricultural and Environmental Ethics* (in press).

Mason, G.J., Cooper, J. and Clarebrough, C. (2001) Frustrations of fur-farmed mink. *Nature* 410, 35–36.

McBride, G., Parer, I.P. and Foenander, F. (1969) The social organization and behaviour of the feral domestic fowl. *Animal Behaviour Monographs* 2, 126–181.

Mendl, M. and Held, S. (2001) Living in groups: an evolutionary perspective. In: Keeling, L.J. and Gonyou, H.W. (eds) *Social Behaviour in Farm Animals*. CAB International, Wallingford, UK, pp. 7–36.

Norrbom, S. (2001) Suckling system in dairy production: experiences and solutions for building design. Diploma work, Swedish University of Agricultural Sciences, Alnarp.

O'Connell, N.E., Beattie, V.E. and Weatherup, R.N. (2002) Influence of feeder type on the performance and behaviour of weaned pigs. *Livestock Production Science* 74, 13–17.

Odén, K. (2003) Fear and aggression in large flocks of laying hens. Effects of sex composition. PhD thesis, Swedish University of Agricultural Sciences, Alnarp.

Olsson, I.A.S. and Keeling, L.J. (2002) The push-door for measuring motivation in hens: laying hens are motivated to perch at night. *Animal Welfare* 11, 11–19.

Pitts, A.D., Weary, D.M., Fraser, D., Pajor, E.A. and Kramer, D.L. (2002) Alternative housing for sows and litters. Part 5. Individual differences in the maternal behaviour of sows. *Applied Animal Behavioural Science* 76, 291–306.

Price, E.O. (1984) Behavioral aspects of animal domestication. *The Quarterly Review of Biology* 59, 1–32.

Price, E.O. (1999) Behavioral development in animals undergoing domestication. *Applied Animal Behaviour Science* 65, 245–271.

Price, E.O. (2002) *Animal Domestication and Behavior*. CAB International, Wallingford, UK.

Quimby, W.F., Sowell, B.F., Bowman, J.G.P., Branine, M.E., Hubbart, M.E. and Sherwood, H.W. (2001) Application of feeding behaviour to predict morbidity in a commercial feedlot. *Canadian Journal of Animal Science* 81, 315–320.

Reid, F. (2002) Integrating layer chickens into a certified organic raspberry and vegetable farm. In: *Proceedings of the 14th IFOAM Organic World Congress, 21–24 August, Victoria, Canada*, p. 80.

Robert, S. and Martineau, G.P. (2001) Effects of repeated cross-fosterings on preweaning behavior and growth performance of piglets and on maternal behavior of sows. *Journal of Animal Science* 79, 88–93.

Robert, S., Matte, J.J., Farmer, C., Girard, C.L. and Martineau, G.P. (1993) High-fibre diets for sows: effects on stereotypies and adjunctive drinking. *Applied Animal Behaviour Science* 37, 297–309.

Rushen, J. (1986) Aversion of sheep to electro-immobilization and physical restraint. *Applied Animal Behaviour Science* 15, 315–324.

Schütz, K. and Jensen, P. (2001) Effects of resource allocation on behavioural strategies: a comparison of red junglefowl (*Gallus gallus*) and two domesticated breeds of poultry. *Ethology* 107, 753–765.

Schwartzkopf-Genswein, K.S., Stookey, J.M., Crowe, T.G. and Genswein, B.M.A. (1998) Comparison of image analysis, exertion force, and behavior measurements for use in the assessment of beef cattle responses to hot-iron and freeze branding. *Journal of Animal Science* 76, 972–979.

Shea-Moore, M.M. (1998) The effect of genotype on behaviour in segregated early-weaned pigs tested in an open field. *Journal of Animal Science* 76 (Suppl. 1), 100.

Sprecher, D., Hostetler, D. and Kaneene, J. (1997) A lameness scoring system that uses posture and gait to predict dairy cattle reproductive performance. *Theriogenology* 47, 1179–1187.

Stangel, G. and Jensen, P. (1991) Behaviour of semi-naturally kept sows and piglets (except suckling) during 10 days post partum. *Applied Animal Behaviour Science* 31, 211–227.

Stolba, A. and Wood-Gush, D.G.M. (1989) The behaviour of pigs in a semi-natural environment. *Animal Production* 48, 419–425.

Stricklin, W.R. (2001) The evolution and domestication of social behaviour. In: Keeling, L.J. and Gonyou, H.W. (eds) *Social Behaviour in Farm Animals*. CAB International, Wallingford, UK, pp. 83–112.

Tannenbaum, J. (1991) Ethics and animal welfare: The inextricable connection. *Journal of American Veterinary Medical Association* 198, 1360–1376.

Thodberg, K., Jensen, K.H., Herskin, M.S. and Jørgensen, E. (1999) Influence of environmental stimuli on nest building and farrowing behaviour in domestic sows. *Applied Animal Behaviour Science* 63, 131–144.

Tucker, C.B., Weary, D.M. and Fraser, D. (2003) Effects of three types of free-stall surfaces on preferences and stall usage by dairy cows. *Journal of Dairy Science* 86 , 521–529.

Weary, D.M. and Fraser, D. (1995) Signalling need: costly signals and animal welfare assessment. *Applied Animal Behaviour Science* 44, 159–169.

Weary, D.M., Pajor, E.A., Fraser, D. and Honkanen, A.-M. (1996a) Sow body movements that crush piglets: a comparison between two types of farrowing accommodation. *Applied Animal Behaviour Science* 49, 149–158.

Weary, D.M., Pajor, E.A., Thompson, B.K. and Fraser, D. (1996b) The relationship between piglet body condition and proximity to the sow: a trade-off between feeding and the risk of mortality by maternal crushing? *Animal Behaviour* 51, 619–624.

7

Applied Ethology: the Basis for Improved Animal Welfare in Organic Farming

Susanne Waiblinger,[1] Johannes Baumgartner,[1] Marthe Kiley-Worthington[2] and Knut Niebuhr[1]

[1]Institute of Animal Husbandry and Animal Welfare, University of Veterinary Medicine Vienna, Veterinärplatz 1, A-1210 Vienna, Austria; [2]ECO Research Centre, Little Ash Eco Farm, Throwleigh, Okehampton, Devon EX20 4QJ, UK

Editors' comments

Ethology is highly relevant to this book because organic farming emphasizes animals' natural behaviour. The authors of this chapter apply our understanding of that behaviour to practical examples and dilemmas. Whereas Chapter 6 emphasized the broader principles of animal behaviour, this chapter discusses the practical consequences of these principles in more detail. What is the 'nature' of an animal, and how can we use our understanding of that question to offer practical guidelines for handling and housing of organic animals? The authors give many examples of appropriate organic living conditions based on our knowledge of the behaviour of each species. They also offer their views on questions that we cannot avoid facing when talking about animal welfare in the future development of organic systems.

Introduction

Consumers associate organic animal husbandry with good animal welfare and 'happy' animals (Sies and Mahlau, 1997). Organic husbandry is seen in opposition to intensive, conventional agriculture with large groups, barren environments and behavioural problems. In fact, the starting point of organic animal husbandry is to give both the animals and the humans concerned a higher quality of life by respecting their nature and

© CAB International 2004. *Animal Health and Welfare in Organic Agriculture* (eds M. Vaarst, S. Roderick, V. Lund and W. Lockeretz)

requirements: the principles of organic animal husbandry postulate that housing systems are adapted to the species-specific needs of the animals, and that the preservation and promotion of animal health result from optimizing the animals' housing, feeding, breeding and care (Sundrum, 1993). These principles are also reflected in organic standards. Consequently, with respect to their behaviour the welfare of the animals in general is better than in conventional systems (e.g. Hörning, 1998).

At present, the focus is often on health problems, because classical treatments of disease may not be compatible with the principles and standards of organic husbandry, but experience with alternative treatments is still limited (as discussed in Chapter 13). However, in a more holistic approach, behaviour is what constitutes the bridge between the animal and its environment, and many health problems are certainly related to deficits in husbandry, management and knowledge. Therefore, further improvements of organic husbandry systems are also needed concerning behavioural requirements. Furthermore, severe problems such as cross-suckling in calves, tail biting in pigs and feather pecking in poultry need to be solved.

Ethology, the study of animal behaviour, has a critical role to play in the further development of organic husbandry that goes hand-in-hand with improved animal welfare. We will briefly discuss the meaning of animal welfare in science, the organic context and public understanding. Natural behaviour plays a central role here, and we outline what natural behaviour means, its role for a life of quality in farm animals and the implications for husbandry. We will then apply this for three main farm animal species – cattle, pigs and laying hens – and at the end summarize the significance of ethology and animal behaviour for organic animal husbandry now and in the future.

Natural Behaviour – the Key for a Life of Quality in Animals

Definitions and expectations regarding animal welfare

Judgements concerning animal welfare have been made on various grounds. For example, Lorz (1973) suggested that good animal welfare involves the animal living in a state of harmony with its environment, both physically and psychologically; that is, as van Putten (1973, 2000) suggested, the animal is able to adapt to or cope with its environment (Broom, 1986). The welfare of an animal can be assessed on a scale from very poor to very good as a result of its ability to cope. Welfare is poor when the individual has difficulty in coping with the conditions encountered (Broom, 1992). Difficulties in coping result in negative emotions, behavioural problems and physical health problems. Good welfare, involving high quality of life for the animal, does not mean only that the animal is able to cope with the environment, that suffering is avoided and biological functioning is enabled, which is what has been discussed the most in applied ethology so far. Rather, good welfare must go beyond this and ensure that the animal has positive emotions and experiences. Only recently has attention been addressed to this aspect of welfare, including how to assess it (Mench, 1998; Knierim, 1998; Désiré *et al.*, 2002).

Organic definitions of animal welfare emphasize the importance of naturalness, i.e. animals should be able to express their species-specific, natural behaviour by being provided with a natural environment or an environment with key features (Kiley-Worthington, 1989; Lund, 2000; Lund and Röcklinsberg, 2001; see also Chapter 5). This is close to consumers' expectations: they believe that good animal welfare standards come as close as possible to nature (Ouédraogo, 2002), and they associate organic farming with pictures of animals in nature (Sies and Mahlau, 1997; Starzinger, 2001). As will be outlined in this chapter, the possibility of performing natural behaviour in fact is linked both to the animal's ability to cope with the environment and to positive experiences, and thus with the quality of life.

Evolution, domestication and natural behaviour

The wild ancestors of our domestic animals adapted to their environment during evolution in their body structures, their physiology and their behaviour (Darwin, 1859). Through the long-lasting evolutionary process, the appropriate body, mind and behaviour for survival and reproduction were selected for and imprinted genetically, building the basic genetic structure for the species-specific behaviour (the animals' 'nature'). But how, when, where and why a certain behaviour is performed – within the possible range of its nature – is the result of local conditions and the individual's experiences (its 'nurture'). It is this flexibility that allows for the great adaptability and variation of behaviour, shown in particular by domestic animals.

For example, the exploration and food selection of the goat in any environment will depend on: (i) her internally controlled hunger; (ii) the nutrients she requires; (iii) her genetically imprinted preferences for browsing rather than grazing; (iv) her ability to find the appropriate food as a result of what she has learned about the environment, and also from others; and (v) her past experience of eating particular plants and how to do it (e.g. avoid the prickles on gorse). Thus, both internal and external stimuli affect the motivation to perform a behaviour, and how it is done. The results of the behaviour or the doing of it, in turn, feed back to the motivation, i.e. to do more of it, or to do something else (Buchholtz, 1993; Jensen and Toates, 1993). In the natural environment, these motivational systems or behavioural control mechanisms are closely linked to the function, e.g. to ensure that the goat provides herself with sufficient nutrients in a difficult environment.

The differences in behaviour between farm animals and their wild counterparts are primarily quantitative rather than qualitative (Price, 1999). Chapter 6 gives some examples of changes in response thresholds and frequency of behaviour. The behavioural repertoire of the wild ancestors still exists in our domesticated animals, many of which can survive in their natural environment (for pigs, e.g. Stolba and Wood-Gush, 1984; Jensen, 1986; for cattle, Reinhardt, 1980a,b; Kiley-Worthington and de la Plain, 1983; for chickens, e.g. McBride *et al.*, 1969; Wood-Gush *et al.*, 1978). Even animals from intensive housing systems very quickly show all behavioural elements when brought into an appropriate environment (e.g. in pigs, Stolba, 1984).

The set of behaviour and the underlying control mechanisms that have evolved in the species and are still possessed by our farm animals can be referred to as their species-specific or 'natural' behaviour. However, definitions of 'natural' behaviour are rare. The behavioural repertoire of a species in a (semi)-natural habitat is often referred to as 'normal' (Stauffacher, 1992), 'natural' (Algers, 1992) or 'species-specific' behaviour. Lund and Röcklinsberg (2001) specifically consider 'natural behaviour' to be 'those sets of control systems developed by the evolutionary process to allow the animal to register and react to internal and external stimuli in order to optimize survival and reproduction'.

Significance of natural behaviour for animal welfare and husbandry

Conditions in husbandry often differ substantially from those in nature. If the evolved species-specific behavioural control systems are disregarded during rearing and husbandry so that the animal no longer can cope successfully with the environment, welfare problems arise (Wechsler, 1993; Fraser *et al.*, 1997). The animal's ability to cope depends on the adaptive capacity that it possesses both genetically as a member of its species and as an individual with its special development and lifetime experiences.

The inability to cope can have two main underlying causes (Fraser *et al.*, 1997). The first is a motivation to perform a behaviour without direct necessity. In domesticated animals kept under the control of humans, the functions of many behaviours have been taken over by the stockman or the housing. For example, providing food supersedes searching for food and sometimes even chewing it. Long walks to water are not necessary because water is provided. Artificial insemination and the absence of mixed groups of both sexes prevents searching for, courting and copulating with a mating partner. Lack of opportunities to perform behaviours can lead to behavioural problems, because the animals are motivated to perform the behaviours anyway – due to the lack of a negative feedback controlling the motivation (Buchholtz, 1993). For instance, if pigs are fed highly concentrated feed in a trough, the function of nutrition is fulfilled, but they still have a strong urge to root and explore (which in nature is necessary to search for food). Preventing them from doing so leads to behavioural problems such as tail biting. When prevented from performing a behaviour when they are highly motivated, animals may not only experience negative feeling or fail to experience positive ones, but they may also suffer physical health problems. Sows prevented from nest building show prolonged parturition that might result in higher stillbirth rate, less vital piglets and higher risk of MMA (mastitis, metritis, agalactia) (Weber and Troxler, 1988; Bertschinger *et al.*, 1994; Plonait, 1997).

The second source of inability to cope is that the environment presents challenges that the animals cannot meet, such as physical constraints that prevent them from performing certain behaviours. Behaviour is somewhat flexible and adaptable, but there are behavioural, physiological and anatomical limits. For instance, cattle naturally graze while walking. In this 'pasture pace', with one front leg in front of the other, the head is closer to the ground compared to standing upright with both forelegs parallel. When eating from a feeding rack while standing, their forelegs must be parallel; this makes it harder for them to reach the ground. If the feeding table is not raised sufficiently (20 cm is recommended in loose housing), the load on the shoulder and forelegs is greater and the animals may become lame and develop lesions at the shoulder (Metzner, 1976; Molz, 1989; Waiblinger *et al.*, 2001).

Another possibility is that a behaviour did not evolve because in the natural habitat it was not necessary. Many pig breeds have hairless, unpigmented skin. They often get sunburned if offered access to an outside yard without shade and a wallowing site. They did not develop behaviour to protect themselves, because wild boars are dark-haired and mainly live in forests.

In contrast, enabling natural behaviour will provide the animals with positive experiences and emotions and a life of quality. Even weakly motivated behaviour may be important to the animal because it is associated with positive feelings of comfort, satisfaction or pleasure or because it contributes to a sense of control, social support or engagement (Mench, 1998).

In general, fulfilling the behavioural requirements nurtures health. Psychological well-being and reduction of stress are important for good immune function and health (e.g. Henry and Stephens, 1977; Kiley-Worthington, 1977). However, there appear to be contradictions. Health depends largely on the hygienic situation and pathogen pressure. A husbandry situation might be hygienically advantageous whilst disregarding behavioural requirements. On the other hand, a free-range environment rich in diverse stimuli that allows the animal to perform its full behavioural repertoire may also present a high risk of parasitic infestation or predation. The farmer must overcome these risks by appropriate management. Good organic agricultural practice demonstrates that such problems can be overcome. For instance, cattle or sheep badly managed on pasture can develop parasitic gastroenteritis leading to production losses. However, good grazing management (such as rotational and multi-species grazing), combined with regular control, can reduce the parasitic infestation to a degree where no treatment is necessary and the animals are not harmed (e.g. Brelin, 1979; Githigia *et al.*, 2001; also Chapter 14).

Besides allowing animals to perform species-specific behaviour, some features of the environment are important for all species and will contribute to the quality of the animals' life. These include the chance to make choices and decisions and have some control over the environment (Dickinson, 1980; Toates, 1987). Animals have expectations and become frustrated if these are not fulfilled. For instance, pigs fed on an irregular, unpredictable schedule showed more aggressive interactions and a chronic stress response (Carlstead, 1986; Barnett and Taylor, 1997). The implication for husbandry is that one must give the animals choices and not decide what is 'better' for them (e.g. to give them permanent access to outside runs, even during bad weather). Furthermore, this means offering them a rich environment with diverse stimuli and letting them predict and control many aspects of their environment, such as by ensuring routines for feeding and milking and by reducing changes in the social structure.

Is there a substitute for nature?

'Natural behaviour' may be a misleading term for housing systems. The question arises: for good welfare, must we offer an environment as close as possible to the natural habitat of the species, or can some elements be substituted?

Respecting the behavioural requirements of the animals does not inevitably mean turning them out into their natural environment or a copy of it, but rather ensuring they have appropriate environmental stimuli (Wechsler *et al.*, 1991). Of course, the more an environment resembles the natural habitat, the easier it is to enable the species-specific behaviour, but it is possible to reduce the natural environment to key features and stimuli, so that animals still can perform the species-specific behaviour and have a good quality of life.

An example is the family pen for pigs (Stolba, 1984). First, the key behaviours and corresponding environmental features were identified by observing pigs in a semi-natural environment. Then, the environment was reduced step-by-step while retaining the relevant features, such as structuring the pen into different areas for rooting, nesting and dunging (Stolba, 1984; Wechsler *et al.*, 1991). Further examples of these key features and of how a knowledge of them is necessary for good welfare will be given in the species-specific sections that follow.

However, is offering key features in an 'artificial' environment enough for organic livestock? Does it comply with the values of organic farming and consumers' expectations? For some features of nature, there is no substitute in a (closed) housing system, e.g. natural light, sun, fresh air and wind, and natural ground. Therefore, organic standards usually require access to an open run or pasture. A difficult challenge is to ensure sufficient environmental diversity and complexity. The animal's active interaction with the environment and dealing with environmental challenges seem to be central for animal welfare (Wemelsfelder and Birke, 1997). In this respect, too, a free-range environment has far more to offer than indoor housing.

In Table 7.1 we summarize the discussion so far. Organic farmers should provide the animals in their care with as high a quality of life as possible. In the following sections, we discuss necessary preconditions for three species as well as possible conflicts and their solutions.

Table 7.1. An outline of different levels of quality of life, how they are characterized with respect to behaviour and health aspects, what system would correspond to it (optimal management assumed) and where current organic husbandry standards fall.

Welfare level	Characteristics	Example in cattle	Standard
Very good	Natural behaviour possible, very few behavioural restrictions; health problems prevented	Beef suckler herds outside on pasture, weaning at 9 months	Beyond standards; greatest approach to natural life
Good	Parts of natural behaviour not possible, but behavioural restrictions not leading to problems; adaptation possible, no sign of distress, no health problems	Beef bulls on deep litter with outside run, partially on pasture and feed with sufficient roughage	Minimum organic standard; natural life only partially attained
Poor	Distinct behavioural restrictions in some functional systems; abnormal behaviour (tongue rolling)	Beef bulls on deep litter, few space, insufficient roughage	Improved conventional systems
Very poor	High restrictions in several functional behavioural systems; health problems and abnormal behaviour (e.g. skin lesions, joint lesions, disturbed lying behaviour, tail tip lesions, tongue rolling)	Beef bulls on fully slatted floors, high stocking density, insufficient roughage	Accepted in many welfare laws, in conventional systems

Askham Bryan College
LIBRARY BOOK

Cattle

Species-specific behaviour

To know about the natural behaviour of any species, and thereby, to be able to design environments to fulfil their needs, it is necessary for ethologists to study wild or feral groups. The ancestor of cattle, the aurochs, *B. primigenius*, became extinct in 1627, but there are several wild close relatives, for example the bison in Europe and northern America (Lott, 1991), the gaur in India and the Cape buffalo in Africa (Prins, 1997). There are few herds of feral cattle (e.g. Hall and Moore, 1986; Lazo, 1994), but cattle have been studied living with little interference from humans (e.g. for *Bos taurus*, Hafez *et al.*, 1963; Kiley-Worthington and de la Plain, 1983; for *Bos indicus*, Reinhardt, 1980a,b). Here we summarize what we know about the behaviour of cattle from these studies and then outline the needs of cattle.

Social structure

Cattle are extremely social, living in mixed herds of different ages and sexes, based on several cow-centred groups (adult cows and their female and subadult male offspring). The herd rarely has more than 50 animals, although occasionally, many groups will join together temporarily for migrations or because of food resources. The herds may occupy specific home ranges. At 2–4 years of age, males leave their group to join a multimale group or live solitary. During the rutting season, they join a cow herd, but some males might stay in the herd during the whole year. The females and some of the bulls will remain with the same group for most of their lives. Consequently, they have stable groups in which they know each other (Randle, 1994). They form long-lasting social bonds, especially between mother and offspring, but also between same-aged animals and between siblings (Reinhardt, 1980a; Reinhardt and Reinhardt, 1981; Kiley-Worthington and de la Plain, 1983).

The social organization of cattle is distinguished by a relatively stable dominance–subordinate relationships, but even more by affiliative behaviour, behaviour that strengthens the cohesion of the group by cooperation and tolerance, rather than competition (Randle, 1994). Individuals have roles within the group that may change, depending on circumstances. Their daily activities are performed synchronously.

Calving and cow–calf interaction

Cows may calve at any time of the year, although there is generally a peak in the spring. The cows separate from the herd to calve, but in some studies of domestic cows, only some really separated from the herd to calve (Kiley-Worthington and de la Plain, 1983; Lidfors and Jensen, 1988). The calf 'lies out' rather than follow the mother in the first week. After 1–3 weeks, the cow rejoins the herd with her calf (Reinhardt, 1980a). In the herd, a calf tends to stay in peer groups rather than follow its mother (Sato *et al.*, 1987). However, the mother repeatedly comes back to the calf, who performs a lot of social behaviour with her, such as playing and social licking (Reinhardt, 1980a).

The cows suckle their calves around eight times per day in the first week, later three to five times per day, with each bout taking around 10 min. They wean them at 8–12 months (Walker, 1962; Reinhardt, 1980a; Kiley-Worthington and de la Plain, 1983; Porzig and Sambraus, 1991). Calves start to graze and ruminate for a remarkable amount of time at 3 weeks of age and regularly graze with the adults at 4–6 months (Arnold and Dudzinski, 1978; Reinhardt, 1980a).

Feeding, locomotion, resting and comfort

Cattle are specialized for living on the edge of forests, eating mainly grasses and ground level herbs and occasionally browsing trees. Consequently they have evolved to have varied diets. They primarily consume grass high in crude fibre, which needs to be chewed thoroughly before it can be digested. Rumination allows them to spend as little as 4–9 h a day eating where they might be in view of predators, and thereafter withdraw to some quiet corner to rechew for another 4–9 h, most of the time (60–80%) while lying. While grazing, they walk slowly, taking 50–80 bites of grass per minute; a dairy cow in total takes around 20,000 bites per day (Porzig and Sambraus, 1991).

Cattle on pasture drink from one to four times daily (Hafez and Bouissou, 1975), at up to 20 l/min, by lowering their muzzles into the water at a 60° angle to keep their nostrils above the water (Metzner, 1976). Cattle mainly move while grazing. They travel an average of 5 km per day going to grazing sites or water (Hafez and Bouissou, 1975), but, depending on the distribution of food and water, the distance can increase to 30 km. Cattle rest around 8 h, most of the time lying. When lying down and getting up, they use a head lunge to take weight from the hindleg. Cattle maintain their skin by self-grooming and reduce irritation by using their tails, scratching and rubbing themselves and shaking their heads.

Implications for organic cattle husbandry and some problems

Social environment and space

Living in groups is a basic characteristic of cattle, a social species, and can only be fulfilled by keep them loose housed. Tied cattle are very restricted in their behaviour, which has negative consequences on health (Bendixen *et al.*, 1988a,b). Regular access to exercise, such as pasturing the animals during summer or regularly offering an outside run, has only a limited beneficial effect (Wiederkehr *et al.*, 2001; Spycher *et al.*, 2002). Consequently, tying systems are not in line with organic principles and should be avoided.

Except in beef suckler herds, the social environment differs from groups in nature with respect to composition (single sex and similar age) and stability (frequent regrouping or new members). Integration of heifers into a milking herd and regrouping can lead to stress responses with reduced lying times, an increase in cortisol and a decrease in milk yield (Hasegawa *et al.*, 1997; Knierim, 1999a,b). Dairy cattle are regrouped frequently to meet the nutritional requirements for their stage of lactation or milk yield. One way of overcoming the conflict between their social and nutritional requirements is to keep cows in one group with access to separate feeding areas. To date, only a few farms are organized in this way. If regrouping is unavoidable, measures to reduce stress should be taken, such as regrouping on pasture, where there is plenty of space and where the natural ground reduces the risk of claw lesions. In horned cows, it seems advantageous to introduce heifers individually to reduce aggression (Menke *et al.*, 2000). Nevertheless, Knierim (1999a,b) found some hints of social support among heifers introduced as a group. This is an area where further ethological research is needed.

The most successful approach is to maximize the herd's stability to allow long-lasting social bonds and social support (Sachser *et al.*, 1998), and to reduce social stress by decreasing the number of regroupings and enhancing the predictability and controllability of the social environment.

Competition can be reduced by a wider distribution and sufficient number of resources such as drinkers, brushes and concentrate feeders. Space allowance is also related to lameness (Leonard *et al.*, 1996) and agonistic interactions (Menke *et al.*, 1999); adequate space helps to avoid long standing periods, which are detrimental to claw health (Galindo and Broom, 1994).

To enable synchronous herd behaviour, at least one feeding and lying place (cubicle) per animal is necessary; otherwise agonistic behaviour increases and daily rhythms change (Wierenga, 1983a,b; Stumpf *et al.*, 2000).The duration of lying during the usual resting times increases when more cubicles are available than cows (Wierenga *et al.*, 1985). Food should be available in a consistent quality all the time to avoid competition. If limited food is offered, animals must be fixed in the feeding rack to ensure that low-ranking animals can eat enough, without disturbances.

Cow–calf interactions

In contrast with beef suckler cattle, a natural social structure rarely exists with dairy cows, so that two major behavioural restrictions are almost universal. First, dairy cows now rarely run with bulls, or in mixed sex and age groups. Second, their calves are taken away shortly after birth, with both mother and calf showing stress reactions: increased vocalization, movement and aggression (Kiley-Worthington, 1995). Though the reaction to separation is stronger after 4–14 days of suckling, later separation has advantages for health and performance advantages (Krohn *et al.*, 1990; Weary and Chua, 2000; Flower and Weary, 2001; for a review, see Krohn, 2001).

Non-nutritive sucking (calves sucking at the pen or bucket), cross-sucking (calves sucking at various body parts of other calves) and inter-sucking (weaned sub-adult or adult cows sucking at the udder of a group-member) are common behavioural problems in artificially reared animals that might cause skin irritations and infections of the navel, the scrotum or the udder (Plath, 1999; Keil, 2000). Different factors contribute to this abnormal behaviour, but the main cause seems to be unfulfilled sucking motivation (Sambraus, 1985), perhaps because the internal signal to stop the sucking motivation (possibly satiety) comes too late compared with the short duration of milk intake in artificial rearing systems (Egle *et al.*, 1999; de Passillé, 2001).

Several measures may reduce cross-sucking, for example:

- Restricting the calves for 10 minutes after drinking milk (Graf *et al.*, 1989).
- Using a gated automatic milk feeder, where calves are protected from displacement from the teat (Weber, 1999; Weber and Wechsler, 2001).
- Using teat buckets instead of open buckets.
- Providing calves with the teat bucket after milk intake.
- Especially, reducing milk flow rate to prolong nutritive suckling (de Passillé, 2001; Jung and Lidfors, 2001; Loberg and Lidfors, 2001a).

However, these measures do not completely prevent cross-sucking, which may continue after weaning. It might be prevented by feeding fulfilling their energetic and behavioural requirements (constant availability of food, long feeding times) after weaning (Keil *et al.*, 2000; Keil and Langhans, 2001). Such problems rarely arise when cows are kept with their calves in an appropriate environment. Therefore, systems where calves can run with their dairy cows should be promoted in organic agriculture. Pilot studies (see Box 7.1) and limited practical experience are encouraging, but also show the need for additional research.

More common on organic dairy farms is to use foster cows for multiple suckling, i.e. one cow suckles two to four calves (e.g. Hudson, 1977; Loberg and Lidfors, 2001b). The calves live in groups, have some contact with adult cows, and can perform natural sucking behaviour. However, they may not all be properly mothered. In the study of Loberg and Lidfors (2001b), foster cows licked the calves very rarely; in beef cows, a high proportion (60%) of calves were only tolerated (that is suckled but not licked), but not adopted by the cow in double-suckling systems in Saler cattle. Consequently, those calves gained less weight than the mother's

Box 7.1. Single suckling.

Pilot studies have been done on organic farms to study the problems and economics of leaving calves with their mothers in a dairy herd running with a bull (Kiley-Worthington, unpublished observations). The milking cows and their calves were kept together as a herd during the day. At night the calves were separated. In the morning, the cows were milked and let out with their calves. The disadvantages of this system were that some of the cows withheld their milk at milking, and milk yields were lower. The advantages of the system were substantial:

- Lower labour costs (once a day milking and very little labour looking after the calves).
- Very little sickness in the calves, no mastitis in the cows.
- Better grown calves that sold at higher prices,
- Running with a bull, the cows were back in oestrus within 2 months of calving, and 90% were pregnant on their first return.
- Better overall economic performance and longer productive life of cows.

The system can be adapted to the producer and the herd. If the cows are very high yielding, they may be milked twice a day, with the calves running with them all the time.

The positive effects on production and the health of the calves and cows are supported by other studies (Boden and Leaver, 1994; Flower and Weary, 2001; Margerison *et al.*, 2003; for a review, see Krohn, 2001). The experience of one conventional farmer documented in the farming press (Hovi, 1998) suggests that such a system can also be used without major problems with large, loose housed herds.

own calf (Le Neindre, 1982). Although in a multiple suckler system the calves might only partly be able to perform and experience natural cow–calf interactions, they are less restricted than with artificial rearing.

The same limitations may apply to multiple suckling in beef herds. Economic constraints might force small organic farms to make more money from each cow. Here extra calves can be bought in, usually from dairy herds. The tendency of the beef cow to discriminate against calves other than her own (Le Neindre, 1982) can be reduced by some measures tested experimentally (Kiley, 1976; Le Neindre and Garel, 1979): introducing the calf as soon as possible after birth, and if possible covering it with amniotic fluid. Also, to ensure that her own calf is not disadvantaged by a stronger and more active introduced calf, her own calf should be made to suck as soon as possible. One way of ensuring that both calves do well is to have them suckle in synchrony for the first few days. The economic performance of cattle managed in this way on organic farms is above that of normal single-suckler beef cows (Kiley-Worthington and Randle, 1997).

Comfort

Self-grooming of all naturally reached body parts is possible only on a non-skid floor. Otherwise animals avoid licking areas that are difficult to reach, and slip during self-grooming, resulting in claw lesions (Sommer and Troxler, 1986). Automatic brushes are popularly used by cows.

Access to an outside run for loose housed cattle outside the pasturing season is not compulsory in all organic standards. However, a permanent accessible outside run, sufficiently sunny during winter and unroofed, is necessary for cattle to perform some of their natural behaviour, such as sun bathing. In Switzerland, cows have been observed to use an outside run especially often on sunny days in autumn, standing at right angles to the sun to get the most radiation (Arnold and Dudzinski, 1978; Krötzl and Hauser, 1997). Cattle also choose to go out in the rain (Krötzl and Hauser, 1997). The yard has positive effects on locomotion and social behaviour by increasing space and by structuring the housing (Menke *et al.*, 2000). It can be fitted with brushes, drinkers and additional hay feeders (Van Caenegem and Krötzl Messerli, 1997).

If cattle are kept on pasture, shelter should be available to protect them from sun during hot periods, and from rain and wind during cold. Trees can serve to give shade and protection in summer. During the winter in temperate and cold areas, free-range cattle need at least a dry lying area with protection against wind (Wallbaum *et al.*, 1997); in wet areas, this can only be offered by a roofed area. Cattle on pasture use a shelter after 2 h of rain (Vandenheede *et al.*, 1994).

Lying and feeding

The head lunge that cattle use to get up takes up to 1.5 m, measured from the carpus to the tip of the muzzle (Boxberger, 1983). The carpal joints serve as a pivot and thereby must carry a high load during lying down and getting up (Boxberger, 1983). Thus, a comfortable lying area for cattle consists of a soft and non-skid surface big enough for them to get up and down and lie unrestricted. A free lying area with deep litter or straw flow pen is the best way of ensuring this. Sufficiently large cubicles with high-quality straw bedding (a straw mattress, which is a compact mattress composed of straw and cow dung) and flexible separations can also provide sufficient comfort (Hörning and Tost, 2001, Hörning, 2002).

Unfortunately, many farms still use cubicles that do not fulfil the cows' requirements, as shown by their lower daily duration of lying, fewer but longer bouts of lying, less comfortable lying positions, abnormal horse-like raising (Buchwalder *et al.*, 2000; Chaplin *et al.*, 2000; Hörning *et al.*, 2001) and more lameness than with deep litter (Somers *et al.*, 2001). Soft lying mats are equivalent to straw mattresses regarding cow behaviour in temperate conditions, but cows develop hock lesions with them (Wechsler *et al.*, 2000) and prefer straw bedding in (cold) winter (Manninen *et al.*, 2002). Sand is often recommended in cubicles to avoid mastitis, but cows prefer straw and soft lying mats (Manninen *et al.*, 2002).

To avoid skin and claw lesions from a heavy load and to allow a relaxed feeding posture, the feeding area should have these three features: a raised feeding table or crib (20–25 cm in loose housing), an inclined feeding rack and food always within reach (Boxberger, 1983; Konggaard, 1983; Hansen *et al.*, 1998; Waiblinger *et al.*, 2001). Drinkers should not be too high – at most 80 cm for adults – so that the cattle can use their normal drinking posture (Metzner, 1976). Water flow must be high enough; lactating cows must have access to a large amount of water at least twice a day. Water troughs are best.

Handling

Another cause of behavioural problems and inconveniences in cattle is bad handling. Research has developed some general rules and knowledge about how to reduce the animals' fear of humans and improve the ease of handling (Grandin, 1989, 1993a,b; Boivin *et al.*, 1992).

All cattle can be taught to be easy to handle and not to be frightened or attack defensively. For example, it is important to provide bulls with an appropriate social environment, to handle them with confidence and to teach them acceptable behaviour as calves. Individually reared bulls were found to threaten and attack humans more than group-reared ones (Price and Wallach, 1990). Positively handled animals are easier to handle, whilst lack of habituation and negative interactions such as shouting and hitting lead to a higher level of fear of humans and more animals attacking humans (Boivin *et al.*, 1992; Grandin, 1993a,b; Hemsworth *et al.*, 2000; Lensink *et al.*, 2001; Waiblinger *et al.*, 2002; for reviews, see Hemsworth and Coleman, 1998; Rushen *et al.*, 1999). Further examples of the importance of how the animals are handled and the human–animal relationship are given in Chapter 10.

Assessing husbandry systems

The restrictions on the different functional behavioural systems (e.g. maternal, social, lying behaviour) in different environments can be assessed, with this assessment used as a guide for improving the quality of cattle's lives. The least behavioural restriction for cattle is with outdoor suckled cattle, as described in Box 7.2. Here, the physical and social environments and the level of stimulation and active interaction with the environment resemble those of natural conditions.

Box 7.2. An example of husbandry close to nature – a beef suckler herd.

Provided that feeding and care are appropriate, the behavioural and physical requirements are best fulfilled in outdoor-suckled cattle allowed to live in stable herds with at least one bull. They can mate and raise their own young, which remain with the herd until finished and sent for beef. In temperate areas, it is necessary to maintain grass for the next year, avoid poaching, and ensure some shelter for the animals. This can be achieved with open-front barns with deep litter straw bedding that they can enter when they wish during bad weather, including a separate creep area for calves. Fodder (hay or silage) is always available, with extra feed as appropriate. The animals choose to go out or in. However, the cycling heifers must be separated from the maternal group to avoid becoming pregnant too young.

Pigs

Species-specific behaviour

Our knowledge of the natural behaviour of pigs is derived from studies on wild boars (Gundlach, 1968; Frädrich, 1974; Meynhardt, 1978; Mauget, 1981; Graves, 1984), feral pigs, and domestic pigs kept in semi-natural environments (Jensen and Wood-Gush, 1984, Stolba and Wood-Gush, 1984, 1989; Petersen *et al.*, 1989; Wechsler *et al.*, 1991).

Social structure

Pigs are gregarious animals. Their basic social unit consists of two to six related sows, their most recent litters and juvenile offspring of previous litters (Graves, 1984). The number of sows depends upon the available resources. Within sow and offspring groups, sows form a stable linear hierarchy, based on age and size (Beilharz and Cox, 1967; Ewbank, 1976). The hierarchy is maintained by subordinates avoiding dominants, rather than by dominants attacking subordinates (Jensen, 1980, 1982). Individuals recognize each other largely by smell, whereas sight is relatively unimportant once a social order has been established (Ewbank *et al.*, 1974). Non-member sows are rarely allowed into a group (Stolba and Wood-Gush, 1989). Juvenile males leave the group at 7–8 months of age and form groups of two or three. Mature males are solitary and nomadic, which permits promiscuous mating (Mauget, 1981).

Pigs live in home ranges of 100–2500 ha, but they are not territorial. The size and use of the home range depends on the available resources and is related to the social organization (Mauget, 1981). Within the home range, maternal groups build communal nests for sleeping, preferably at the border of forest or bush habitats. Pigs leave the nest site to defecate

(Stolba and Wood-Gush, 1984). Most suidae species are diurnal (Frädrich, 1965; Gundlach, 1968; van Putten, 1978a; Mauget, 1981).

Their time budget is controlled mainly by the search for food. Domestic pigs in a semi-natural enclosure were most active for some hours in the morning and the late afternoon to early evening, resting in the middle of the day and during the night (Wood-Gush *et al.*, 1990).

During nursing, the rhythm becomes polyphasic. Pigs in a group prefer to eat, suckle and rest in synchrony (Hsia and Wood-Gush, 1984).

Pigs have many vocalizations, varying in frequency, tone and magnitude. They communicate by means of grunts, squeals, snarls and snorts. The function of only a few of these is known, such as contact grunts, warning calls, sow lactation grunts, begging calls of piglets (described in detail in Chapter 6) and mating songs (Grauvogl, 1958; Klingholz *et al.*, 1979; Weary and Fraser, 1995). Smells from faeces, urine and secretions of the metacarpal, lachrymal, salivary and preputial glands are thought to be important for familiarity, dominance and reproduction.

Reproduction, nursing and weaning

Wilds boar and feral pigs have pronounced seasonal reproductive periods, one or two matings per year in most studies (Mauget, 1981), whereas domestic pigs breed more or less year round. Reproduction is synchronized within a sow group. A new boar joins the maternal group for each mating season and courts the sows in heat. During oestrus, sows actively search for boars and stay close.

A few days before farrowing, a pregnant sow leaves the maternal group to search for a suitable nest site. Farrowing nests will be built on the periphery of the group's home range. Nest building begins about 16–20 h before farrowing. The initial phase of nest building, the excavation of a hollow, is thought to be mainly internally controlled, whereas the second phase, the arrangement of nesting material, depends on environmental stimuli (Jensen, 1993). Nest building is mostly finished 2–4 h before farrowing (Gundlach, 1968; Jensen, 1993).

During farrowing, the sow mostly lies still. The newborn piglets usually find the teats within less than 30 min, and for several hours they sample different teats and ingest colostrum (Fraser *et al.*, 1995). During the first few hours, the typical cyclical pattern of nursing develops, with nursing intervals of 40–60 min (Castren *et al.*, 1993). The newborn piglets frequently fight to get control of highly productive teats, and a stable teat order develops within the first week. Thereafter, the pigs tend to suckle at the same teat for the rest of the lactation (Fraser, 1975; Martys, 1982). After farrowing, the sow and litter stay in or near the nest for about 10 days. This allows them to form a strong bond. The sow and her piglets recognize each other by smells and vocal cues (Jensen and Redbo, 1987).

After this isolation period, the sow and litter rejoin the primary maternal group (Stangel and Jensen, 1991). Reintegration involves much social activity but little overt fighting, and the social activity stabilizes about 8 weeks postpartum (Petersen *et al.*, 1989). Piglets from several litters are reared together. Social bonds among litter mates remain stronger than other bonds in the group (Newberry and Wood-Gush, 1986).

Piglets also develop dominance relationships among themselves that do not correlate with the teat order (Martys, 1991). Play amongst piglets begins within the first few days after birth and peaks between weeks 2 and 6 (Newberry and Wood-Gush, 1986; Blackshaw *et al.*, 1997). Under free-range conditions, some piglets tend to suckle from mothers other than their own (Jensen and Stangel, 1992). Whereas some piglets suckle opportunistically on several mothers, others get completely adopted and integrated into another litter (Newberry and Wood-Gush, 1986; Goetz and Troxler, 1993). Weaning in free-ranging domestic pigs is a gradual process that begins 2–3 weeks after birth but is not finished until the pigs are on average 13–17 weeks, a weaning age similar to that of wild boars (Gundlach, 1968, Newberry and Wood-Gush, 1985; Jensen, 1995). It is characterized by a gradual decrease of suckling frequency and more sucklings with the sow standing (Jensen and Recén, 1989; Jensen and Stangel, 1992).

Feeding and temperature regulation

Pigs are omnivorous. The diet of wild boars and feral pigs is based on plants (grass, roots, fruit, berries, seeds), but animals such as worms, frogs and rodents may also be important (Hansen and Karstad, 1959). Foraging behaviour is closely linked to exploratory behaviour, for which the pig has a highly sensitive and well-adapted snout. Domestic pigs in a semi-natural enclosure have been noted to spend 6–8 h per day searching for food by rooting, grazing and browsing (Wood-Gush *et al.*, 1990).

Since pigs – wild and domestic alike – have very limited sweating and panting abilities; they rely on wallowing and skin wetting for cooling in hot weather (Baldwin and Ingram, 1967). Pigs outside their thermoneutral zone adjust their lying posture to dissipate excessive heat or to limit its loss. Pigs huddle in cold weather, while a resting group will spread out when it is hot.

In summary, pigs have a highly complex behavioural repertoire. They live in stable social groups within a home range, their activities show a variable diurnal biphasic rhythm and their exploratory behaviour is strongly developed. These facts must be considered in designing and managing housing systems for pigs.

Implications for the husbandry of pigs

Grouping

Apart from adult boars and sows around farrowing, pigs are social animals and therefore should be kept in groups. Their natural groupings are small and stable (Gundlach, 1968; Frädrich, 1974). In commercial production, they are usually kept in larger groups and repeatedly mixed with unfamiliar individuals. It is common to group piglets from different litters after weaning and to regroup unfamiliar pigs by weight or sex during the growing and finishing phases. Problems arise from disrupted social bonds and the repeated need to rebuild a hierarchy under restricted conditions.

Thus, mixing of pigs should be avoided whenever possible. The frequency and intensity of fighting after mixing can be reduced by enriching the pen (Schaefer *et al.*, 1990; Petersen *et al.*, 1995), by providing food *ad libitum*, and by grouping after dark (Barnett *et al.*, 1996). Regrouping young piglets results in less aggression than with older ones (Jensen, 1994). Newly weaned piglets have more problems coping with unfamiliar housing than coping with unfamiliar piglets (Puppe *et al.*, 1997). Playing music or other sound provides no improvement in conditions for piglets during weaning (Cloutier *et al.*, 2000).

According to the Council Regulation (EC) No. 1804/1999, sows in organic farms must be kept in groups, except in the last stage of pregnancy and during suckling. There are many group housing systems for dry sows that differ regarding grouping system, group size, feeding system, flooring and bedding. As a consequence it is not possible to make a general statement about welfare in group housing (Edwards, 2000). A 'static' (stable) grouping system minimizes disturbances of the group by limiting changes in its composition. Once a group is formed after weaning or at the beginning of pregnancy, no other sows are added and the group disbands only at farrowing. The 'dynamic' system, in contrast, regularly adds recently bred animals and removes those approaching farrowing.

'Dynamic' groups regularly result in post-regrouping aggression, but allow farms to operate with only a few larger groups. Problems with social behaviour can occur if the group has more than 100 animals. With small groups, on the other hand, space may be very limited, impairing locomotion and not allowing the animals to avoid each other (IGN-Working Group 'Group Housing of Dry Sows', 2000). Dynamic groupings can benefit by providing well-defined areas within the pen that new groups can claim as their own when being integrated into the main group (van Putten and van de Burgwal, 1990). Fighting at grouping can be reduced by regrouping sows that have already been kept in the same group before. A 'grouping arena', with adequate flooring surface and enough space, where sows can establish a hierarchy, minimizes the risk of injuries (Deininger, 1998; Deininger *et al.*, 2002). Edwards *et al.* (1993) used a central suspended barrier in a mixing pen and found that it reduced fighting. The impact of large groups on social organization and aggression is poorly understood. Whilst retaining the ability to discriminate between pen mates and foreign pigs, sows from large groups display a marked reduction in aggressive tendency towards foreign individuals (Turner *et al.*, 2001).

Since much less space is available in commercial pig production than in nature, great attention must be paid to giving each animal enough space for its behavioural requirements. Pigs prefer to separate their dunging/activity area from the lying area, choosing to lie in the most thermally comfortable and undisturbed areas, and excreting in areas that are cold, wet or draughty. They need space for social interactions or to withdraw from or be out of sight of pen mates (McGlone and Curtis, 1985). Insufficient space increases agonistic interactions and the incidence of body lesions (Weng *et al.*, 1998), and reduces resting time (Ewbank and Bryant, 1969), mating rate (Hemsworth *et al.*, 1986) and weight gain and feed conversion efficiency (Brumm, 1996).

Feeding

Pigs should be allowed to eat simultaneously. If their feed is restricted, each pig should have sufficient space at the trough. *Ad libitum* feeding of weaning, growing and finishing pigs has become increasingly common because it enables the housing of large groups, which saves costs. Over the last few years, several types of feeders have been developed, making it difficult to give a general recommendation about the number of pigs per trough and the number of spaces provided per group. For a tube feeder, there should be no more than ten pigs per trough (Kircher, 2001). Nielsen *et al.* (1995a) found that groups of five, ten, 15 or 20 pigs fed with single-space feeders showed no differences in the number of attempts to displace other pigs from the feeder or in production variables. Pigs in groups of 20 made fewer but longer visits to the feeder, and ate more and faster than pigs in smaller groups. Pigs in groups of ten given access to a four-space trough visited the feeder much more frequently and for shorter duration than pigs with access to a single-space feeder, but there were no differences in production variables (Nielsen *et al.*, 1995b). If pigs are fed restrictively on a ration basis, the feeding space should allow all animals to feed simultaneously to limit competition.

Dry sows typically are fed simultaneously in individual feeding stalls, or they are fed sequentially from a single or a few stalls, as in electronic feeding stations. When feed is restricted and competition for feed is not well controlled, dominant sows become fat and subordinates become thin. This causes reproductive problems and low levels of milk production. Precise rationing of each sow without aggression can only be guaranteed by individually confining the animals at feeding. When sows are fed in electronic feeding stations, aggression and feeder occupation following the start of the feeding cycle are lower with one feeding cycle per day than with two (Weber *et al.*, 1993).

Providing straw as bedding material and starting the cycle in the evening for overnight feeding may reduce aggression and feeder occupation following the start of the cycle (Jensen *et al.*, 2000). Another strategy to control competition is to feed sows *ad libitum* with a high-fibre diet, which prevents them from getting fat. Sugarbeet pulp without molasses may be used to feed sows *ad libitum* during gestation without reducing productivity. However, food intake may be too high to make *ad libitum* feeding of pregnant sows an attractive option (Whittaker *et al.*, 2000). In organic farming, fresh or dry fodder or silage must be added to pigs' daily rations (Council Regulation (EC) No. 1804/1999). Grower pigs with access to roughage showed less aggression (Olsen *et al.*, 2002).

Flooring

To prevent injury, floors must not be slippery. The roughness of the surface is important for regulating hoof growth. Smooth surfaces and deep litter systems can cause excessive hoof growth, leading to lameness (Geyer, 1979). The floor in the lying area has to be clean, dry and comfortable. The thermal characteristics of flooring materials must be related to the thermal requirements of the pigs and the ambient temperature. In hot conditions, being cooled by the floor may be more important to the pigs than physical comfort or insulation. If pigs are kept on perforated or slatted floors, the size of the slots, the surface roughness and the edge design must be taken into account. To reduce the risk of claw injuries, bar width and slot space between bars of perforated floors need to be adjusted to the size of the pigs' claws (Geyer, 1979; Mulitze, 1989). Early-weaned piglets prefer an insulated floor to a barren expanded metal or wire floor. Floors with a high ratio of solid to slotted area are preferred (Marx and Schuster, 1986).

Bedded flooring not only improves comfort, but also allows manipulatory and investigatory activities. Also, in the case of straw, it may provide dietary fibre and allow pigs to express feeding behaviour. Pigs provided with bedding (mainly straw) are reported to be more active and to exhibit increased rooting and exploratory behaviour than pigs housed on bare flooring without bedding (Fraser *et al.*, 1991; Arey and Franklin, 1995; Beattie *et al.*, 1995; Guy *et al.*, 2002). Straw bedding also reduces destructive behaviour directed at pen mates, such as tail biting, ear biting and belly nosing (Troxler, 1981; Fraser *et al.*, 1991; Beattie *et al.*, 1995; Day *et al.*, 2002). Straw or other destructible materials for investigation and manipulation should be provided whenever possible; it can be bedding material, but need not be.

The interest of pigs in manipulating artificial objects decreases with familiarity (Heizmann *et al.*, 1988). Rooting appears to be a need that is performed regardless of feeding level or nutritional feedback. This suggests that a suitable rooting substrate should be provided even for pigs that are fed *ad libitum* (Beattie and O'Connell, 2002). In preference to straw, growing pigs favour substrates that are similar in texture to soil, such as peat, mushroom compost and sawdust (Beattie *et al.*, 1997). Commercial pigs kept outdoors are often given nose rings to inhibit rooting and minimize pasture damage (see Chapter 8).

Temperature regulation

The thermoneutral zone of a pig varies with its age, size and nutritional status (Mount, 1960, 1968). Draughts, ambient temperature, floor type, bedding and the design of the lying area all affect the pig's thermoregulation. Straw on the floor helps to maintain the temperature close to the thermoneutral zone. For finishing pigs the ambient temperature in a lying area with straw bedding should not drop below 9°C, whereas without bedding, the minimum is 17°C (Mayer, 1999). At a temperature above thermoneutrality, pigs cannot dissipate heat fast enough to regulate their body temperature, and respond with reduced activity, modified lying behaviour and wallowing. As a result, the lying area may become much dirtier, especially on a concrete floor (McKinnon *et al.*, 1989). Therefore, above 18°C, pigs should be provided with showers and outdoor runs (Mayer, 1999).

A wallow is another effective way for pigs to cool down. Wallowing also plays a role in skin and hair care of pigs (van Putten, 1978a). Olsen *et al.* (2001) reported that pigs used a wallow for lying and oral behaviour within the whole temperature range (–4 to +24° C), but this behaviour lasted longer above 15°C. Providing a wallow can cause problems with internal parasites if it is not cleaned regularly (Simantke, 2000). If pigs are exposed to sunshine for too long in summer, they often burn their skin. For pigs kept outside in areas without natural shade, a shelter must be provided.

Farrowing and nursing

Special emphasis should be given to the housing of farrowing and lactating sows. Under commercial conditions, sows are usually moved to the farrowing accommodation 3–7 days before the expected farrowing and are penned individually. To enable them to turn around, to perform nest-building behaviour and to separate their dunging area from their lying area, sows must not be confined in farrowing crates. The minimum space to allow undisturbed behaviour at the nest and to prevent piglets from being crushed is 7.5 m^2 (Schmid, 1992, 1993). These findings correspond with the Council Regulation (EC) No. 1804/1999, according to which the minimum indoor surface area for farrowing sows in organic farming is 7.5 m^2.

A loose-housing farrowing pen must have a lying/nest area with bedding and an activity/defecating area. Cronin *et al.* (1998) concluded that the width of the nest area affects sow and piglet behaviour that may be relevant to piglet survival. Anti-crushing rails around the walls or inward-slipping bars to limit the area where the sow can lie can help reduce crushing of piglets. Providing straw as nest-building material presumably favours piglet survival by affecting the timing and quality of nest building, reducing the duration of the first part of farrowing, and reducing nest building and postural changes during farrowing (Thodberg *et al.*, 1999).

Another important factor in the design of a farrowing pen is the type of floor, which affects the choice of the nest site and nest-building behaviour. Newborn piglets often develop abrasions on their front legs from contact with the floor during suckling activity (Furniss *et al.*, 1986).

The sow and piglets are separated much earlier under commercial conditions than in nature. Several studies have shown that weaning may be a source of distress, causing an increase in vocalizations, aggression, plasma cortisol concentration and digestive disorders. Weaning piglets at 7 weeks of age, as required by the Council Regulation (EC) No. 1804/1999, is better than weaning earlier, but is still thought to be a problem for the piglets.

The temperature requirements of lactating sows and piglets are very different. The thermoneutral zone is likely to be around 34°C for newborn piglets and 25–30°C for 4–6 kg piglets, whereas for sows, it is about 15°C (Mount, 1968). This makes it necessary to match the ambient temperature to the sows' requirements and provide a well-designed nest with straw bedding for the piglets, which can be heated with either infrared or underground heaters. Over the last few years several 'get-away pens' have been developed, such as the Schmid pen (Schmid, 1992, 1993), the FAT pen (Weber, 1996, 2000) and the Werribee pen (Cronin *et al.*, 1996, 1998). These systems usually do not have access to an outdoor run, which must be provided to all animals in organic farming according to Council Regulation (EC) No. 1804/1999. However, little research has been done about the effects of an outdoor run on the behaviour and health of lactating sows and suckling piglets.

Group housing

An interesting alternative to keeping sows and piglets in farrowing pens for the whole lactation is to move them to a group pen 10–14 days after farrowing, where 3–15 sows and their litters are kept together until weaning (Algers, 1991). This system, often referred to as 'multi-suckling units', is reported to reduce aggression among sows and piglets and allow a better design of the pen than an individual farrowing pen, especially if an outdoor run must be provided.

However, cross-suckling (presence of alien piglets at the udder during milk ejection) may be a problem. Cross-suckling results in more fighting amongst piglets (Pedersen *et al.*, 1998), missed nursings, and nursings without milk ejection (Arey and Sancha, 1996), and reduced milk intake and weight gain (Puppe and Tuchscherer, 1995). Cross-suckling declines with decreasing group size (Fraser and Broom, 1990), and with decreasing variation in the age and size of the litters (Brodmann *et al.*, 1995).

In group farrowing systems, the sows are already grouped before farrowing and give birth in small compartments within the group pen; therefore they are never moved during the lactation (van Putten and van de Burgwal, 1989; Goetz and Troxler, 1995a,b; Arey and Sancha, 1996). The mortality of piglets may be higher in these systems.

An ethological approach that aims to satisfy the motivations and behavioural requirements of domestic pigs is the 'family pen system' (Stolba and Wood-Gush, 1989; Wechsler *et al.*, 1991). It is a combined breeding, rearing and fattening system. The sows are kept in stable groups throughout production. The piglets are weaned naturally by their mother and are not removed from the family group until they have reached market weight. By this time the sows, which are served during lactation, are ready to farrow again, and the cycle is complete. Arey and Sancha (1996) reported that piglet mortality was not significantly different from farrowing crates. The pen is an adaptation of the features that released normal behaviour in a semi-natural enclosure. It contains nest areas, activity areas and rooting areas. Wechsler (1996) reported that lactational oestrus occurred in 54% of the sows before the piglets were 7 weeks old; it is a problem if not all sows of a group can be mated.

Poultry (Laying Hens)

Domesticated species of poultry currently kept in organic agriculture are mainly terrestrial birds of the order galliformes, the most important being chickens and turkeys. Others, such as guinea fowl and quail, play a minor role. Also important are aquatic species of the order anseriformes, such as ducks, muscovy ducks and geese. Here we concentrate on chickens, especially laying hens.

Species-specific behaviour

The ancestor of the domestic chicken, the red jungle fowl, *G. gallus* (with several subspecies) lives from India to Malaysia at the edges of densely covered forest areas (Fumihito *et al.*, 1996). Because it is difficult to study in its natural habitat, we also draw on studies of unconfined flocks of red jungle fowl in zoos (e.g. Collias *et al.*, 1966; Dawkins, 1989) or in pens (Fölsch, 1981b; Kruijt, 1964). We also present information on feral domestic chickens studied on two islands off Australia (McBride *et al.*, 1969) and in Scotland (Duncan *et al.*, 1978); these showed behavioural traits qualitatively consistent with those of the red jungle fowl.

Social structure

Chickens are highly social. They form comparatively stable groups (flocks of 6–30 at San Diego Zoo according to Collias *et al.*, 1966) that centre their lives around a roosting site (tree) in a territory or home range of few hectares with watering places and less densely covered or open feeding sites (Collias and Collias, 1967; Wood-Gush *et al.*, 1978). However, McBride *et al.* (1969) report that feral fowl have fixed, defended territories only during the breeding season, with overlapping home ranges around the roosting place the rest of the year.

Within the flock a dominance hierarchy is established. Breeding flocks are usually composed of four to six females and a dominant male, sometimes with a few subordinate males (Collias and Collias, 1967; McBride *et al.*, 1969). Females form an independent dominance hierarchy (Collias and Collias, 1996) and are rather tolerant of females from other flocks (McBride *et al.*, 1969). In domesticated chickens, status is affected by age, breed, comb size and colour, and body weight (Mench and Keeling, 2001). All behaviours (e.g. walking, preening, feeding, resting) are highly synchronized (Savory *et al.*, 1978; Mench and Keeling, 2001).

Reproduction

Red jungle fowl breed seasonally, from March to May (Collias and Collias, 1967). Before laying, hens separate from the group and start inspecting possible nesting sites (Fölsch, 1981b). By giving a nesting call the hen attracts a male who will accompany her and show her possible nests (Fölsch, 1981b; McBride *et al.*, 1969). Nests are located on the ground, covered by bushes or other structures (Collias and Collias, 1967; Duncan *et al.*, 1978). When a nest is finally chosen, the hen will scratch and turn to create a suitable shallow depression. The nest also will have a few leaves and feathers. Whilst sitting on the nest the hen throws loose material on her back or places it along her body (McBride *et al.*, 1969; Fölsch, 1981a).

Eggs are usually laid in the morning. After laying the egg, the hen quietly moves away from the nest and gives a cackle, which attracts a male to accompany her back to the flock (Fölsch, 1981a). During this occasion mating occurs (McBride *et al.*, 1969). Red jungle fowl usually have a clutch of five to ten eggs (Collias and Collias, 1967). After the whole clutch is laid, the hen starts brooding, briefly leaving the nest only once a day (Duncan *et al.*, 1978; Fölsch, 1981b).

The chicks of a clutch hatch at nearly the same time. Chicks are nidifugous and are led out of the nest by the hen within 36 h after the first chick is hatched (Fölsch, 1981b). Except for social and reproductive behaviour, most of the behavioural repertoire of red jungle fowl chicks (e.g. locomotion, feeding, drinking and comfort behaviour) is completely developed within the first 2 weeks (Kruijt, 1964). Domestic chickens already dust bathe by the third day. They fly as early as 4–5 weeks (Collias and Collias, 1967), and Wood-Gush *et al.* (1978) observed a brood at age 7 weeks roosting in a tree 7 m above the ground. Broods are left by the hen at 5–8 weeks (Wood-Gush *et al.*, 1978; Collias and Collias, 1996). Wood-Gush *et al.* (1978) observed broods of feral domestic chickens staying together in a group and integrating into the flock at age 4–5 months.

Feeding, movement and body care

Chickens are omnivorous. Red jungle fowl eat a great variety of items, including seeds, insects (caterpillars, termites, etc.), spiders, snails, leaves, grasses and fruits (Collias and Collias, 1967). McBride *et al.* (1969) observed their population of feral domestic chickens also eating carrion. Laying hens even can be observed hunting small animals, such as mice and frogs. Chicks seem to depend mainly on animal sources of protein, such as invertebrates (Collias and Collias, 1967; Savory *et al.*, 1978). Red jungle fowl and feral domestic chickens spend half to two-thirds of the day in feeding and foraging behaviour, such as walking, scratching and pecking (Collias and Collias, 1967; Savory *et al.*, 1978; Dawkins, 1989). Red jungle fowl usually get water from waterholes (Collias and Collias, 1967).

Chickens are very active. Locomotion is closely related to foraging, and red jungle fowl might spend more than 60% of the daytime walking even when fed *ad libitum* (Dawkins, 1989). Nevertheless, they generally move within a small area, about 140 m in diameter (Collias and Collias, 1967). Similarly McBride *et al.* (1969) report average distances of 60–150 m between roosting trees of dominant males. Red jungle fowl and feral chickens fly very little, mostly to access or leave the roosting tree or to escape. Both red jungle fowl and feral chickens roost on branches of high trees during both night and day (Collias and Collias, 1967; Wood-Gush *et al.*, 1978).

Comfort behaviour such as preening and dust bathing are important for maintaining the plumage in good conditions. Dawkins (1989) found red jungle fowl spending more than 10% of the daytime preening. Besides egg laying, dust bathing is the chicken's most complex behaviour, but so far it has been studied mainly under experimental conditions (Vestergaard *et al.*, 1997).

To summarize, chickens have a highly complex behavioural

repertoire. Their behaviour is arranged in a distinctive daily pattern with peaks of activity in the morning and afternoon. Typically, they start preening before daylight, leave their night roost, go to search for food, (lay an egg in domestic hens), rest at midday, search for food again and start roosting again before sunset. Both their behaviour and daily activity pattern must be taken into account in designing and managing a hen house.

Implications for the husbandry of laying hens

Grouping

In today's organic agriculture, hens are usually kept in groups of several hundred to a thousand. Compared to the social environment usually experienced by their ancestors, this presents a considerable challenge. As described earlier, in red jungle fowl, a flock usually is much smaller, allowing individual recognition and the formation of a dominance hierarchy (Pagel and Dawkins, 1997). Whether the critical group size for individual recognition is actually 100, as reported by Guhl (1953), is open to debate (Mench and Keeling, 2001).

Although there have been reports of sub-groups forming in laying hens (Bölter, 1987) or individual recognition of at least some birds (Odén *et al.*, 2000), hens probably do not form a dominance hierarchy in larger flocks (Mench and Keeling, 2001). In these flocks, hens may classify other hens by features such as comb or body size (Pagel and Dawkins, 1997) and give way to those perceived as higher in status. Hens in large flocks show little aggression, even when mixed with unfamiliar flocks (Hughes *et al.*, 1997), although in single flocks agonistic behaviour can be a problem (Gunnarsson *et al.*, 1995). Cocks seem to reduce agonistic interactions between females (Odén *et al.*, 1999). Even though commercial flocks are bigger than 'natural' red jungle fowl groups, hens seem to cope well with the larger size. Furthermore, it is very difficult to design and manage alternative systems for groups of fewer than 100 birds that take into account most of the behavioural requirements of laying hens. This does not mean, however, that group size is unimportant; for example, groups of more than 500 birds seem to use outside runs less (Hirt *et al.*, 2000; Niebuhr *et al.*, 2001).

Feeding

Feeders should be evenly distributed and easily accessible, and linear feeders should be equipped with perches over the feeding space to prevent hens from roosting in the trough. Attention should be paid to avoid having linear feeders act as barriers for the birds. This can be overcome by raising the feeding equipment in combination with raised perches or walks. Nevertheless, the birds must be carefully observed, especially at the start of laying, as birds not used to raised areas or perches may have trouble getting to the feeding trough. As many feeding systems do not truly feed *ad libitum*, because of the diurnal rhythm of food intake in laying hens, feed should be plentiful, especially in late daytime hours.

Even when fed mash, laying hens spend less time feeding than red jungle fowl or feral hens spend foraging and feeding. Therefore, a scratching area with adequate litter is important for the welfare of laying hens, including to prevent feather pecking and cannibalism (Appleby *et al.*, 1992; see also Chapter 8). From practical evidence, long straw seems to be the most suitable material for foraging, scratching and pecking. It remains attractive longer when given in intervals as bales, and can be manipulated and torn into smaller pieces, of which at least a part is also swallowed. Also, giving hay, silage and other materials (e.g. pecking stones), as well as scattering grain, allows foraging and keeps the hens busy.

The outdoor run also provides excellent foraging possibilities, as long as it is open for the hens during daytime, covered with vegetation, and easy to access. However, it cannot replace an indoor scratching area lit 16 h a day. In outdoor runs, hens can take in a lot of fresh plant material; Hughes and Dun (1983) found a daily intake as high as 30 g dry matter per hen. Free range areas, which now are compulsory for laying hens in organic farming, generally should have a maximum distance of 100–150 m from the hen house, since more distant areas are hardly used (Niebuhr *et al.*, 2001). This maximum distance corresponds closely to the distances covered by red jungle fowl in their natural habitat.

The run also should be equipped with cover, such as trees and bushes. Apart from hiding places, cover provides shade and thus roosting places at midday. To prevent excessive vegetation use and nutrient loading, a rotation system must be used (at least two runs of 4 m^2 each), thereby also reducing the risk of parasitic infection. Particularly during bad weather and in winter, a covered run ('bad weather run', 'winter garden') is an excellent addition to deep litter or aviary systems with access to a free-range area. It provides an extra area for scratching and pecking all year round and thus reduces stocking density in the hen house. Furthermore, it allows sun bathing and can be equipped with boxes of sand for dust bathing, thereby decreasing the amount of dust produced in the hen house.

As feeding and drinking are closely associated (Fölsch, 1981a), drinkers should be close to the feeders. Given a choice, hens generally prefer open water surfaces, as in bell or cup drinkers, rather than nipple drinkers. Drinking from nipples is not a natural behaviour and must be learned, but switching from nipple to bell drinkers also may cause problems (Appleby *et al.*, 1992). Rearing facilities should therefore be equipped with drinkers of various types, or at least the type that the hens will later find in the hen house.

Body care, nesting and sleeping

Apart from physical space, comfort behaviour mainly requires two facilities within a hen house: raised areas or perches for preening, and litter for dust bathing. Although free-range hens can dust bathe in the run, a separate large box with sand should be present in the hen house or the 'winter garden'. Sand seems to be preferred to wood shavings or straw (Sanotra *et al.*, 1995), but remains attractive only if the boxes are refilled regularly.

Perches are not only a preferred site for laying hens to preen, but also the most important facility for roosting and sleeping, especially at night (Blokhuis, 1984). Higher perches are generally preferred (e.g. Olsson and Keeling, 2000). Series of perches should not be positioned steeper than 45°, as the birds may have problems in descending (Lambe *et al.*, 1997). In adult laying hens, raised perches reduce the number of birds on the floor and lead to less agonistic interaction (Cordiner and Savory, 2001). Perches higher than 70 cm could also reduce feather damage (Wechsler and Huber-Eicher, 1998). Raised perches are especially important during early rearing, as rearing without perches leads to poorer spatial ability later in life (Fröhlich, 1991; Gunnarsson *et al.*, 2000). It also increases the probability of feather pecking (Huber-Eicher and Audigé, 1999), cannibalism (Fröhlich, 1991; Gunnarsson *et al.*, 1999) and floor eggs in adult hens (Appleby *et al.*, 1988a; Gunnarsson *et al.*, 1999).

Floor eggs are an important problem in alternative systems, as they are often dirty or broken, which further encourages egg eating (Appleby *et al.*, 1992). To prevent floor laying, nests should be easy accessible but separated from the areas of activity. Nests with litter are preferred over rollaway nests (Appleby *et al.*, 1988b). Apart from rearing factors, nest site selection and floor laying are influenced by age, genotype, social structure, layout of the hen house and nests, number of nests (or area) per hen, and management (Bauer, 1995). Dark and enclosed sites in the littered area should be avoided (Appleby *et al.*, 1992). Raised nests should have at least two perches or a slatted area in front to let the hens inspect them.

As with other species, facilities for laying hens should be structured into functional areas corresponding to functional behavioural systems such as feeding, egg laying, drinking, roosting and foraging. Use of each area should be limited to a certain behavioural system, e.g. areas intended for roosting should not be equipped with feeders, and scratching areas should be sufficiently lighted to prevent egg laying in the litter and allow explorative behaviour.

Conclusions for Organic Animal Husbandry – Current Knowledge, Current Systems and Future Requirements

Although ethology, the study of animal behaviour, has made considerable progress towards understanding what animal welfare is and how animals must be kept to have a high quality of life, in many systems, major elements of natural behaviour cannot be performed, leading to welfare problems. But how far we should go in offering the animals a 'natural' environment is still a matter for debate. Whilst a tree that is used for rubbing is easily replaced by a brush, there are no artificial replacements for calves or piglets. Thus, is it acceptable to separate the calf from the cow or the piglet from the sow much earlier than would happen in nature? Is it acceptable to frequently separate individual animals that have bonded together, exposing them to stress and reducing the possibilities of social support? Of course, we always will have to interfere with 'natural behaviour', but we have the responsibility of balancing our own interests with those of the animals. We need to think much more about where the boundaries are to be drawn so that both the animals and the humans have lives of quality and can mutually enrich each other's.

A step towards enabling all natural behaviour was made many years ago in pigs, with the Stolba family pen system (described earlier), where the pigs are provided not only with all relevant features of the physical environment, but also a social structure similar to nature's. However, although the first version was developed two decades ago, little research has been done on this system. In dairy cows, only recently have efforts been undertaken to avoid early separation of calves and cows, which is accepted by most organic standards. Pilot work in this area and some practical experience are promising. However, to make it acceptable economically, contact between cow and calf is allowed only part of the time or only for the first 3 weeks of life. Some farmers have already gone further, integrating calves and young stock into the dairy herd, thereby approaching a natural social structure.

There are many other areas where further research is needed to solve existing problems or to move towards a higher quality of life, e.g. keeping ducks, geese and hens as egg layers in free-range systems, raising sows and fattening pigs outdoors, using males rather than artificial insemination, keeping males with the herds, and not castrating males.

Taking seriously the values of naturalness and the holistic view of the organic movement means that it is not enough that the animals can fulfil their needs in a minimal fashion. A major challenge for organic husbandry is not only to optimize current systems by using existing ethological knowledge, but also to develop more innovative systems in which all aspects of natural behaviour are taken into account. Beyond avoiding suffering, such systems make positive experiences an important part of the animal's life. It is not important to 'prove' that a sow needs to root or a hen needs to dust bathe, because these are natural behaviours belonging to the animal's species-specific nature. Rather, researchers and farmers should concentrate on answering more detailed questions to help use our understanding of animals' nature (e.g. Kiley-Worthington, 1993). This will make respect for and knowledge of animal behaviour the basis for a sustainable organic agriculture.

References

Algers, B. (1991) Group housing of farrowing sows – health aspects of a new system. In: *Proceedings VII International Congress on Animal Hygiene, Leipzig*, p. 851.

Algers, B. (1992) Natürliches Verhalten – ein natürlicher Begriff? *Berliner Münchner Tierärztliche Wochenschrift* 105, 372–374.

Appleby, M.C., Duncan, I.J.H. and McRae, H.E. (1988a) Perching and floor laying by domestic hens: experimental results and their commercial application. *British Poultry Science* 29, 351–357.

Appleby, M.C., Hogart, G.S. and Hughes, B.O. (1988b) Nest box design and nesting material in a deep litter house for laying hens. *British Poultry Science* 29, 215–222.

Appleby, M.C., Hughes, B.O. and Elson, H.A. (1992) *Poultry Production Systems. Behaviour, Management and Welfare*. CAB International, Wallingford, UK.

Arey, D.S. and Franklin, M.F. (1995) Effects of straw and unfamiliarity on fighting between newly mixed growing pigs. *Applied Animal Behaviour Science* 45, 23–30.

Arey, D.S. and Sancha, E.S. (1996) Behaviour and productivity of sows and piglets in a family system and in crates. *Applied Animal Behaviour Science* 50, 135–145.

Arnold, G.W. and Dudzinski, M.L. (1978) *Ethology of Free-ranging Domestic Animals*. Elsevier, Amsterdam.

Baldwin, B.A. and Ingram, D.L. (1967) Behavioural thermoregulation in pigs. *Physiology and Behaviour* 2, 15–21.

Barnett, J.L. and Taylor, I.A. (1997) Sequential versus concurrent feeding on acute and chronic stress responses in pigs. In: *Fifth International Livestock Environment Symposium*. American Society of Agricultural Engineers, St Joseph, Michigan, pp. 607–612.

Barnett, J.L., Cronin, G.M., McCallum, T.H., Newman, E.A. and Hennessy, D.P. (1996) Effects of grouping unfamiliar adult pigs after dark, after treatment with Amperozide and by using pens with stalls, on aggression, skin lesions and plasma cortisol concentrations. *Applied Animal Behaviour Science* 50, 121–133.

Bauer, T. (1995) Ergebnisse von Untersuchungen zum Nestwahlverhalten von Legehennen in alternativen Haltungssystemen. Thesis, Humboldt Universität, Berlin.

Beattie, V.E. and O'Connell, N.E. (2002) Relationship between rooting behaviour and foraging in growing pigs. *Animal Welfare* 11, 295–303.

Beattie, V.E., Walker, N. and Sneddon, I.A. (1995) Effects of environmental enrichment on behaviour and productivity of growing pigs. *Animal Welfare* 4, 207–220.

Beattie, V.E., Walker, N. and Sneddon, I.A. (1997) Preference testing of substrates by growing pigs. *Animal Welfare* 7, 27–34.
Beilharz, R.G. and Cox, D.F. (1967) Social dominance in swine. *Animal Behaviour* 15, 117–22.
Bendixen, P.H., Vilson, B., Ekesbo, I. and Åstrand, D.B. (1988a) Disease frequency in dairy cows in Sweden V. Mastitis. *Preventive Veterinary Medicine* 5, 263–274.
Bendixen, P.H., Vilson, B., Ekesbo, I. and Åstrand, D.B. (1988b) Disease frequency in dairy cows in Sweden VI. Tramped teat. *Preventive Veterinary Medicine* 6, 25.
Bertschinger, H.U., Drossaert van Dusseldorp, P. and Troxler, J. (1994) Incidence of intramammary infection and of mastitis in sows kept in four different farrowing systems. In: *Proceedings of the 8th International Congress on Animal Hygiene, 12–16 September 1994*. International Society for Animal Hygiene, St Paul, Minnesota, pp. 93–96.
Blackshaw, J.K., Swain, A.J., Thomas, F.J.M. and Gillies, K.J. (1997) The development of playful behaviour in piglets from birth to weaning in three farrowing environments. *Applied Animal Behaviour Science* 55, 37–49.
Blokhuis, H.J. (1984) Rest in poultry. *Applied Animal Behaviour Science* 12, 289–303.
Boden, R.F. and Leaver, J.D. (1994) A dual purpose cattle system combining milk and beef production. *Animal Production* 58, 463–464.
Boivin, X., LeNeindre, P., Chupin, J.M., Garel, J.P. and Trillat, C. (1992) Influence of breed and early management on ease of handling and open-field behaviour of cattle. *Applied Animal Behaviour Science* 32, 313–323.
Bölter, U. (1987) Felduntersuchungen zum Sozialverhalten von Hühnern in der Auslauf- und Volierenhaltung. Thesis, Justus-Liebig-Universität Giessen, Germany.
Boxberger, J. (1983) *Wichtige Verhaltensparameter von Kühen als Grundlage zur Verbesserung der Stalleinrichtung*. Forschungsbericht Agrartechnik der MEG, Weihenstephan.
Brelin, B. (1979) Mixed grazing with sheep and cattle compared with single grazing. *Swedish Journal of Agricultural Research* 9, 112–120.
Briedermann, L. (1971) Ermittlung zur Aktivitätsperiodik des mitteleuropäischen Wildschweines (Sus s. scrofa L.). Zool. *Garten N.F., Leipzig* 40, 302–327.
Brodmann, N., Wechsler, B. and Rist, M. (1995) Strategien von fremdsaugenden Ferkeln bei der Gruppenhaltung von ferkelführenden Sauen. In: *Aktuelle Arbeiten zur artgemäßen Tierhaltung 1994. KTBL-Schrift* 370, 237–246.
Broom, D.M. (1986) Indicators of poor welfare. *British Veterinary Journal* 142, 524–526.
Broom, D.M. (1992) Animal welfare: its scientific measurement and current relevance to animal husbandry in Europe. In: Phillips, C. and Piggins, D. (eds) *Farm Animals and the Environment*. CAB International, Wallingford, UK, pp. 245–254.
Brumm, M.C. (1996) Effect of space allowance on barrow performance to 136 kilograms body weight. *Journal of Animal Science* 74, 745–749.
Buchholtz, C. (1993) Das Handlungsbereitschaftsmodell – ein Konzept zur Beurteilung und Bewertung von Verhaltensstörungen. In: Martin, G. (ed.) *Leiden und Verhaltensstörungen bei Tieren. Tierhaltung* 23, Birkhäuser Verlag, Basel, pp. 93–109.
Buchwalder, T., Wechsler, B., Hauser, R., Schaub, J. and Friedli, K. (2000)

Liegeplatzqualität für Kühe im Boxenlaufstall im Test. *Agrarforschung* 7, 292–296.

Carlstead, K. (1986) Predictability of feeding: its effect in agonistic behaviour and growth in grower pigs. *Applied Animal Behaviour Science* 16, 25–38.

Castren, H., Algers, B., de Passillé, A.M., Rushen, J. and Uvnas-Moberg, K. (1993) Early milk ejection, prolonged parturition and periparturient oxytocin release in the pig. *Animal Production* 57, 465–471.

Chaplin, S.J., Tierney, G., Stockwell, C., Logue, D.N. and Kelly, M. (2000) An evaluation of mattresses and mats in two dairy units. *Applied Animal Behaviour Science* 66, 263–272.

Cloutier, S., Weary, D.M. and Fraser, D. (2000) Can ambient sound reduce distress in piglets during weaning and restraint? *Journal of Applied Animal Welfare Science* 3, Vol. 2, 107–116.

Collias, N.E. and Collias, E.C. (1967) A field study of the Red Jungle Fowl in north-central India. *The Condor* 69, 360–386.

Collias, N.E. and Collias, E.C. (1996) Social organization of a red junglefowl, *Gallus gallus,* population related to evolution theory. *Animal Behaviour* 51, 1137–1354.

Collias, N.E., Collias, E.C., Hunsaker, D. and Minning, L. (1966) Locality fixation, mobility and social organization within an unconfined population of Red Jungle Fowl. *Animal Behaviour* 14, 550–559.

Cordiner, L.S. and Savory, C.J. (2001) Use of perches and nestboxes by laying hens in relation to social status, based on examination of consistency of ranking orders and frequency of interaction. *Applied Animal Behaviour Science* 71, 305–317.

Cronin, G.M., Simpson, G.J. and Hemsworth, P.H. (1996) The effects of the gestation and farrowing environments on sow and piglet behaviour and piglet survival and growth in early lactation. *Applied Animal Behaviour Science* 46, 175–192.

Cronin, G.M., Dunsmore, B. and Leeson, E. (1998) The effects of farrowing nest size and width on sow and piglet behaviour and piglet survival. *Applied Animal Behaviour Science* 60, 331–345.

Darwin, C. (1859) *The Origin of Species by Means of Natural Selection, or the Preservation of Favoured Races in the Struggle of Life.* John Murray, London.

Dawkins, M.S. (1989) Time budgets in red junglefowl as a baseline for the assessment of welfare in domestic fowl. *Applied Animal Behaviour Science* 24, 77–80.

Day, J.E.L., Burfoot, A., Docking, C.M., Whittaker, X., Spoolder, H.A.M. and Edwards, S.A. (2002) The effects of prior experience of straw and the level of straw provision on the behaviour of growing pigs. *Applied Animal Behaviour Science* 76, 189–202.

Deininger, E. (1998) Beeinflussung der aggressiven Auseinandersetzungen beim Gruppieren von abgesetzen Sauen. *FAT-Schrift* 49, Tänikon.

Deininger, E., Friedli, K. and Troxler, J. (2002) Gruppieren von Sauen nach dem Absetzen. *Tierärztliche Umschau* 57, 234–238.

de Passillé, A.M. (2001) Sucking motivation and related problems in calves. *Applied Animal Behaviour Science* 72, 175–187.

Désiré, L., Boissy, A. and Veissier, I. (2002) Emotions in farm animals: a new approach to animal welfare in applied ethology. *Behaviour Proceeding* 60, 165–180.

Dickinson A. (1980) *Contemporary Animal Learning Theory*. Cambridge University Press, Cambridge.

Duncan, I.J.H., Savory, C.J. and Wood-Gush, D.G.M. (1978) Observations on the reproductive behaviour of domestic fowl in the wild. *Applied Animal Ethology* 4, 29–42.

EC Council Regulation (1999) Council Regulation (EC) No 1804/1999 of 19 July 1999 supplementing Regulation (EEC) No. 2092/91 on organic production of agricultural products and indications referring thereto on agricultural products and foodstuffs to include livestock production (OJ L 222 of 24.8.1999). *Official Journal L 083, 04/04/2000 P. 0035 – 0035 EN.*

Edwards, S. (2000) Alternative housing for sows: System studies or component analyses? In: Blokhuis, H.J., Ekkel, E.D. and Wechsler, B. (eds) *Improving Health and Welfare in Animal Production*. EAAP Publication No. 102. Wageningen, pp. 99–107.

Edwards, S.A., Mauchline, S. and Stewart, A.H. (1993) Designing pens to minimise the aggression when sows are mixed. *Farm Building Progress* 113, 20–23.

Egle, B., Meier, K., Richter, T. and Von Borell, E.H. (1999) Gegenseitiges Besaugen von Kälbern unter dem Einfluss von Glucosezufütterung. In: *Aktuelle Arbeiten zur artgemäßen Tierhaltung 1998. KTBL-Schrift* 382, 137–145.

Ewbank, R. (1976) Social hierarchy in suckling and fattening pigs: a review. *Livestock Production Science* 3, 363–372.

Ewbank, R. and Bryant, M.J. (1969) The effects of population density upon the behaviour and economic performance of fattening pigs. *Farm Building Progress* 18, 14–15.

Ewbank, R., Meese, G.B. and Cox, J.E. (1974) Individual recognition and the dominance hierarchy in the domestic pig. The role of sight. *Animal Behaviour* 22, 473–480.

Flower, F.C. and Weary, D.M. (2001) Effects of early separation on the dairy cow and calf: 2. Separation at 1 day and 2 weeks after birth. *Applied Animal Behaviour Science* 70, 275–284.

Fölsch, D.W. (1981a) Das Verhalten von Legehennen in unterschiedlichen Haltungssystemen unter Berücksichtigung der Aufzuchtmethoden. In: Fölsch, D.W. and Vestergaard, K. (eds) *The Behaviour of Fowl. The Normal Behaviour and the Effect of Different Housing Systems and Rearing Methods. Tierhaltung* 12, Birkhäuser Verlag, Basel, pp. 9–114.

Fölsch, D.W. (1981b) Die Veranlagung zum Brutverhalten und zur Aufzucht von Leghorn-Hybriden und Bankiva-Hühnern. In: Fölsch, D.W. and Vestergaard, K. (eds) *The Behaviour of Fowl. The Normal Behaviour and the Effect of Different Housing Systems and Rearing Methods. Tierhaltung* 12, Birkhäuser Verlag, Basel, pp. 133–143.

Frädrich, H. (1965) Zur Biologie und Ethologie des Warzenschweines (Phacochoerus aethiopicus) unter Berücksichtigung des Verhaltens anderer Suiden. *Zeitschrift für Tierpsychologie* 22, 328–393.

Frädrich, H. (1974) A comparison of behaviour in the suidae. In: *The Behaviour of Ungulates and Its Relation to Management*. IUCN Pub. 24, 133–143, Morgues.

Fraser, D. (1975) The 'teat order' of suckling pigs. II. Fighting during suckling and the effects of clipping the eye teeth. *Journal of Agricultural Science, Cambridge* 84, 393–399.

Fraser, A.F. and Broom, D.M. (1990) Feeding. In: *Farm Animal Behaviour and Welfare*, 3rd edn. CAB International, Wallingford, UK, pp. 79–98.

Fraser, D., Phillips, P.A., Thompson, B.K. and Tennessen, T. (1991) Effect of straw on the behaviour of growing pigs. *Applied Animal Behaviour Science* 30, 307–318.

Fraser, D., Phillips, P.A., Thompson, B.K., Pajor, E.A., Weary, D.M. and Braithwaite, L.A. (1995) Behavioural aspects of piglet survival and growth. In: Varley, M.A. (ed.) *The Neonatal Pig – Development and Survival*. CAB International, Wallingford, UK, pp. 287–312.

Fraser, D., Weary, D.M., Pajor, E.A. and Milligan, B.N. (1997) A scientific conception of animal welfare that reflects ethical concerns. *Animal Welfare* 6, 187–205.

Fröhlich, E.K.F. (1991) Zur Bedeutung erhöhter Sitzstangen und räumlicher Enge während der Aufzucht von Legehennen. In: *Aktuelle Arbeiten zur artgemäßen Tierhaltung 1990. KTBL-Schrift* 344, 36–46.

Fumihito, A., Miyake, T., Takada, M., Shingu, R., Endo, T., Gojobori, T., Kondo, N. and Ohno, S. (1996) Monophyletic origin and unique dispersal patterns of domestic fowls. *Proceedings of the National Academy of Sciences, USA* 93, 6792–6795.

Furniss, S.J., Edwards, S.A., Lightfoot, A.L. and Spechter, H.H. (1986) The effect of floor type in farrowing pens on pig injury. I. Leg and teat damage of suckling piglets. *British Veterinary Journal* 142, 434–440.

Galindo, F. and Broom, D.M. (1994) How does social behaviour of dairy cows affect the occurrence of lameness? *Applied Animal Behaviour Science* 41, 272–273.

Geyer, H. (1979) Morphologie und Wachstum der Schweineklaue. Grundlagen für Stallbodengestaltung und Klauenpathologie. *Schweizer Archiv für Tierheilkunde* 121, 275–293.

Githigia, S.M., Thamsborg, S.M. and Larsen, M. (2001) Effectiveness of grazing management in controlling gastrointestinal nematodes in weaner lambs on pasture in Denmark. *Veterinary Parasitology* 99, 15–27.

Goetz, M. and Troxler, J. (1993) Farrowing and nursing in the group. In: Collins, E. and Boon, C. (eds) *Livestock Environment IV*. American Society of Agricultural Engineers, St Joseph, Michigan, p. 159.

Goetz, M. and Troxler, J. (1995a) Sauen in Gruppen während der Geburt und Säugezeit. *FAT-Schrift* 50, Tänikon, Switzerland.

Goetz, M. and Troxler, J. (1995b) Group housing of sows during farrowing and lactation. *Transaction of the ASAE* 38, 1495–1500.

Graf, B., Verhagen, N. and Sambraus, H.H. (1989) Reduzierung des Ersatzsaugens bei künstlich aufgezogenen Kälbern durch Fixierung nach dem Tränken oder Verlängerung der Saugzeit. *Züchtungskunde* 61, 384–400.

Grandin. T. (1989) Behavioural principles of livestock handling. *The Professional Animal Scientist* 5, 1–11.

Grandin, T. (1993a) Behavioural principles of cattle handling under extensive conditions. In: Grandin, T. (ed.) *Livestock Handling and Transport*. CAB International, Wallingford, UK, pp. 43–58.

Grandin, T. (1993b) Handling facilities and restraint of range cattle. In: Grandin, T. (ed.) *Livestock Handling and Transport*. CAB International, Wallingford, UK.

Grauvogl, A. (1958) Über das Verhalten der Hausschweine mit besonderer

Berücksichtigung des Fortpflanzungsverhaltens. Thesis, Freie Universität Berlin.

Graves, H.B. (1984) Behaviour and ecology of wild and feral swine (*Sus scrofa*). *Journal of Animal Science* 58, 482–492.

Guhl, A.M. (1953) Social behaviour of the domestic fowl. *Kansas Agricultural Experimental Station Technical Bulletin* 73, 1–48.

Gundlach, H. (1968) Brutfürsorge, Brutpflege, Verhaltensontogenese und Tagesperiodik beim Europäischen Wildschwein (*Sus scrofa* L.). *Zeitschrift für Tierpsychologie* 25, 955–995.

Gunnarsson, S., Odén, K., Algers, B., Svedberg, J. and Keeling, L. (1995) *Poultry Health and Behaviour in a Tiered System for Loose Housed Layers*. Report 35, Department of Animal Hygiene, Swedish University of Agricultural Sciences, Skara.

Gunnarsson, S., Keeling, L.J. and Svedberg, J. (1999) Effect of rearing factors on the prevalence of floor eggs, cloacal cannibalism and feather pecking in commercial flocks of loose housed laying hens. *British Poultry Science* 40, 12–18.

Gunnarsson, S., Yngvesson, J., Keeling, L.J. and Forkman, B. (2000) Rearing without early access to perches impairs the spatial skills of laying hens. *Applied Animal Behaviour Science* 67, 217–228.

Guy, J.H., Rowlinson, P., Chadwick, J.P. and Ellis, M. (2002) Behaviour of two genotypes of growing-finishing pig in three different housing systems. *Applied Animal Behaviour Science* 75, 193–206.

Hafez, E.S.E. and Bouissou, M.F. (1975) The behaviour of cattle. In: Hafez, E.S.E. (ed.) *The Behaviour of Domestic Animals*. Ballière Tindall, London, pp. 203–245.

Hafez, E.S.E, Schein, M.W and Ewbank, R. (1963) The behaviour of cattle. In: Hafez, E.S.E (ed.) *The Behaviour of Domestic Animals*. Ballière Tindall, London, pp. 247–296.

Hall, S.J.G. and Moore, G.F. (1986) Feral cattle of Swona, Orkney Islands. *Mammal Review* 16, 89–96.

Hansen, R.P. and Karstad, L. (1959) Feral swine in the south-eastern United States. *Journal of Wildlife Management* 23, 64–74.

Hansen, K., Pallesen, C.N. and Chastain, J.P. (1998) Dairy cow pressure on self-locking feed barriers. *Proceedings of the 4th International Dairy Housing Conference*. St Louis, Missouri, pp. 312–319.

Hasegawa, N., Nishiwaki, A., Sugawara, K. and Ito, I. (1997) The effects of social exchange between two groups of lactating primiparous heifers on milk production, dominance order, behavior and adrenocortical response. *Applied Animal Behaviour Science* 51, 15–27.

Heizmann, V., Hauser, C. and Mann, M. (1988) Zum Erkundungs- und Spielverhalten juveniler Hausschweine in der Stallhaltung. In: *Aktuelle Arbeiten zur artgemäßen Tierhaltung 1987. KTBL-Schrift* 323, 243–265.

Hemsworth, P.H. and Coleman, G.J. (1998) *Human–Livestock Interactions: the Stockperson and the Productivity and Welfare of Intensively Farmed Animals*. CAB International, Wallingford, UK.

Hemsworth, P.H., Barnett, J.L., Hansen, C. and Winfield, C.G. (1986) Effects of social environment on welfare status and sexual behaviour of female pigs. II. Effects of space allowance. *Applied Animal Behaviour Science* 16, 259–267.

Hemsworth, P.H., Coleman, G.J., Barnett, J.L. and Borg, S. (2000) Relationships between human-animal interactions and productivity of commercial dairy cows. *Journal of Animal Science* 78, 2821–2831.

Henry, J.P. and Stephens, P.M. (1977) *Stress, Health, and the Social Environment.* Springer, London.

Hirt, H., Hördegen, P. and Zeltner, E. (2000) Laying hen husbandry: group size and use of hen-runs. In: Alföldi, T., Lockeretz, W. and Niggli, U. (eds) *IFOAM 2000 – The World Grows Organic. Proceedings of the 13th International IFOAM Scientific Conference,* Basel, 28–31 August 2000. VDF, Zürich, p. 363.

Hörning, B. (1998) Tiergerechtheit und Tiergesundheit in ökologisch wirtschaftenden Betrieben. *Deutsche Tierärztliche Wochenschrift* 105, 313–321.

Hörning, B. (2002) Vergleich von Milchviehlaufstallsystemen unter nutztierethologischen und wirtschaftlichen Gesichtspunkten. Habilitation, Universität Gesamthochschule Kassel, Witzenhausen, Germany.

Hörning, B. and Tost, J. (2001) Lying behaviour of dairy cows in different loose housing system. In: *Proceedings of the International Symposium, Animal Welfare Considerations in Livestock Housing Systems.* 2nd Technical Section of the International Commission of Agricultural Engineering (C.I.G.R.), Zielona Góra, Poland, pp. 229–238.

Hörning, B., Zeitlmann, C. and Tost, J. (2001) Unterschiede im Verhalten von Milchkühen im Liegebereich verschiedener Laufstallsysteme. In: *Aktuelle Arbeiten zur artgemäßen Tierhaltung 2000. KTBL-Schrift* 403, 153–162.

Hovi, M. (1998) A new approach to calf rearing. *Organic Farming* 60, 18–19.

Hsia, L.C. and Wood-Gush, D.G.M. (1984) The relationship between social facilitation and feeding behaviour in pigs. *Applied Animal Ethology* 8, 410.

Huber-Eicher, B. and Audigè, L. (1999) Analysis of risk factors for the occurrence of feather pecking in laying hen growers. *British Poultry Science* 40, 599–604.

Hudson. S.J. (1977) Multi-fostering of calves on to nurse cows at birth. *Applied Animal Ethology* 3, 57–63.

Hughes, B.O. and Dun, P. (1983) Production and behaviour of laying domestic fowls in outside pens. *Applied Animal Ethology* 11, 201.

Hughes, B.O., Carmichael, N.L., Walker, A.W. and Grigor, P.N. (1997) Low incidence of aggression in large flocks of laying hens. *Applied Animal Behaviour Science* 54, 215–234.

IGN-Working Group 'Group Housing of Dry Sows' (2000) *Report on the IGN-Workshop Group Housing of Dry Sows, 9–11 September 1998.* Bundesanstalt für Alpenländische Landwirtschaft.

Jensen, P. (1980) An ethogram of social interaction pattern in group-housed dry sows. *Applied Animal Ethology* 6, 341–350.

Jensen, P. (1982) An analysis of agonistic interaction patterns in group-housed dry sows: aggression regulation through an 'avoidance order'. *Applied Animal Ethology* 9, 47–61.

Jensen, P. (1986) Observations on the maternal behaviour of free-ranging domestic pigs. *Applied Animal Behaviour Science* 16, 131–142.

Jensen, P. (1993) Nest building in domestic sows: the role of external stimuli. *Animal Behaviour* 45, 351–358.

Jensen, P. (1994) Fighting between unacquainted pigs. Effects of age and of individual reaction pattern. *Applied Animal Behaviour Science* 41, 37–52.

Jensen, P. (1995) The weaning process of free-ranging domestic pigs: within-and between-litter variations. *Ethology* 100, 14–25.

Jensen, P. and Recén, B. (1989) When to wean – observations from free-ranging domestic pigs. *Applied Animal Behaviour Science* 23, 49–60.
Jensen, P. and Redbo, I. (1987) Behaviour during nest leaving in free-ranging domestic pig. *Applied Animal Behaviour Science* 18, 355–362.
Jensen, P. and Stangel, G. (1992) Behaviour of piglets during weaning in a semi-natural enclosure. *Applied Animal Behaviour Science* 33, 227–238.
Jensen, P. and Toates, F.M. (1993) Who needs 'behavioural needs'? Motivational aspects of the needs of animals. *Applied Animal Behaviour Science* 37, 161–181.
Jensen, P. and Wood-Gush, D.G.M. (1984) Social interactions in a group of free-ranging sows. *Applied Animal Behaviour Science* 12, 327–337.
Jensen, K.H., Sørensen, L.S., Bertelsen, D., Pedersen, A.R., Jørgensen, E., Nielsen, N.P. and Vestergaard, K.S. (2000) Management factors affecting activity and aggression in dynamic group housing systems with electronic sow feeding: a field trial. *Animal Science* 71, 535–545.
Jones, G. F. (1998) Genetic aspects of domestication, common breeds and their origin. In: Ruvinsky, A. and Rothschild, M.F. (eds) *The Genetics of the Pig*. CAB International, Wallingford, UK, pp. 17–50.
Jung, J. and Lidfors, L. (2001) Effects of milk, milk flow and access to a rubber teat on cross-sucking and non-nutritive sucking in dairy calves. *Applied Animal Behaviour Science* 72, 201–213.
Keil, N.M. (2000) Development of intersucking in dairy heifers and cows. Dissertation, Swiss Federal Institute of Technology, Zürich.
Keil, N.M. and Langhans, W. (2001) The development of intersucking in dairy calves around weaning. *Applied Animal Behaviour Science* 72, 295–308.
Keil, N.M., Audigé, L. and Langhans, W. (2000) Factors associated with intersucking in Swiss dairy heifers. *Preventive Veterinary Medicine* 45, 305–323.
Kiley, M. (1976) Fostering and adoption in beef cattle. *British Cattle Breeders Digest* 31, 42–55.
Kiley-Worthington, M. (1977) *Behavioural Problems of Agricultural Animals.* Oriel Press, Stocksfield, UK.
Kiley-Worthington, M. (1989) Ecological, ethological and ethically sound environments for animals: toward symbiosis. *Journal of Agricultural Ethics* 2, 232–247.
Kiley-Worthington, M. (1993) *Eco-Agriculture, Food First Farming*. Souvenir Press, London.
Kiley-Worthington, M. (1995) The future of the dairy industry. Can we overcome the environmental, ethological and ethical problems? If so how? In: *3rd International Dairy Housing Conference, Orlando, Florida*, pp. 524–532.
Kiley-Worthington, M. and de la Plain, S. (1983) *The Behaviour of Beef Suckler Cattle*. Birkhäuser Verlag, Basel.
Kiley-Worthington, M. and Randle, H.D. (1997) The practicalities and economics of ethologically and ecologically raised double suckled beef. *Biological Agriculture and Horticulture* 16, 381–393.
Kircher, A. (2001) Untersuchung zum Tier-Fressplatzverhältnis bei der Fütterung von Aufzuchtferkeln und Mastschweinen an Rohrbreiautomaten unter dem Aspekt der Tiergerechtheit. *FAT-Schrift* 53, Tänikon.
Klingholz. F., Siegert, C. and Meynhardt, H. (1979) Die akustische Kommunikation des Europäischen Wildschweines (*Sus scrofa* L.). *Der Zoologische Garten* 49, 277–303.
Knierim, U. (1998) Wissenschaftliche Untersuchungsmethoden zur Beurteilung

der Tiergerechtheit. In: *Beurteilung der Tiergerechtheit von Haltungssystemen. KTBL-Schrift* 377, 40–48.

Knierim, U. (1999a) Das Verhalten von Färsen bei der Einzel- oder Gruppeneinführung in die Milchviehherde. In: *Aktuelle Arbeiten zur artgerechten Tierhaltung 1998. KTBL-Schrift* 382, 115–120.

Knierem (1999b) Social and resting behaviour of heifers after single or group introduction to the dairy herd. In: Veissier, I. and Boissy, A. (eds) *Proceedings of the 32nd International Congress of the ISAE, Clermont-Ferrand*, p. 153.

Konggaard, S.P. (1983) Feeding conditions in relation to welfare for cows in loose-housing systems. In: Baxter, S.H., Baxter, M.R. and MacCormack, J.A.D. (eds) *Farm Animal Housing and Welfare*. Martin Nijhoff Publishers, Boston, Massachusetts, pp. 272–282.

Krohn, C.C. (2001) Effects of different suckling systems on milk production, udder health, reproduction, calf growth and some behavioural aspects in high producing dairy cows – a review. *Applied Animal Behaviour Science* 72, 271–280.

Krohn, C.C., Jonasen, B. and Munksgaard, L. (1990) Cow–calf relations III. The effect of 6–8 weeks suckling on behaviour of the cow, milk production and udder health and reproduction. Report from the National Institute of Animal Science, Fredriksberg, Denmark.

Krötzl, H. and Hauser, R. (1997) Ethologische Grundlagen zum Platzbedarf, zur Gestaltung und zum Betrieb von Laufhöfen bei Kühen im Laufstall. *Agrartechnische Forschung* 3, 141.

Kruijt, J.P. (1964) Ontogeny of social behaviour in Burmese red junglefowl. *Behaviour,* Supplement XII. Brill, Leiden, The Netherlands.

Lambe, N.R., Scott, G.B. and Hitchcock, D. (1997) Behaviour of laying hens negotiating perches at different heights. *Animal Welfare* 6, 29–41.

Lazo, A. (1994) Social segregation and the maintenance of social stability in a feral cattle population. *Animal Behaviour* 48, 1133–1141.

Le Neindre, P. (1982) Cow–calf relationships: the effect of management systems. In: Signoret, P. (ed.) *Welfare and Husbandry of Calves*. Martin Nijhoff Publishers, Boston, Massachusetts, pp. 24–35.

Le Neindre, P. and Garel, J.-P. (1979) Adoption of foster calves by Salers cows several days after calving. *Annales de Zootechnie* 28, 231–234.

Lensink, J., Fernandez, X., Cozzi, G., Florand, L. and Veissier, I. (2001) The influence of farmers' behavior on calves' reactions to transport and quality of veal meat. *Journal of Animal Science* 79, 642–652.

Leonard, F.C., O'Connell, J.M. and O'Farrell, K.J. (1996) Effect of overcrowding on claw health in first-calved Friesian heifers. *British Veterinary Journal* 152, 459–472.

Lidfors, L. and Jensen, P. (1988) Behaviour of free-ranging beef cows and calves. *Applied Animal Behaviour Science* 20, 237–247.

Loberg, J. and Lidfors, L. (2001a) Effect of milkflow rate and presence of a floating nipple on abnormal sucking between dairy calves. *Applied Animal Behaviour Science* 72, 189–199.

Loberg, J. and Lidfors, L. (2001b) Effect of stage of lactation and breed on dairy cows' acceptance of foster calves. *Applied Animal Behaviour Science* 74, 97–108.

Lorz, A. (1973) *Tierschutzgesetz. Kommentar*. Beck, Munich.

Lott, D.F. (1991) American bison sociobiology. *Applied Animal Behaviour Science* 29, 135–145.

Lund, V. (2000) Is there such a thing as 'organic' animal welfare? In: Hovi, M. and Trujillo, R.G. (eds) *Diversity of Livestock Systems and Definition of Animal Welfare. Proceedings of the Second NAHWOA Workshop, Córdoba, 8–11 January 2000*. University of Reading, Reading, pp. 151–160.

Lund, V. and Röcklinsberg, H. (2001) Outlining a conception of animal welfare for organic farming systems. *Journal of Agricultural and Environmental Ethics* 14, 391–424.

Manninen, E., de Passillé, A.M., Rushen, J., Norring, M. and Saloniemi, H. (2002) Preferences of dairy cows kept in unheated buildings for different kind of cubicle flooring. *Applied Animal Behaviour Science* 75, 281–292.

Margerison, J.K., Phillips, C.J.C. and Preston, T.R. (2003) Effect of restricted suckling in dairy cows on milk production, reproduction and calf production. *Journal of Animal Science* (in press).

Martys, M. (1982) Gehegebeobachtungen zur Geburts- und Reproduktionsbiologie des Europäischen Wildschweines (Sus scrofa L.). *Zeitschrift für Säugetierkunde* 47, 100–113.

Martys, M.F. (1991) Ontogenie und Funktion der Saugordnung und Rangordnung beim Europäischen Wildschwein, *Sus scrofa* L. *Sitzung der Tagung über Wildschweine und Pekaris im Zoo Berlin vom 12. bis 15. Juli 1990*. Frädrich-Jubiläumsband, Bongo, Berlin, 18, 219–232.

Marx, D. and Schuster, H. (1986) Ethologische Wahlversuche mit frühabgesetzten Ferkeln während der Flatdeckhaltung. *Deutsche tierärztliche Wochenschrift* 93, 65–104.

Mauget, R. (1981) Behavioural and reproductive strategies in wild forms of *Sus scrofa* (European wild boar and feral pigs). In: Sybesma, W. (ed.) *The Welfare of Pigs*. Martinus Nijhoff, The Hague.

Mayer, C. (1999) Stallklima, ethologische uns klimatische Untersuchungen zur Tiergerechtheit unterschiedlicher Haltungssysteme in der Schweinemast. *FAT-Schrift* 50.

McBride, G., Parer, I.P. and Foenander, F. (1969) The social organization and behaviour of the feral domestic fowl. *Animal Behaviour Monographs* 2, 126–181.

McGlone, J.J. and Curtis, S.E. (1985) Behaviour and performance of weanling pigs in pens equipped with hide areas. *Journal of Animal Science* 60, 20–24.

McKinnon, A.J., Edwards, S.A., Stephens, D.B. and Walters, D.E. (1989) Behaviour of groups of weaner pigs in three different housing systems. *British Veterinary Journal* 145, 367–372.

Mench, J.A. (1998) Thirty years after Brambell: whither animal welfare science? *Journal of Applied Animal Welfare Science* 1, 91–102.

Mench, J.A. and Keeling, L.J. (2001) The social behaviour of domestic birds. In: Keeling, L.J. and Gonyou, H.W. (eds) *Social Behaviour in Farm Animals*. CAB International, Wallingford, UK, pp. 177–209.

Menke, C., Waiblinger, S., Fölsch, D.W. and Wiepkema, P.R. (1999) Social behaviour and injuries of horned cows in loose housing systems. *Animal Welfare* 8, 243–258.

Menke, C., Waiblinger, S. and Fölsch, D.W. (2000) Die Bedeutung von Managementmaßnahmen im Laufstall für das Sozialverhalten von Milchkühen. *Deutsche tierärztliche Wochenschrift* 107, 262–268.

Metzner, R. (1976) Kennwerte tiergerechter Versorgungseinrichtungen des Kurzstandes für Fleckviehkühe. Dissertation, TU München, Germany.

Meynhardt, H. (1978) *Schwarzwild-Report*. Neumann-Neudamm, Melsungen.
Molz, C. (1989) Beziehung zwischen haltungstechnischen Faktoren und Schäden beim Milchvieh in Boxenlaufställen. Dissertation, LMU München, Germany.
Mount, L.E. (1960) The influence of huddling and body size on the metabolic rate of the young pig. *Journal of Agriculture Science* 55, 101–105.
Mount, L.E. (1968) *The Climatic Physiology of the Pig*. Edward Arnold, London.
Mulitze, P. (1989) Die Bestimmung der Trittsicherheit perforierter Stallfußböden für die Schweinehaltung. Dissertation, Fachbereich Agrarwissenschaften, Justus-Liebig-Universität Gießen, Germany.
Newberry, R.C. and Wood-Gush, D.G.M. (1985) The suckling behaviour of domestic pigs in a semi-natural environment. *Behaviour* 95, 11–25.
Newberry, R.C. and Wood-Gush, D.G.M. (1986) Social relationship of piglets in a semi-natural environment. *Animal Behaviour* 34, 1311–1318.
Newberry, R.C. and Wood-Gush, D.G.M. (1988) Development of some behaviour patterns in piglets under semi-natural conditions. *Animal Production* 46, 103–09.
Niebuhr, K., Harlander-Matauschek, A. and Troxler, J. (2001) *Untersuchungen zum Einfluß der Gruppengröße und der Größe der Auslauföffnungen auf die Auslaufnutzung bei Legehennen in Freilandhaltung*. Endbericht zu Handen des Ministeriums für Land- und Forstwirtschaft, Umwelt und Wasserwirtschaft. Institut für Tierhaltung und Tierschutz, Veterinärmedizinische Universität Wien, Vienna.
Nielsen, B.L., Lawrence, A.B. and Whittemore, C.T. (1995a) Effect of group size on feeding behaviour, social behaviour, and performance of growing pigs using single-space feeders. *Livestock Production Science* 44, 73–85.
Nielsen, B.L., Lawrence, A.B. and Whittemore, C.T. (1995b) Feeding behaviour of growing pigs using single or multi-space feeders. *Applied Animal Behaviour Science* 47, 235–246.
Odén, K., Vestergaard, K.S. and Algers, B. (1999) Agonistic behaviour and feather pecking in single-sexed and mixed flocks of laying hens. *Applied Animal Behaviour Science* 62, 219–231.
Odén, K., Vestergaard, K.S. and Algers, B. (2000) Space use and agonistic behaviour in relation to sex composition in large flocks of laying hens. *Applied Animal Behaviour Science* 67, 307–320.
Olsen, A.W., Dybkjær, L. and Simonsen, H.B. (2001) Behaviour of growing pigs kept in pens with outdoor runs: II. Temperature regulatory behaviour, comfort behaviour and dunging preferences. *Livestock Production Science* 69, 265–278.
Olsen, A.W., Simonsen, H.B. and Dybkjaer, L. (2002) Effect of access to roughage and shelter on selected behavioural indicators of welfare in pigs housed in a complex environment. *Animal Welfare* 11, 75–87.
Olsson, I.A.S. and Keeling, L.J. (2000) Night time roosting in laying hens and the effect of thwarting access to perches. *Applied Animal Behaviour Science* 68, 243–256.
Ouédraogo, A.P. (2002) Consumers' concern about animal welfare and the impact on food choice: social and ethical conflicts. INPL-Nancy, 84–86.
Pagel, M. and Dawkins, M.S. (1997) Peck orders and group size in laying hens: 'future contracts' for non-aggression. *Behavioural Processes* 40, 13–25.
Pedersen, L.J., Studnitz, M., Jensen, K.H. and Giersing, A.M. (1998) Suckling

behaviour of piglets in relation to accessibility to the sow and the presence of foreign litters. *Applied Animal Behaviour Science* 58, 267–279.

Petersen, H.V., Vetergaard, K. and Jensen, P. (1989) Integration of piglets into social groups of free-ranging domestic pigs. *Applied Animal Behaviour Science* 23, 223–236.

Petersen, V., Simonsen, H.B. and Lawson, L.G. (1995) The effect of environmental stimulation on the development of behaviour of pigs. *Applied Animal Behaviour Science* 45, 215–224.

Plath, U. (1999) Beurteilung verschiedener Tränketechniken und Betreuungsmaßnahmen hinsichtlich ihrer Auswirkungen auf die oralen Aktivitäten, den Gesundheitszustand un die Mastleistung über zwei bis acht Wochen alter Mastkälber in Gruppenhaltung. Dissertation, Tierärztliche Hochschule, Hannover, Germany.

Plonait, H. (1997) Perinatale Sterblichkeit. In: Plonait, H. and Birckhardt, K. (eds) *Lehrbuch der Schweinekrankheiten*. Parey, Berlin, pp. 503–507.

Porzig, E. and Sambraus, H.H. (1991) *Nahrungsaufnahmeverhalten landwirtschaftlicher Nutztiere*. Deutscher Landwirtschaftsverlag Berlin GmbH, Berlin.

Price, E.O. (1999) Behavioral development in animals undergoing domestication. *Applied Animal Behaviour Science* 65, 245–271.

Price, E.O. and Wallach, S.J.R. (1990) Physical isolation of hand-reared Hereford bulls increases their aggressiveness toward humans. *Applied Animal Behaviour Science* 27, 263–267.

Prins, H.H.T. (1997) *Ecology and Behaviour of the African Buffalo*. Chapman and Hall, London.

Puppe, B. and Tuchscherer, M. (1995) Zum Saugverhalten und zur Entwicklung biochemischer Parameter bei Ferkeln unter den Bedingungen einer Gruppenhaltung säugender Sauen: Erste Ergebnisse. *Berliner und Münchner Tierärztliche Wochenschrift* 108, 161–166.

Puppe, B., Tuchscherer, M. and Tuchscherer, A. (1997) The effect of housing conditions and the social environment immediately after weaning on the agonistic behaviour, neutrophil/lymphocyte ratio, and plasma glucose level in pigs. *Livestock Production Science* 48, 157–164.

Randle, H. (1994) Adoption and personality in cattle. PhD thesis, University of Exeter, UK.

Reinhardt, V. (1980a) *Untersuchungen zum Sozialverhalten des Rindes. Tierhaltung* 10, Birkhäuser Verlag, Basel.

Reinhardt, V. (1980b) Social behaviour of *Bos indicus*. In: Wodzicka-Tomaszewska, M., Edey, T.M. and Lynch, J.J. (eds) *Reviews in Rural Science IV. Behaviour in Relation to Reproduction, Management and Welfare of Farm Animals*. University of New England, pp. 153–156.

Reinhardt, V. and Reinhardt, A. (1981) Cohesive relationships in a cattle herd (*Bos indicus*). *Behaviour* 77, 121–151.

Rushen, J., Taylor, A. and de Passillé, A. (1999) Domestic animals' fear of humans and its effect on their welfare. *Applied Animal Behaviour Science* 65, 285–303.

Sachser, N., Dürschlag, M. and Hirzel, D. (1998) Social relationships and the management of stress. *Psychoneuroendocrinology* 23, 891–904.

Sambraus, H.H. (1985) Cross suckling of bucket reared calves that were tied for varying times after feeding. *Applied Animal Behaviour Science* 13, 179.

Sanotra, G.S., Vestergaard, K.S., Agger, J.F. and Lawson, L.G. (1995) The relative preference for feathers, straw, wood-shavings and sand for dustbathing, pecking and scratching in domestic chicks. *Applied Animal Behaviour Science* 43, 263–277.

Sato, S., Wood-Gush, D.G.M. and Wetherill, G. (1987) Observations of creche behaviour in suckler calves. *Behaviour Proceeding* 15, 126–142.

Savory, C.J., Wood-Gush, D.G.M. and Duncan, I.J.H. (1978) Feeding behaviour in a population of domestic fowl in the wild. *Applied Animal Ethology* 4, 13–27.

Schaefer, A.L., Salomons, M.O., Tong, A.K.W., Sather, A.P. and Lepage, P. (1990) The effect of environmental enrichment on aggression in newly weaned pigs. *Applied Animal Behaviour Science* 27, 41–52.

Schmid, H. (1992) Arttypische Strukturierung der Abferkelbucht. In: *Aktuelle Arbeiten zur artgemäßen Tierhaltung 1991. KTBL-Schrift* 351, 27–36.

Schmid, H. (1993) Ethological design of a practicable farrowing pen. In: Nichelmann, M., Wierenga, H.K. and Braun, S. (eds) *Proceedings of the International Congress on Applied Ethology.* KTBL Darmstadt, Berlin, pp. 238–241.

Sies, S. and Mahlau, G. (1997) *Das Image der Landwirtschaft – Ergebnisse von Assoziationstests.* Lehrstuhl für Agrarmarketing, Institut für Agrarökonomie der Universität Kiel, Germany.

Simantke, C. (2000) *Ökologische Schweinehaltung: Haltungssysteme und Baulösungen.* Bioland, Mainz, Germany.

Somers, J.G.C.J., Noordhuizen-Stassen, E.N., Frankena, K. and Metz, J.H.M. (2001) Epidemiological study on claw disorders in dairy cattle: impact of floor systems. In: *Proceedings of the 12th International Symposium on Lameness in Ruminants, 9th–13th January 2002, Orlando, Florida, USA*, pp. 350–354.

Sommer, T. and Troxler, J. (1986) Ethologische und veterinärmedizinische Beurteilungskriterien in Bezug auf die Tiergerechtheit von Loch- und Spaltenböden für Milchvieh. In: *Aktuelle Arbeiten zur artgemäßen Tierhaltung 1985. KTBL-Schrift* 311, 73–85.

Spycher, B., Regula, G., Wechsler, B. and Danuser, J. (2002) Gesundheit und Wohlergehen von Milchkühen in verschiedenen Haltungsprogrammen. *Schweizer Archiv für Tierheilkunde* 144, 519–530.

Stangel, G. and Jensen, P. (1991) Behaviour of semi-naturally kept sows and piglets (except suckling) during 10 days postpartum. *Applied Animal Behaviour Science* 31, 211–227.

Starzinger, B. (2001) 'Bio'-Blicke – Von der Stadt auf's Land. *Freiland-Journal* 4, 14–15.

Stauffacher, M. (1992) Ethologische Grundlagen zur Beurteilung der Tiergerechtheit von Haltungssystemen für landwirtschaftliche Nutztiere und Labortiere. *Schweizerisches Archiv für Tierheilkunde* 134, 115–125.

Stolba, A. (1984) Verhaltensmuster von Hausschweinen in einem Freigehege: Bemerkungen zum Film. In: *Aktuelle Arbeiten zur artgemässen Tierhaltung 1983. KTBL-Schrift* 299, 106–115.

Stolba, A. and Wood-Gush, D.G.M. (1984) The identification of behavioural key features and their incorporation into a housing design for pigs. *Annales de Recherches Vétérinaires* 15, 287–298.

Stolba, A. and Wood-Gush, D.G.M. (1989) The behaviour of pigs in a semi-natural environment. *Animal Production* 48, 419–425.

Stumpf, S., Nydegger, F., Wechsler, B. and Beyer, S. (2000) Untersuchungen zum Tier-Fressplatzverhältnis und zur Fressplatzgestaltung bei der Selbstfütterung von Milchkühen am Fahrsilo. In: *Aktuelle Arbeiten zur artgemäßen Tierhaltung 1999. KTBL-Schrift* 391, 103–110.

Sundrum, A. (1993) Tierschutznormen in der ökologischen Nutztierhaltung und Möglichkeiten zu ihrer Kontrolle. *Deutsche tierärztliche Wochenschrift* 100, 41–88.

Thodberg, K., Jensen, K.H., Herskin, M.S. and Jørgensen, E. (1999) Influence of environmental stimuli on nest building and farrowing behaviour in domestic sows. *Applied Animal Behaviour Science* 63, 131–144.

Toates, F. (1987) The relevance of models of motivation and learning to animal welfare. In: Wiepkema, P.R. and van Adrichem, P.W.M. (eds) *Biology of Stress in Farm Animals: An Integrative Approach.* Martin Nijhoff Publishers, Dordrecht, The Netherlands, pp. 153–186.

Troxler, J. (1981) Beurteilung zweier Haltungssysteme für Absetzferkel. In: *Aktuelle Arbeiten zur artgemäßen Tierhaltung 1980. KTBL-Schrift* 264, 151–164.

Turner, S.P., Horgan, G.W. and Edwards, S.A. (2001) Effect of social group size on aggressive behaviour between unacquainted domestic pigs. *Applied Animal Behaviour Science* 74, 203–215.

Van Caenegem, L. and Krötzl Messerli, H. (1997) Der Laufhof für den Milchvieh-Laufstall. *FAT-Bericht* 493, Tänikon, Switzerland.

van Putten, G. (1973) Enkele aspecten van het gedrag van varkens. *Proceedings* Varkensstudiedag, 10.5., Gent, Belgien. Wessanen, Wormerveer, The Netherlands, pp. 43–46.

van Putten, G. (1978a) Schwein. In: Sambraus, H.-H. (ed.) *Nutztierethologie.* Ulmer Verlag, Stuttgart, Germany, pp. 168–213.

van Putten, G. (1978b) Comfort behaviour in pigs: informative for their well-being. In: *Proceedings of the 28th Annual Meeting, European Association for Animal Production, Brussels, Belgium, August 1977*, pp. 70–76.

van Putten, G. (2000) An ethological definition of animal welfare with special emphasis on pig behaviour. In: Hovi, M. and Trujillo, R.G. (eds) *Diversity of Livestock Systems and Definition of Animal Welfare. Proceedings of the Second NAHWOA Workshop, Córdoba, 8–11 January 2000.* University of Reading, Reading, UK, pp. 120–134.

van Putten, G. and van de Burgwal, J.A. (1989) Tiergerechte Gruppenhaltung im Abferkelstall. In: *Aktuelle Arbeiten zur artgemäßen Tierhaltung 1988. KTBL-Schrift* 336, 93–108.

van Putten, G. and van de Burgwal, J.A. (1990) Vulva biting in group-housed sows: preliminary report. *Applied Animal Behaviour Science* 26, 181–186.

Vandenheede, M., Shehi, R., Nicks, B., Canart, B., Dufrasne, I., Biston, R. and Lecomte, P. (1994) Influence de la pluie sur l'utilisation d'un abri par des taurillons au paturage. *Annales de Medecine Veterinaire* 138, 91–94.

Vestergaard, K.S., Skadhauge, E. and Lawson, L.G. (1997) The stress of not being able to perform dustbathing in laying hens. *Physiology and Behavior* 62, 413–419.

Waiblinger, S., Reichmann, V., Troxler, J., Dreiseitel, H., Haller J., and Windischbauer, G. (2001) Einfluß eines Vorrückfressgitters auf Druckbelastungen und Schäden an den Schultern von Milchkühen. In: *Bau,*

Technik und Umwelt in der landwirtschaftlichen Nutztierhaltung. Institut für Agrartechnik der Universität Hohenheim, pp. 462–465.

Waiblinger, S., Menke, C. and Coleman, G. (2002) The relationship between attitudes, personal characteristics and behaviour of stockpeople and subsequent behaviour and production of dairy cows. *Applied Animal Behaviour Science* 79, 195–219.

Walker, B.E. (1962) Suckling and grazing behaviour of beef heifers and calves. *New Zealand Journal of Agricultural Research* 5, 331–338.

Wallbaum, F., Wassmuth, R. and Langholz, H.-J. (1997) Outdoor wintering of suckler cows in low mountain range. In: Arendonk, J.A.M. (ed.) *Outdoor Wintering of Suckler Cows in Low Mountain Range.* Book of abstracts of the 48th annual meeting of the European Association for Animal Production, Vienna, 25–28.8.97, pp. 226–226.

Weary, D.M. and Chua, B. (2000) Effects of early separation on the dairy cow and calf 1.separation at 6 h, 1 day and 4 days after birth. *Applied Animal Behaviour Science* 69, 177–188.

Weary, D.M. and Fraser, D. (1995) Signalling need: costly signals and animal welfare assessment. *Applied Animal Behaviour Science* 44, 149–158.

Weber, R. (1996) Neue Abferkelbuchten ohne Fixation der Muttersau. Wenig höhere Investitionen, praxisüblicher Zeitbedarf. *FAT-Bericht* 481, Tänikon, Switzerland.

Weber, R. (1999) Reduzierung des Besaugens von Artgenossen bei Kälbern durch Verwendung eines verschliessbaren Tränkestandes. In: *Aktuelle Arbeiten zur artgemäßen Tierhaltung 1998. KTBL-Schrift* 382, 146–152.

Weber, R. (2000) Alternative housing systems for farrowing and lactating sows. In: Blokhuis, H.J., Ekkel, E.D. and Wechsler, B. (eds) *Improving Health and Welfare in Animal Production.* EAAP Publication No. 102. Wageningen, pp. 109–115.

Weber, R. and Troxler, J. (1988) Die Bedeutung der Zeitdauer der Geburt in verschiedenen Abferkelbuchten zur Beurteilung auf Tiergerechtheit. In: *Aktuelle Arbeiten zur artgemäßen Tierhaltung 1987. KTBL-Schrift* 323, 172–184.

Weber, R. and Wechsler, B. (2001) Reduction in cross-sucking in calves by the use of a modified automatic teat feeder. *Applied Animal Behaviour Science* 72, 215–223.

Weber, R., Friedli, K. and Winterling, C. (1993) Einfluß der Abruffütterung auf Aggressionen zwischen Sauen. In: *Aktuelle Arbeiten zur artgemäßen Tierhaltung 1992. KTBL-Schrift* 356, 155–166.

Wechsler, B. (1993) Verhaltensstörungen und Wohlbefinden – ethologische Überlegungen. In: Martin, G. (ed.) *Leiden und Verhaltensstörungen bei Tieren. Tierhaltung* 23, Birkhäuser Verlag, Basel, pp. 50–64.

Wechsler, B. (1996) Rearing pigs in species-specific family groups. *Animal Welfare* 5, 25–35.

Wechsler, B. and Huber-Eicher, B. (1998) The effect of foraging material and perch height on feather pecking and feather damage in laying hens. *Applied Animal Behaviour Science* 58, 131–141.

Wechsler, B., Schmid, H. and Moser, H. (1991) *Der Stolba-Familienstall für Hausschweine. Ein tiergerechtes Haltungssystem für Zucht- und Mastschweine. Tierhaltung* 22, Birkhäuser Verlag, Basel.

Wechsler, B., Schaub, J., Friedli, K. and Hauser, R. (2000) Behaviour and leg injuries in dairy cows kept in cubicle systems with straw bedding or soft lying mats. *Applied Animal Behaviour Science* 69, 189–197.

Wemelsfelder, F. and Birke, L. (1997) Environmental challenge. In: Appleby, M.C. and Hughes, B.O. (eds) *Animal Welfare*. CAB International, Wallingford, UK, pp. 35–47.

Weng, R.C., Edwards, S.A. and English, P.R. (1998) Behaviour, social interactions and lesion score of group-housed sows in relation to floor space allowance. *Applied Animal Behaviour Science* 59, 307–316.

Whittaker, X., Edwards, S.A., Spoolder, H.A.M., Corning, S. and Lawrence, A.B. (2000) The performance of group-housed sows offered a high fibre diet ad libitum. *Animal Science* 70, 85–93.

Wiederkehr, T., Friedli, K. and Wechsler, B. (2001) Einfluss von regelmäßigem Auslauf auf das Vorkommen und den Schweregrad von Sprunggelenksschäden bei Milchvieh im Anbindestall. In: *Aktuelle Arbeiten zur artgemäßen Tierhaltung 2000. KTBL-Schrift* 403, 163–170.

Wierenga, H.K. (1983a) Auswirkungen einer Beschrankung der Liegeplätze im Laufstall auf das Verhalten von Milchkühen. *Tierzüchter* 11, 473–475.

Wierenga, H.K. (1983b) The influence of the space for walking and lying in a cubicle system on the behaviour of dairy cattle. In: Baxter, S.H., Baxter, M.R. and MacCormack, J.A.D. (eds) *Farm Animal Housing and Welfare*. Martin Nijhoff Publishers, Boston, Massachusetts, pp. 171–179.

Wierenga, H.K., Metz, J.H.M. and Hopster, H. (1985) The effect of extra space on the behaviour of dairy cows kept in a cubicle house. In: Zayan, R (ed.) *Social Space for Domestic Animals*, pp. 160–169.

Wood-Gush, D.G.M., Duncan, I.J.H. and Savory, C.J. (1978) Observations on the social behaviour of domestic fowl in the wild. *Biology of Behaviour* 3, 193–205.

Wood-Gush, D.M.G., Jensen, P. and Algers, B. (1990) Behaviour of pigs in a novel semi-natural environment. *Biology of Behaviour* 15, 62–73.

Askham Bryan College
LIBRARY BOOK

8

Mutilations in Organic Animal Husbandry: Dilemmas Involving Animal Welfare, Humans and Environmental Protection

Christoph Menke,[1] Susanne Waiblinger,[1] Merete Studnitz[2] and Monique Bestman[3]

[1]Institute of Animal Husbandry and Animal Welfare, Veterinarian University of Vienna, 1040 Vienna, Austria; [2]Department of Animal Health and Welfare, Research Centre Foulum, PO Box 50, DK-8830 Tjele, Denmark; [3]Louis Bolk Institute, Hoofdstraat 24, NL-3972 LA Driebergen, The Netherlands

Editors' comments

Animal welfare is just one of many interests that must be considered on an organic farm, and it may compete with others, such as environmental protection, human safety and adequate production. This forces us to make choices based on the values we cherish in life, our basic ethical view. Hence, our choices may depend on whether we are rooted, for example, in the biodynamic or the ecological approach to organic farming. The authors of this chapter discuss dilemmas presented by three mutilations done either in the interest of the environment or to reduce the adverse consequences of intensive animal keeping: dehorning in cows, nose-ringing in pigs and beak-trimming in hens. Scientific, philosophical, ethical and tradition-based arguments for and against these mutilations are presented, the aim being to illustrate the complexity of these value-laden questions, for which there are no simple 'right' or 'wrong' answers. This leads to consideration of issues regarding how we keep animals in European organic production systems today and in the future, issues that should be discussed not only among people working daily with animals, but also in the larger organic movement, including consumers, policy makers and other stakeholders.

© CAB International 2004. *Animal Health and Welfare in Organic Agriculture* (eds M. Vaarst, S. Roderick, V. Lund and W. Lockeretz)

Introduction

Among the main aims of organic animal husbandry is good animal welfare, achieved by respecting the animals' behavioural and physiological needs (Sundrum, 1993; see also the introduction to this book and Chapter 5). Many consumers also expect that the organic animal products they purchase are reared under high standards of animal welfare (Sies and Mahlau, 1997). As discussed in Chapter 5, several authors have argued for a broad concept of animal welfare in organic farming that takes into account the values of the organic movement. Consequently, achieving good animal welfare in an organic sense does not only mean preventing anxiety, suffering and pain; it goes far beyond that, and includes allowing the animals to perform their natural, species-specific behaviour by providing them with an adequate environment (see Chapters 5 and 7). It also means respecting the integrity of the animals (for a definition of integrity see Chapter 5, where the 'biocentric' framework is discussed).

Farm animal species have two main behavioural characteristics in common: they are social species, searching for contact with their companions, and they spend a lot of time feeding and searching for food. To fulfil these behavioural characteristics and facilitate natural behaviour, and thereby establish a basis for improved welfare, organic farmers in general keep the animals in groups and in free-range conditions. These principles and practices are reflected in organic regulations, e.g. the EC Council regulation No. 1804/1999 and the IFOAM standards (Chapter 4; Schmid, 2000).

Living in groups and in free-range conditions not only enables social contact among the animals but also stimulates locomotion and enhances the animals' health, such as udder health in cows (Washburn *et al.*, 2002). However, social groups of farm animals in today's production systems typically differ from free-living groups of their wild ancestors with regard to group composition, size, space, stability over time and food provided. Thus, negative effects on animal welfare could arise from agonistic or abnormal social interactions. Furthermore, under free-range conditions, the animals' actions or excretions can harm the environment. To prevent these negative effects, mutilations are common, including in organic farming. Dehorning of cattle, beak trimming of poultry and nose ringing of pigs all are controversial examples that raise critical, challenging discussions about organic animal husbandry.

All mutilations entail anxiety, suffering, pain and, in some cases, sensory deprivation (e.g. dehorning: Taschke, 1995; beak trimming: Duncan *et al.*, 1989). They may change the species-specific behaviour (e.g. dehorning: Sambraus *et al.*, 1979; nose ringing: Horrell *et al.*, 2000), and they obviously compromise the integrity of the animals. Various studies have shown that, under certain circumstances, it is possible to keep animals more or less successfully without mutilations. The dilemma in this is that other characteristics and goals of organic farming may lead to other priorities that might favour nose ringing, for example. This chapter discusses the ethical and practical consequences of these mutilations, the possible practical solutions and the result of studies in this field, with the focus on dilemmas involving different interests in organic farming.

Dehorning of Dairy Cows – Creating a Dilemma in Organic Agriculture?

In Germany, nearly 40% of dairy cows are kept in loose housing systems. In 1991, Menke (1996), after an intensive search in Germany, found only 100 farms with horned dairy cows in loose housing. More recently, however, Hörning (1997) found that 10% of conventional dairy farms with loose housing systems had horned cows, whilst on organic farms the figure was 27%. These results suggest that in recent years, with increasing use of loose housing systems in organic farming and the prohibition of dehorning by the biodynamic farming organization Demeter, the proportion of horned dairy cows in loose housing systems has noticeably increased.

Still, dehorning is common with loose housing on organic farms in Germany. This might apply even more in other European countries, such as the UK, Denmark or The Netherlands, where dairy cows generally are loose-housed in larger herds and where farms with a long tradition of dehorning are now converting to organic farming.

Reasons for dehorning

Horned dairy cows can be considered to be dangerous both for other cows in the herd (injuries and stress for subdominant animals), and for stockpersons. Although studies on this topic are rare, some of these, based on single herds, support these opinions. Vaarst *et al.* (1991, 1992) reported on an organic herd in Denmark where a common reason for culling was severe horn traumas; some cows could not give birth to more calves because of vaginal horn traumas. Oester (1977) found a higher level of avoidance behaviour among horned dairy cows in one herd than among dehorned ones, and inferred that cows should be dehorned. Graf (1974) found that dehorned cows reduce their individual distances (Fraser and Broom, 1990), which would allow a lower space per dehorned cow in the stable. Other authors pointed out the problem of horns on fattening cattle, which cause skin injuries during transport, leading to losses in the value of the leather (Meischke *et al.*, 1974; Shaw *et al.*, 1976; Ernst, 1977; Kretzmann *et al.*, 1985).

Reasons for keeping the horns

Arguments against dehorning arise from three considerations: the procedure itself (pain, suffering); the social functions of the horns; and the importance of horns in biodynamic agriculture, which is practised by some organic farmers.

Pain

Dehorning of cattle is a painful invasion involving opening the skullcap (Taschke, 1995; Graf *et al.*, 1996; Faulkner and Weary, 2000). Calves and cows show distinct pain and defence behaviour reactions during dehorning. Afterwards they shake their heads, wag their tails and walk backwards. Salivary cortisol levels increase markedly during and after dehorning (Taschke and Fölsch, 1997; Graf and Senn, 1999). Taschke (1995) found suppurations of the wound in 46% of the calves after 1 week, 30% after 2 weeks and 5% after 3 weeks. Finding no evidence that the sensation of pain was lower in younger calves, Taschke urged that dehorning always be done with anaesthesia, no matter how young the animal. Nevertheless, in most European countries, dehorning without anaesthesia is allowed until age 3 months. In some other countries, anaesthesia is compulsory when dehorning (e.g. in Denmark). Dehorning of calves or cows is combined with a deletion of nerves, which might result in the development of neuromas and chronic pain (Taschke, 1995), especially in adults.

Social functions

The gene for hornlessness is dominant, but it has not asserted itself in evolution. It can be argued that horns have a biological benefit under free-living conditions. Horns play a role in social behaviour. They act as organs for display, and in a pair of cows the horns may serve to assess not only body size, but also the opponent's fighting ability (Espmark, 1964; Geist, 1966; Walther, 1966). The presentation of horns influences the rank of individual cows (Bouissou, 1972; Kimstedt, 1974), and can reduce the frequency of agonistic behaviour (Graf, 1974; Fries, 1978). In hornless (dehorned or polled) herds of dairy cows, weight is the main physical characteristic influencing rank (Wagnon *et al.*, 1966; Dickson *et al.*, 1967; Fries, 1978), whereas in horned herds the age and size of the horns are the main influences (Schein and Fohrman, 1955; Brantas, 1968; Süss, 1973; Sambraus and Osterkorn, 1974; Reinhardt, 1980). This is important with regard to social stability in herds where cows are allowed to grow old, and where the groups are relatively stable (e.g. in Switzerland, Austria and southern Germany). Sambraus *et al.* (1979) observed that in such herds the herd structure was more stable, because older animals were less often challenged by younger animals because of their big horns, despite their reduced physical strength.

Biodynamics

In biodynamic farming, horns are regarded as being related to the digestion of cattle; one sign of this relationship is seen in the co-development of the rumen and the horns, with both organ systems being undeveloped at birth. Also, according to biodynamic principles, horns have specific functions, e.g. they are necessary for the special preparations used by biodynamic practitioners (Steiner, 1924).

A study of the behaviour of horned dairy cows

To evaluate the situation of horned dairy cows in loose housing systems in practice, Menke *et al.* (1999) investigated 35 farms in Switzerland and Germany. The herds differed greatly in agonistic social behaviour, defined as 'chasing and pushing away', and in injuries. Injuries caused by horns ranged from two to 63 per cow, with more than 90% of the injuries being superficial hair grazes. In 77% of the herds the frequency of injuries was less than 16 per cow (Fig. 8.1). There was a close correlation between injuries and the degree of agonistic social behaviour. In 77% of the herds, chase-away behaviour was seen a maximum of 0.035 times per hour and push-away behaviour a maximum of 0.31 times per hour per cow. Comparable studies on social behaviour of dehorned cows showed similar or higher values for push-away and chase away behaviour: 2.64 on pasture (Graf, 1974); 0.34 on pasture (Andreae *et al.*, 1985); 0.9 in loose housing (Kondo and Hurnik, 1990); and 2.39 in loose housing with yard and pasture (Jonasen, 1991). However, the results also show that some farms (herd numbers 12, 25, 21, 31 and 34) had considerable problems in keeping horned dairy cows in loose housing.

Factors influencing the behaviour of horned cows

The study in Switzerland and Germany, using stepwise regression analysis, identified several farming conditions affecting social behaviour and injuries (Menke, 1996; Menke *et al.*, 1999). These factors can be divided into characteristics of the animals, housing, management and human–animal relationship.

Herd size

In large herds, agonistic behaviour was more common and social licking less common than in small herds. It may be hypothesized that this is a consequence of the higher frequency that cows meet in the crowded space and the possible lack of mutual recognition, exacerbated by more frequent introduction of new cows into the herd (Frey and Berchtold,

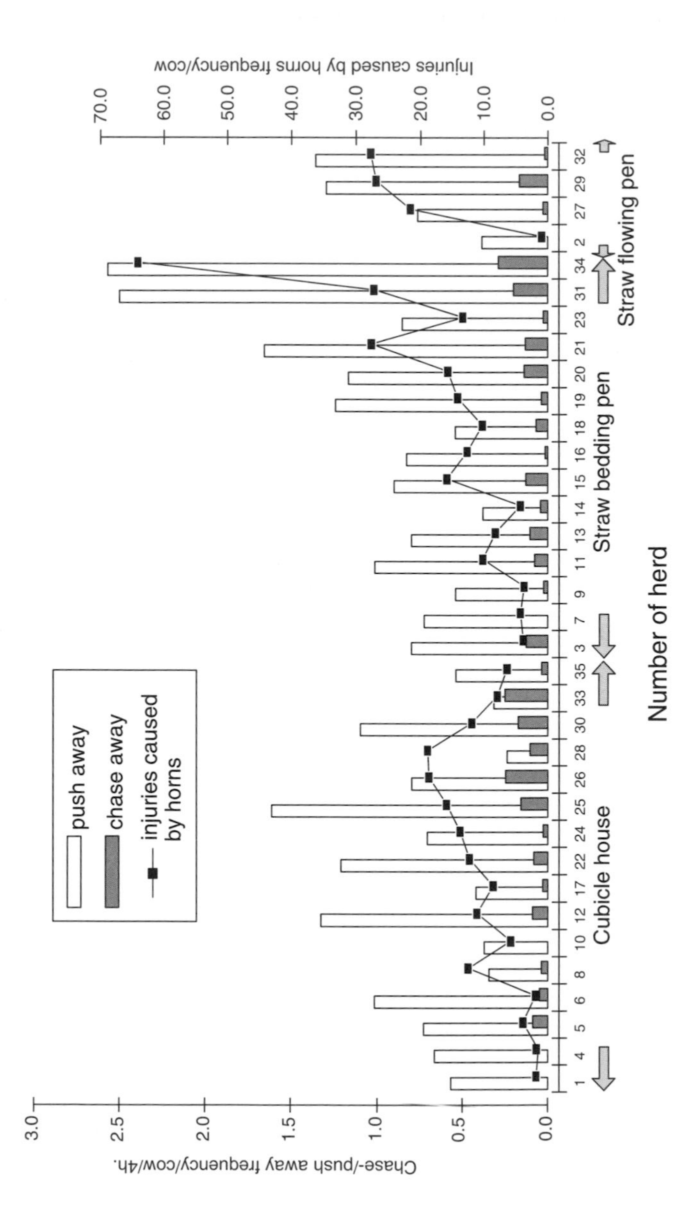

Fig. 8.1. Comparison of dairy herds regarding chase-away and push-away behaviour (per 4 h), and injuries caused by horns.

1983; Arave *et al.*, 1984). These conditions reduce the herd's social stability (more chasing and pushing away). However, in the largest herd (92 horned cows, herd 35) injuries were infrequent. Herd size was shown not to be the most important factor in this system.

Housing conditions

Housing should offer the subdominant animals the opportunity to avoid or withdraw from dominant ones. Housing designs that do not allow this create stress in horned as well as dehorned animals, and can cause severe injuries in horned herds. For example, Waiblinger *et al.* (2001) showed that feeding racks with a bar and locking device above the neck are not suited for horned cows, because the cows do not have enough time to get free of the opened feeding rack to withdraw from higher ranking cows. The total amount of space allocated to animals has been shown to be related to both agonistic behaviour and skin injuries (Lippitz *et al.*, 1973; Czakó, 1978; Metz and Wierenga, 1984; Andreae *et al.*, 1985; Menke *et al.*, 1999; Baars and Brands, 2000).

Herd management

How the herd was managed had a greater effect on agonistic behaviour or injuries than did housing factors (Menke *et al.*, 1999). On German/Swiss farms where high levels of problem solving were evident, agonistic behaviour and skin injuries were infrequent. Farmers who identify and prevent situations that promote competition can reduce negative effects in their herds. Examples included feeding *ad libitum* feeds of constant quality, home rearing of replacement cows, reducing the time that dry or newly calved cows are separated from the herd, temporarily removing restless oestrus cows from the herd, and rounding the tips of the horns (Menke and Waiblinger, 1999; Waiblinger *et al.*, 2001).

The human–animal relationship

The intensity and quality of human–animal contact is strongly associated with social behaviour and injuries (Menke, 1996; Waiblinger, 1996; Waiblinger and Menke, 1999). The intensity/quality of contact and the frequency of the milkers' friendly interactions were highly correlated with management factors related to social behaviour (Table 8.1; Waiblinger, 1996; Menke *et al.*, 1999). This suggests that a close contact enhances the understanding and knowledge of both the individual animals and the herd structure, and thereby better enables the stockperson to recognize, prevent or solve problems, as already suggested by Seabrook (1984).

Furthermore, a stockperson's interactions with cows and their decisions regarding management or housing have a common basis, their

Table 8.1. Spearman rank correlations between management and human–animal relationship.

	Friendly interaction during milking	Intensity/quality of contact
Problem solving	0.44[b]	0.63[c]
Management social behaviour	0.54[a]	0.56[b]

[a]*P*<0.05; [b]*P*<0.01; [c]*P*<0.001.

attitudes towards the animals. Besides personality, attitude is the major concept in explaining the behaviour of humans (Ajzen and Fishbein, 1980; Schiefele, 1990). In general, attitudes are consistent with behaviour (Schiefele, 1990). Accordingly, stockpersons' attitudes are important determinants of their behaviour and consequently the human–animal relationship, animal welfare, and behaviour and yield of the animals, as discussed in Chapter 10. When keeping horned cows in loose housing, the attitudes towards this system, towards the cows and towards the horns are important for the farmer's behaviour and decisions. It may be the case that farmers who see the horns as related to the integrity of the cow will adapt their management accordingly. In contrast, farmers who do not view horns as important to the integrity of the animals will see problems with horns, and instead of searching for other solutions may simply decide on dehorning (for an example see Menke and Waiblinger, 1999).

Human welfare and horns: a conflict?

The risk of accidents with horns is a frequently used as an argument for dehorning; it is the most difficult to oppose, as human welfare is involved. Without a doubt, cattle are large animals and can be dangerous, and horns present an additional risk of injury, but injuries by horns make up only a small proportion of all accidents with animals, occurring mainly in connection with tying and untying the animals in tied stables (Trachsler, 1993; Waiblinger and Menke, 2002). None of the stockpersons in the study of Menke *et al.* (1999) had been seriously hurt by a horned cow in loose housing, though some herds had existed for 20 years. The examples given above indicate that good human–animal relationships are crucial in minimizing the risk of accidents with cows in general, and with horned cows in particular (Fig. 8.2).

Breeding for polled animals

One possibility for solving the problem of the painful dehorning process is to breed for polled animals. Some cows have been polled for many years (e.g. some Hereford lines and Aberdeen Angus). In a few modern

Fig. 8.2. A good human–animal relationship is important in reducing the risk of accidents with cows, horned or dehorned.

breeds, polledness has been systematically introduced since the 1980s. For example, in the Norwegian Red Cattle (NRF), 25% of the calves were polled in 2000–2002. Polledness is a dominant trait in cattle, and investigations so far have not shown any negative correlations with regard to milk yield or disease problems. By using polled sires, it is possible to change the phenotype from horned to polled over several generations. It is relevant to consider the possibilities for introducing polled breeds into organic farming, and what consequences it will have for the individual animals, the herds and organic farming in general.

A dilemma: horns, dehorning and organic agriculture

Keeping horned dairy cows in loose housing is possible without unacceptable risks for animal and human welfare if the farmer implements high levels of management and suitable housing identified above. Housing conditions must allow the animals to satisfy basic behavioural needs and provide for sufficient handling.

Horns play an important part in the social behaviour of cattle. The dilemma connected with dehorning arises when the horn is regarded as part of the integrity of the animal and an expression of its individuality. Dehorning can be considered as taking away part of what makes a cow a cow, including part of its expression and possibilities for communicating. It may change the cows' social behaviour and the social structure and stability of the herd.

Without horns, they still can interact and create a social structure, of course.

All cows, whether or not they have horns, should be provided with an environment where they can perform their natural behaviour. Because of the traumas caused by horns, horned cows require more space, in case of deficiencies in husbandry conditions, than is recommended for dehorned cows. Meeting this requirement may be beneficial for animal welfare because it reduces stress, pain and suffering, horns or no horns. If, in contrast, dehorning makes it possible to give the animals a more restricted environment and a more stressful herd dynamic, it is an alarming sign of a lack of understanding of the ideas of organic farming. In such cases, dehorning is merely a means of adapting the animals to the husbandry system rather than offering them an adequate environment to perform their species-specific behaviour. In essence, this can be said to contradict the principles of organic husbandry and the 'organic' concepts of animal welfare (see Chapter 5; Alrøe *et al.*, 2001).

Keeping horned dairy cows requires skill and an understanding of the animals. Growing experience due to increasing numbers of farmers keeping horned cattle in loose housing can be a basis for an exchange among farmers and further developments. The advice of the organic farming organizations should impart necessary knowledge now available to farmers who prefer to have horned cattle.

The dilemma of horns and dehorning is clearly related to the broader question of what we do to animals that we have taken into our household and have domesticated for thousands of years. The introduction of polled animals raises the interesting question of whether we should expect a species to remain fixed or to develop further over time, as the species changes its role in the world. This dilemma will be discussed in the last part of this chapter along with other kinds of mutilations.

Feather Pecking in Laying Hens

Feather pecking is one of the main welfare problems in poultry keeping, especially in non-cage systems where feather-pecking birds have many potential victims at their disposal. Studies in Switzerland, Great Britain and The Netherlands have shown that respectively 50% (Häne *et al.*, 2000), 55% (Green *et al.*, 2000) and 75% (Bestman, 2000) of laying flocks in alternative systems have problems with feather pecking. Of these, only those in the Dutch flocks had not received any beak treatment, which might explain their more frequent feather pecking. Feather pecking can be viewed both as a symptom of reduced animal welfare and a cause of a further decrease in welfare. It also has economic consequences: hens with fewer feathers need more energy to stay warm, up to 27% more according to Tauson and Svensson (1980).

Characteristics of feather pecking

Feather pecking is abnormal behaviour; wild birds in their natural environment do not do it. Feather pecking can be gentle or severe. Gentle feather pecking does not cause damage and is done by most birds in a group. Severe feather pecking is responsible for the damage, but is done by only a small portion of the group (9% in a study by Keeling, 1994). Feather pecking is not aggressive behaviour. Laying hens can start it because of several factors, such as feed deficiencies, absence or poor quality of litter, boredom, and insufficient possibilities to perform their natural behaviour. The common factor that links these diverse factors might be stress.

Often feather pecking and (cloacal) cannibalism are mentioned as a single problem. In The Netherlands survey (unpublished results), it appeared that feather pecking often occurs under specific housing or management conditions, whilst cannibalism is related to factors such as excessively large eggs at the onset of production, or diseases with symptoms such as bloody diarrhoea. Outbreaks of cannibalism are unpredictable (Appleby and Hughes, 1991). Damage caused by feather pecking can be serious even where no cannibalism is seen (Keeling, 1994).

Reasons for beak trimming

To prevent the damage caused by feather pecking, in conventional poultry systems, beaks are partially amputated, or trimmed. In The Netherlands, Belgium, France, Spain, Portugal and Italy, about one-third of both living and dead tissue of the upper and lower beak is removed; in Switzerland and Germany it is less, but living tissue still is removed (Th.G.C.M. Fiks-van Niekerk, The Netherlands, 2002, personal communication). Because a hot knife is used, there is hardly any bleeding afterwards. After this treatment, the birds show less beak-related behaviour, such as eating, drinking and preening (Duncan *et al.*, 1989), probably to avoid pain; this can last up to 6 weeks. Neuromas are formed in the stump and nervous activity is demonstrated that probably can be interpreted by the nervous system as pain. These examples are evidence that chickens suffer chronically after beak trimming (reviewed by Hughes and Gentle, 1995).

Feather pecking can be a big problem in organic poultry keeping, especially on recently converted farms. Some organic farmers may therefore want to have their hens' beaks trimmed. They perceive feather pecking as a greater source of suffering than beak trimming. Additionally, feather damage may lead to larger economic loss. Finally, some farmers have said that they feel uncomfortable having their featherless creatures visible to everybody in their outdoor run on an organic farm where animal welfare is emphasized.

The EU regulations for organic animal husbandry state that trimming of beaks must not be carried out systematically. One can interpret this as a licence to trim beaks as long as it is not done systematically. Sometimes feather pecking in a commercial flock starts at age 25 weeks, or even later. To trim beaks at this age would be very cruel (Duncan *et al.*, 1989), but it can be claimed that it is not being done systematically. Beak trimming before feather pecking starts, before 3 weeks of age, when consequences such as chronic pain are less probable (Gentle *et al.*, 1995), makes the measure systematic and preventive, and therefore prohibited. The aim of this rule is probably to prevent suffering caused by systematic beak trimming, but it allows a cruel solution in cases where everything has gone wrong.

Factors influencing success in keeping beaks intact

If beak trimming is not an option, what other options are there to prevent feather pecking? Several studies have focused on risk factors related to feather pecking. A survey of 36 organic flocks on 25 Dutch organic farms (Bestman, 2000) showed that reduced feather pecking was associated with several factors: farmers having more experience with organic poultry; hens arriving on the farm younger than just before the onset of laying; hen houses laid out with different well-defined functional units for different behavioural needs; more grain scattered daily; and widespread daily use of the outdoor run. The last factor was the strongest, which makes it unfortunate that the EU regulation is not very demanding in this regard, with requirement such as: 'at least 70% of the flock should use the outdoor run under optimum conditions (dry weather, end of the day)' or 'at least 10% of the outdoor surface within 20 meters from the stable should be covered with natural or artificial shelter of at least one meter high under which the hens can walk'.

Risk factors for feather pecking identified in other on-farm studies include: stocking density greater than 10 birds per m^2 during rearing; lack of access to elevated perches during rearing (Huber-Eicher and Audigé, 1999); fewer than 50% of the birds from a flock using the outdoor run; diet changed three or more times during lay; inspections done by one person; no loose litter left by the end of lay; hen house colder than 20°C; lights turned up when the flock was inspected; and bell drinkers used (Green *et al.*, 2000).

The results suggest that significant differences in system design and labour inputs are needed more than in conventional farming. All these factors are under the influence of the farmer, and on many farms in the Dutch study (Bestman, 2000), there were many factors to improve rather than applying beak trimming. For example, although an outdoor run is obligatory, on a majority of the farms, only 20–30% of the hens used it. Either the birds did not learn to range during rearing or the outdoor run was an open field, which is highly unsuitable for a bush bird such as a chicken. In Chapter 13, a case study from a poultry farm in The Netherlands gives a good example of problem-solving related to feather pecking.

A dilemma: feather pecking, beak trimming and organic agriculture

Feather pecking is related to stress, boredom and inappropriate conditions. It is obvious that beak trimming merely reduces the symptoms of an inappropriate environment without solving the problems. Recall the studies of flocks that respectively do (Green *et al.*, 2000; Häne *et al.*, 2000) and do not (Bestman, 2000) use beak trimming, where the proportion of problem flocks was 50–55% of the former versus 75% of the latter. This shows that beak trimming does not solve the problem. If one adheres to the philosophy behind organic farming, the farm and its management should be adapted to the nature of the animals, not vice versa.

If beak trimming is not an option, this will clearly have consequences for the development of organic poultry keeping. Nowadays, there is growing consumer interest in organic products, and as a result, the bigger supermarkets are interested in selling them. Their policy is to buy as cheaply as possible; because they buy very large quantities, they also have the power to demand cheap products.

With eggs, this leads to huge, automated organic farms with up to 10,000 hens. These farms teeter on the edge of the EU regulations concerning organic farming, and it is appropriate to question whether they live up to the ideas of organic farming. They certainly do not fit the image that consumers have when buying organic eggs, and not surprisingly, such farms have the worst health and welfare problems and the fewest possibilities of solving them. Pressure from the big supermarkets leads to this increase in scale and to systems that have nothing to do with the original organic philosophy. This does not show respect for the animals, the farmers or the consumers. Moreover, this same pressure for cheap bulk production years ago led to intensive animal husbandry that we now want to get rid of. Well-motivated farmers with smaller flocks produce eggs that are a bit more expensive but can be considered more 'organic'. In The Netherlands, however, the organic market has become dominated by eggs from huge, low-cost farms, leaving empty buildings on the small farms.

Keeping organic poultry without feather pecking is not easy, and one should realize that before starting. One should regard it as a challenge to work with these highly demanding animals and to do it right. Poultry keeping can be done only by good stockpersons who are willing to look at the nature of their animals. One of the best organic poultry farms in The Netherlands has 4000 hens at most, which arrive at the age of at most

6 weeks. From a young age the hens are stimulated to go outside in an outdoor run covered with maize (to the hens this looks like a forest), and they receive more than 20 g of grain per hen per day scattered in the litter area.

It can be concluded that the problem of feather pecking and considerations of beak trimming illustrate that organic poultry husbandry demands a lot from the flock manager and demands appropriate choices not only on the level of the individual farm, but throughout the food system.

Nose Ringing of Sows

Rooting is an important behavioural pattern of pigs (Hughes and Duncan, 1988). However, rooting of sows in a restricted area can be very damaging to the environment. Therefore, outdoor sows often are nose-ringed to prevent them from uprooting the grass. This is done to maintain the pasture intact, both to prevent nutrient losses and to ensure feeding.

There are two main types of nose rings. One is a plain ring made of steel, 4 cm in diameter. The ring is inserted in the nasal septum with a pair of tongs made specifically for this purpose. Another type (prohibited in Denmark) consists of four or five copper rings inserted in the upper rim of the nose. The copper rings often fall off and the sows must be re-ringed.

Nose ringing is allowed in organic herds in most European countries. However, in indoor conventional production, there is growing awareness that the 'amputation' of parts of a pig to prevent undesirable behaviour is troublesome. Thus in Denmark, routine trimming of teeth and docking of tails have been prohibited. In organic systems, 'amputation' of the sows' possibility of rooting is allowed, but in many countries nose ringing is under discussion. In Norway, the Council for Animal Ethics (Rådet for dyreetik – Norge, 1998) has advised against the use of nose rings on the grounds that they prevent the pig's natural behaviour by inducing pain when the pig tries to root. In the UK, nose ringing is prohibited in organic farming; in Switzerland, some stores do not accept products from pigs that were nose-ringed; in Sweden, nose ringing of sows is prohibited in general.

Especially among organic farmers, there is discussion of whether the requirement of natural behaviour and good welfare can be met when the sows are prevented from rooting. Rooting is considered to be an important behavioural activity in sows, so that preventing them from rooting by nose ringing might reduce their welfare.

On the other hand, providing sows with nose rings will keep the pasture intact. In that way it is believed to result in the most efficient use of the sows' manure and thereby prevent pollution. The intact pasture also gives the sows the possibility of grazing for their own nourishment. All these purposes are very important issues in organic farming, where the animals and the environment must be seen as one system, benefiting from each other. Keeping sows outdoors in organic farming presents the dilemma of having to choose between letting the sows perform their natural behaviour versus optimal nutrient utilization, prevention of pollution and conservation of resources.

Characterization of rooting behaviour

The pig's snout is a very effective organ for foraging as well as for exploring its environment, its fellow pigs, other animals and humans (Adler and Simonsen, 1980). The pig's behaviour in connection with foraging and exploring may be divided into several elements: rooting, sniffing, chewing and manipulating. Foraging and exploration may also be supplemented by the pig scraping its forelegs (Stolba and Wood-Gush, 1989). Furthermore, rooting is observed in connection with nest building, either for resting or for farrowing.

Foraging and exploration are related behavioural categories in how they function and how they are performed, and for both purposes the pig will root. Pigs spend 25–60% of their waking time foraging and exploring (Blasetti *et al.*, 1988; Edwards *et al.*, 1993; Tober, 1996), with rooting accounting for 10–20% of waking time (Stolba and Wood-Gush, 1989; Wood-Gush *et al.*, 1990; Horrell and A´Ness, 1996; Berger *et al.*, 1997; Guilloux *et al.*, 1998; Andresen, 2000). Pigs spend more time rooting when they are hungry, but satisfied pigs also root (Edwards *et al.*, 1993; Young *et al.*, 1994; Day *et al.*, 1995, 1996).

The problems connected with nose ringing: pain and natural behaviour

Problems with nose ringing fall into two categories. One is the pain involved in the procedure, which is done when the sow is mature, and may be repeated if she loses her ring. The other is its influence on species-specific behaviour connected with rooting.

Knowledge about the painfulness of the procedure is lacking. Obviously, the actual intervention is painful, and the pain may last for hours and perhaps even for days. Also, the pain is what prevents the sow from rooting, but little or nothing is known about the pain associated with wearing a nose ring.

The plain ring effectively prevents the sow from rooting, and presumably also influences other functional activities. Thus, Horrell *et al.* (2000) found that ringing impedes grazing, exploration and nest building. It may also affect other behavioural activities involving the snout, such as social behaviour, as suggested by Stolba and Wood-Gush (1989), and sexual behaviour, as suggested by Feenstra (1976). Adverse effects of nose rings on behaviour other than foraging and exploring have not been studied sufficiently.

Is rooting a behavioural need?

The pig's need to explore new stimuli seems to have priority over hunger (Day *et al.*, 1995), and as mentioned, exploration includes rooting, sniffing, manipulating and chewing. In wildlife, feed is distributed by chance. Therefore, the ability to learn about available resources and to maintain this knowledge constitutes a sensible evolutionary strategy and is most likely the explanation of the animal's drive to explore. If a pig is in a barren environment that does not offer the possibility to explore, the absence of relevant stimuli can lead to abnormal behaviour (Stolba and Wood-Gush, 1980; Wood-Gush, 1983; Wood-Gush and Vestergaard, 1989; Petersen *et al.*, 1995). It is obvious that pigs need stimuli to a certain extent, and that the stimuli will be obtained by exploration.

In studies with nose-ringed and unringed gilts, the two groups spent the same time exploring, but in different ways (Studnitz and Jensen, 2002; Studnitz *et al.*, 2003). Whereas the unringed gilts rooted, the ringed gilts chewed, sniffed or grazed more. Rooting may be replaced by movements with the mouth, snout or head that provide the most input (Dailey and McGlone, 1997; Horrell *et al.*, 1998; Studnitz and Jensen, 2002).

However, it is generally assumed that pigs have a specific need to root (Hughes and Duncan, 1988; Horrell *et al.*, 2001), which is precisely the behaviour that is prevented by nose ringing. If a pig does need to root, deprivation of the chance to do so will lead to frustration (Dawkins, 1988), and the pig's behaviour will be influenced. However, in the studies by Studnitz *et al.* (2003) and Studnitz and Jensen (2002) mentioned above, the sows were sufficiently stimulated that they did not show signs of frustration or abnormal behaviour, only a change in the kind of exploratory behaviour.

Furthermore, if rooting is a need, the motivation to root will accumulate if the pig is prevented from rooting, and when finally given the opportunity, the pig will root more than if it had not been deprived (Taylor and Friend, 1986). Studnitz and Jensen (unpublished observations) found that when gilts were prevented from rooting for different lengths of time, when given the option to root, the longer they had been deprived of rooting the more they rooted. However, this sign of accumulated motivation for rooting was observed only at the beginning of the experiment. Later, no accumulated motivation for rooting was noted; instead, they appeared motivated to change to another environment.

Studies in this field suggest that although pigs prefer to explore by rooting, and although rooting has been believed to be behavioural need, the means – rooting behaviour – is less important than the goal – sensory input. The basic natural behaviour exists in these animals, but its purpose can be achieved other than by rooting. There is still a debate over whether this fulfils the basic needs of the animals.

Nose ringing seems not to prevent fulfilment of a behavioural need, but ringing of outdoor sows entails considerable changes to the sows' behaviour. If welfare is defined by whether or not the sows are subjected to suffering, the welfare of the ringed sows is not greatly affected. However, if welfare is defined as having positive experiences and meeting their need to perform species-specific behaviour, then avoiding nose ringing of the sows improves their welfare.

A dilemma: nose ringing, environmental care and organic agriculture

On the one hand, with respect to rooting, the question seems to be one of giving the sows the chance to perform their natural behaviour, rather than a question of welfare. However, questions concerning adverse effects of nose rings on behaviour other than foraging and exploring have not been answered sufficiently.

On the other hand, new research has shown that even though ringed sows do not uproot the grass, the feeding area gets heavily nitrogen loaded (Eriksen and Kristensen, 2001). Consequently, better growth may be expected in these spots the following year because of the large nitrogen input, but this has not been demonstrated (Eriksen *et al.*, 2001). The conclusion on the environmental side must be that nose ringing as it is used today does not properly protect the environment. Therefore, it is not a question of having to choose between enhancing animal welfare (understood as performing natural behaviour) or protecting the environment; rather, it is a question of finding alternatives that do both.

Alternatives to nose ringing in organic agriculture?

Some organic farmers still want to provide their sows with nose rings because they believe that soil preparation will be impeded when the sows can root; some also believe that nose ringing prevents piglet mortality by preventing the sows from digging in the huts. Therefore new production systems without nose ringing must offer reliable alternatives that take many factors into account.

Different approaches can be followed. In outdoor pig production, the pigs' rooting behaviour can be considered as something to be *taken advantage of* rather than *minimized*. Some of the following solutions could be seen as alternative ways of protecting the environment:

- By letting the sows out on a very attractive crop and possibly by feeding them whole beetroots, it may be possible to stimulate them to such an extent that they will root less.

- Rooting behaviour may be used as part of the crop rotation. The pigs can clear the land for cultivation and eat the waste crop (Andresen, 2000).
- Finally, an alternative could be seasonal production with piglets born outdoors in the spring and slaughtered in the autumn, with the dry sows kept indoors during the winter.

In all systems, it is very important to move the feeding places and huts several times during the year, and it would be an advantage to have fewer sows per hectare and move them frequently.

Final Discussion

Mutilations are not wanted in organic farming: they are regarded as restrictions on the performance of natural behaviour and access to a natural life, which are essential for good animal welfare in organic animal husbandry. The main principle of organic husbandry is to adapt the system to the nature of the animal. Mutilations, in contrast, are a way of artificially adapting the animal to the production system. They only take away the symptoms of an inadequate system, and may cause welfare problems themselves.

We have discussed three kinds of mutilations in relation to the dilemmas they present for organic animal husbandry. Beak trimming is prohibited in the EU regulation and is regarded as a direct obstacle to important natural behaviour, as well as a cause of constant pain in the animals when using their beaks. Nose ringing of pigs is prohibited by the organic farming regulations of some countries but accepted by others because of environmental concerns arising from having many animals in a limited area. Nose ringing hinders some elements of natural behaviour, which the pigs seem to compensate for in other ways, e.g. replacing rooting with grazing. Dehorning is generally accepted, because cows seem to form stable social groups even without horns; dehorning of calves is forbidden only in biodynamic farming. Dehorning is nevertheless a kind of mutilation.

All three kinds of mutilations are connected with pain and other adverse effects on the animals. The well-being of the animals is reduced and their behaviour patterns are restricted, especially right after the mutilation. Long-lasting effects on both pain and behaviour are also seen. Beak-trimmed hens and nose-ringed pigs suffer from sensory deprivation, i.e. the lack of species-typical sensory input, and from difficulties in performing some kinds of behaviour. Dehorning and beak trimming can result in the development of neuromas and pain.

The dilemmas connected to mutilations arise because of interests and concerns other than those of single animals and their natural behaviour.

It is a big challenge – one that perhaps cannot be solved within the framework of today's organic farming – to take care of all valid interests. We conclude by discussing the dilemmas for all three animal species, pointing to future choices. (For more discussion of future challenges in organic animal husbandry, see Chapter 17.)

Environmental care and the whole farming system

Important goals for organic farming include care of the environment and close integration of all elements of the farm into a well-functioning whole. Organic products should be produced with minimal harm to the environment, and at the same time with a high level of animal welfare, meaning that the animals must have access to outdoors and other characteristics of a natural life. The possible contradiction between the two will influence the handling of pigs in outdoor areas. As noted above, the presence of pigs in itself may have a severe environmental impact, even if they are ringed. Therefore, the core of the dilemma may not be the nose ring itself, but rather the whole way we produce pigs. Resolving such dilemmas seems challenging, and must take account of the discussion of animal welfare, species-specific behaviour and integrity of the animals. Regarding pigs, for example, it would be relevant to consider rooting behaviour as a means of soil preparation rather than behaviour that should be prevented.

Human management

The main factors in successfully keeping unmutilated animals are the humans on the farm and their attitude and relationship to the animals. Human attitude influences every level of husbandry, from decisions on housing to management and direct handling. Managing the animals requires more effort and skills than just performing a quick amputation. Therefore, farmers who convert from conventional to organic production must convert not only their farms, but also their minds. Those who choose to keep horned cows, for example, or hens with intact beaks, will not be successful unless there is harmony between their basic attitude and the system they chose. Farmers need sufficient information and must be prepared to search for it, including knowledge obtained from on-farm studies and other farmers. For all three examples we have discussed, the key factors identified as preventing negative effects are under the farmers' influence (see also Chapter 10). The farmer has to accept the animals in their integrity, with their special needs and possible problems.

The consumers' role: to pay the price

Keeping animals without mutilation can add to production costs. The question is how to get consumers to accept a reasonable price. Consumers may have high expectations about how organic animals are kept, including that they do not get mutilated, but they are not always willing to pay a higher price. In some cases, they are given a wrong picture of the farm (e.g. dairy products promoted with pictures of horned cows even when the cows actually have no horns). Consumers sometimes lack knowledge about problems such as feather pecking and beak trimming in conventional free-range hens, and the measures necessary to avoid them. This may contribute to their unreadiness to pay a higher price for organic products produced without mutilations.

Biodynamic farming: a farming system with direct interest in horns

In biodynamic farming, another factor may influence the discussion about horned cattle: the need for horns from cows on the whole farm for production of biodynamic preparations. Also, biodynamic farming shows respect for the horns as an important, functional organ and a characteristic of the animals. This leads to a preference for horned cows, which again reflects priorities on the level of both the individual animal and the whole farm.

Breeding towards organic goals: potential and limitations

The possibilities for setting and reaching breeding goals specifically relevant to organic farming also need further investigation (see Chapter 16). These possibilities include influencing the level of feather pecking in hens through breeding, and breeding cattle for polledness or to influence their temperament. Such possibilities are relevant to the discussion of future organic animal husbandry systems. Breeding of polled animals can raise questions as to what is the characteristics of the species 'cattle', and whether it can be a part of the development of breeds and species to develop polled animals. It is easier to have hornless cattle, and developing polled breeds bears the risk of accepting systems that allow less space or poorer living conditions and do not give animals an opportunity for natural behaviour.

Breeding cannot solve the problems of inappropriate housing and management. It might ease things, but as discussed above, both feather pecking in hens and high level of agonistic interaction and social stress in cattle (horned, dehorned or polled) have multiple causes. The main way to avoid disease and behaviour problems and to avoid the need for mutilation in organic farming will be to keep the animals according to the nature of the species, and the overall challenge for organic animal husbandry must be to construct systems that allow the animals to perform natural behaviour.

References

Adler, H.C. and Simonsen, H.B. (1980) *Husdyrs adfærd*. 1. Det Kgl. Danske Landhusholdningsselskab, Copenhagen.

Ajzen, I. and Fishbein, M. (1980) *Understanding Attitudes and Predicting Social Behavior*. Prentice-Hall, Englewood Cliffs, New Jersey.

Alrøe, H.F., Vaarst, M. and Kristensen, E.S. (2001). Does organic farming face distinctive livestock welfare issues? A conceptual analysis. *Journal of Agricultural and Environmental Ethics* 14, 274–299.

Andreae, U., Beneke B. and Smidt, D. (1985) Ethologische Erhebungen über den Raumbedarf und Raumnutzung bei Jungrindern und Milchkühen. *Landbauforschung Völkenrode* Sh.75, 58–84.

Andresen, N. (2000) The foraging pig, resource utilisation, interaction, performance and behaviour of pigs in cropping systems. PhD thesis, Agraria 227, Swedish University of Agricultural Sciences, Uppsala.

Appleby, M.C. and Hughes, B.O. (1991) Welfare of laying hens in cages and alternative systems: environmental, physical and behavioural aspects. *World Poultry Science Journal* 47, 109–128.

Arave, C.W., Hurnik, J.F. and Friend, T.H. (1984) Some observations on the role of behavior in cattle production and future research needs. *Applied Animal Ethology* 11, 413–421.

Baars, T. and Brands, L. (2000) *Een koppel koeien is nog geen kudde*. Louis Bolk Instituut, Driebergen, The Netherlands.

Berger, F., Dagorn, J., Denmat, M.Le., Quillien, J.P., Vaudelet, J.C. and Signoret, J.P. (1997) Perinatal losses in outdoor pig breeding. A survey of factors influencing piglet mortality. *Annales Zootechnie* 46, 321–329.

Bestman, M. (2000) The role of management and housing in the prevention of feather pecking in laying hens. In: Hovi, M. and Bouilhol, M. (eds) *Human–Animal Relationship: Stockmanship and Housing in Organic Livestock Systems. Proceedings of the Third NAHWOA Workshop, Clermont-Ferrand, 21–24 October 2000*. University of Reading, Reading, pp. 21–24.

Blasetti, A., Boitani, L., Riviello, M.C. and Visalberghi, E. (1988) Activity budgets and use of enclosed space by wild boars (*Sus scrofa*) in captivity. *Zoo Biology* 7, 69–79.

Bouissou, M.F. (1972) Influence of body weight and presence of horns on social rank in domestic cattle. *Animal Behaviour* 20, 474–477.

Brantas, G.C. (1968) On the dominance order in Friesian–Dutch dairy cows. *Zeitschrift für Tierzüchtung und Züchtungsbiologie* 84, 127–151.

Czakó, J. (1978) The effect of space on the behaviour and production on dairy cows in large-scale systems. In: *1st World Congress on Applied Ethology of Farm Animals, Madrid*, pp. 275–283.

Dailey, J.W. and McGlone, J.J. (1997) Oral/nasal/facial and other behaviours of sows kept individually outdoors on pasture, soil or indoors in gestation crates. *Applied Animal Behaviour Science* 52, 25–43.

Dawkins, M.S. (1988) Behavioral deprivation: a central problem in animal welfare. *Applied Animal Behaviour Science* 20, 209–225.

Day, J.E.L., Kyriazakis, I. and Lawrence, A.B. (1995) The effect of food deprivation on the expression of foraging and exploratory behaviour in the growing pig. *Applied Animal Behaviour Science* 42, 193–206.

Day, J.E.L., Kyriazakis, I. and Lawrence, A.B. (1996) An investigation into the causation of chewing behaviour in growing pigs: the role of exploration and feeding motivation. *Applied Animal Behaviour Science* 48, 47–59.

Dickson, D.P., Barr, G.R. and Wieckert, D.A. (1967) Social relationship of dairy cows in a feed lot. *Behaviour* 29, 195–203.

Duncan, I.J.H., Gillian, S.S., Seawright, E. and Breward, J. (1989) Behavioural consequences of partial beak amputation (beak trimming) in poultry. *British Poultry Science* 30, 479–488.

Edwards, S.A., Atkinson, K.A. and Lawrence, A.B. (1993) The effect of food level and type on behaviour of outdoor sows. In: Nichelmann, M., Wierenga, H.K. and Braun, S. (eds) *Proceedings of the 27th International Congress of the International Society for Applied Ethology*, 93, Berlin, pp. 501–503.

Eriksen, J. and Kristensen, K. (2001) Nutrient excretion by outdoor pigs: a case study of distribution, utilization and potential for environmental impact. *Soil Use and Management* 17, 21–29.

Eriksen, J., Petersen, S.O. and Sommer, S.G. (2001) [Environmental charging outdoor sows. In Danish]. Miljøbelastning ved søer på friland. Intern Rapport, Danish Institute of Agricultural Sciences, 145, pp. 30–34.

Ernst, A.J. (1977) Dehorning beef cattle. *Queensland Agriculture Journal* 103, 439–442.

Espmark, Y. (1964) Studies of dominance subordination relationships in a group of semi domestic reindeer. *Animal Behaviour* 12, 420–426.

Faulkner, P.M. and Weary, D.M. (2000) Reducing pain after dehorning in dairy calves. *Journal of Dairy Science* 83, 2037–2041.

Feenstra, A. (1976) Svins adfærd. *Beretning fra Statens Husdyrbrugsforsøg* 444, 1–57.

Fraser, A.F. and Broom, D.M. (1990) *Farm Animal Behaviour and Welfare*. Ballière Tindall, London.

Frey, R. and Berchtold, M. (1983) Analyse vorzeitiger Ausmerzungen. *Zuchthygiene* 18, 203–208.

Fries, B. (1978) Ursachen und Auswirkungen des Sozialgeschehens in einer hornlosen Rinderherde. Dissertation, München, Germany.

Geist, V. (1966) Evolution of horn-like organs. *Behaviour* 27, 178–214.

Gentle, M.J., Thorp, B.H. and Hughes, B.O. (1995) Anatomical consequences of partial beak amputation (beak trimming) in turkeys. *Research in Veterinary Science* 58, 158–162.

Graf, B. (1974) Aktivitäten von enthornten und nicht enthornten Milchkühen auf der Weide. Diplomarbeit, Institut für Tierproduktion, ETH, Zürich.

Graf, B. and Senn, M. (1999) Behavioural and physiological responses of calves to dehorning by heat cauterisation with or without local anaesthesia. *Applied Animal Behaviour Science* 62, 153–171.

Graf, B., Trachsler, U., Steiger, M. and Senn, M. (1996) Zur Belastung von Kälbern bei der Enthornung. *Agrarforschung* 3, 247–250.

Green, L.E., Lewis, K., Kimpton, A. and Nicol, C.J. (2000) A cross sectional study of the prevalence of feather pecking in laying hens in alternative systems and its associations with management and disease. *Veterinary Record* 147, 233–238.
Guilloux, A., Berger, F., Bellanger, D., Cossée, B. and Meunier-Salaün, M.C. (1998) Comportement de pâture chez les truies logées en plein air. *Journées Recherche Porcine en France* 30, 189–194.
Häne, M., Huber-Eicher, B. and Fröhlich, E. (2000) Survey of laying hen husbandry in Switzerland. *World's Poultry Science Journal* 56, 21–31.
Hörning, B. (1997) Tiergerechtheit und Verfahrenstechnik eingestreuter Milchviehlaufställe in der Praxis. Dissertation, Gesamthochschule Kassel, Witzenhausen, Germany.
Horrell, I. and A'Ness, P. (1996) The nature, purpose and development of rooting in pigs. *Proceedings of the 30th International Congress of the International Society for Applied Ethology, 14–17 August 1996, Guelph, Ontario, Canada*, p. 100.
Horrell, I., A'Ness, P. and Edwards, S.A. (1998) Nasal-ringing in pigs: the impact of food restriction and environmental enrichment. In: *Proceedings of the 31th International Congress of the ISAE, Prague, Czech Republic*, 97, 76.
Horrell, I., A'Ness, P., Edwards, S.A. and Riddoch, I. (2000) Nose-rings influence feeding efficiency in pigs. *Animal Science* 71, 259–264.
Horrell, I., A'Ness, P., Edwards, S.A. and Eddison, J. (2001) The use of nose-ringing in pigs: consequences for rooting, other functional activities, and welfare. *Animal Welfare* 10, 3–22.
Huber-Eicher, B. and Audigé, L. (1999) Analysis of risk factors for the occurrence of feather pecking in laying hen growers. *British Poultry Science* 40, 599–604.
Hughes, B. and Duncan, I.J.H. (1988) The notion of ethological need, models of motivation and animal welfare. *Animal Behaviour* 36, 1696–1707.
Hughes, B.O. and Gentle, M.J. (1995) Beak trimming of poultry, its implications for welfare. *World's Poultry Science Journal* 51, 51–61.
Jonasen, B. (1991) Social behaviour of dairy cows in a stimulus-rich environment. Internal Report: National Institute of Animal Science, Department of Research in Cattle and Sheep, Copenhagen.
Keeling, L.J. (1994) Feather pecking – who in the group does it, how often and under what circumstances? *Proceedings of the 9th European Poultry Conference*, Glasgow, UK.
Kimstedt, W.M. (1974) Untersuchungen über die Rangordnung beim Hausrind in Abhängigkeit von der Enthornung. Dissertation, Universität Gießen.
Kondo, S. and Hurnik, J.F. (1990) Stabilisation of social hierarchy in dairy cows. *Applied Animal Behaviour Science* 27, 287–297.
Kretzmann, P.M., Wallace, H.G. and Weaver, D.B. (1985) The role of hornless cattle in beef carcass bruising at Cato Rico abattoir. *Journal South African Veterinary Association* 56, 199–200.
Lippitz, O., Kaiser, R. and Klug, F. (1973) Untersuchungen zum Verhalten von Milchkühen im Boxenlaufstall bei unterschiedlichem Tier/Freßplatz-Verhältnis und ständig freiem Zugang zur Krippe. *Tierzucht* 27, 522–525.
Meischke, H.R.C., Ramsay, W.R. and Shaw, F.D. (1974) The effect of horns on bruising in cattle. *Australia Veterinary Journal* 50, 432–434.
Menke, C. (1996) Laufstallhaltung mit behornten Milchkühen. Dissertation, Institut für Nutztierwissenschaften, ETH, Zürich, Switzerland.

Menke, C. and Waiblinger, S. (1999) Behornte Kühe im Laufstall – gewußt wie. Landwirtschaftliche Beratungszentrale Lindau, Lindau, Switzerland.

Menke, C., Waiblinger, S., Fölsch, D.W. and Wiepkema, P.R. (1999) Social behaviour and injuries of horned cows in loose housing systems. *Animal Welfare* 8, 243–258.

Metz, J.H.M. and Wierenga, H.K. (1984) Spatial requirements and lying behaviour of cows in loose housing systems. In: Unshelm, J., v. Putten, G. and Zeeb, K. (eds) *Proceeding of International Congress of Applied Ethology of Farm Animals, Kiel, August 1–4*, pp. 179–183.

Oester, H. (1977) Auswirkungen der Enthornung bei Schweizer Braunvieh Milchkühen im Tiefstreulaufstall auf allgemeine und spezielle Aktivitäten. Diplomarbeit, Zoologisches Institut, Bern, Switzerland.

Petersen, V., Simonsen, H.B. and Lawson, L.G. (1995) The effect of environmental stimulation on the development of behaviour in pigs. *Applied Animal Behaviour Science* 45, 215–224.

Rådet for Dyreetikk – Norge. (1998) *Uttalelse om Trynering til Gris.* www.org.nlh/etikkutvalget.

Reinhardt, V. (1980) *Untersuchungen zum Sozialverhalten des Rindes.* Reihe Tierhaltung 10, Birkhäuser Verlag, Basel.

Sambraus, H.H. and Osterkorn, K. (1974) Die soziale Stabilität in einer Rinderherde. *Zeitschrift für Tierpsychologie* 35, 418–424.

Sambraus, H.H., Fries, B. and Osterkorn, K. (1979) Das Sozialgeschehen in einer Herde hornloser Hochleistungsrinder. *Zeitschrift für Züchtungsbiologie* 95, 81–88.

Schein, M.W. and Fohrman, M.H. (1955) Social dominance relationships in a herd of dairy cattle. *British Journal of Animal Behaviour* 3, 45–55.

Schiefele, U. (1990) *Einstellung, Selbstkonsistenz und Verhalten.* Verlag für Psychologie-Dr.C.J. Hogrefe Verlag, Göttingen, Germany.

Schmid, O. (2000) Comparison of European organic livestock standards with national and international standards – problems of common standards development and future areas of interest. In: Hovi, M. and Trujillo, R.G. (eds) *Diversity of Livestock Systems and Definition of Animal Welfare. Proceedings of the Second NAHWOA Workshop, Córdoba, 8–11 January 2000*, pp. 63–75.

Seabrook M.F. (1984) The psychological interaction between the stockman and his animals and its influence on performance of pigs and dairy cows. *Veterinary Record* 115, 84–87.

Shaw, F.D., Baxter, R.I. and Ramsay, W.R. (1976) The contribution of horned cattle to carcass bruising. *Veterinary Record* 98, 255–257.

Sies, S. and Mahlau, G. (1997) Das Image der Landwirtschaft – Ergebnisse von Assoziationstests. Lehrstuhl für Agrarmarketing, Institut für Agrarökonomie der Universität Kiel, Germany.

Steiner, R. (1924) *Geisteswissenschaftliche Grundlagen zum Gedeihen der Landwirtschaft.* Landwirtschaftlicher Kurs, 6. Auflage, GA 327, Dornach, Switzerland.

Stolba, A. and Wood-Gush, D.G.M. (1980) Arousal and exploration in growing pigs in different environments. *Applied Animal Ethology* 6, 382–383.

Stolba, A. and Wood-Gush, D.G.M. (1989) The behaviour of pigs in a seminatural environment. *Animal Production* 48, 419–425.

Studnitz, M. and Jensen, K.H. (2002) Expression of rooting motivation in gilts following different lengths of deprivation. *Applied Animal Behaviour Science* 76, 203–213.

Studnitz, M. and Jensen, K.H. (submitted) Explorative behaviour in outdoor nose ringed and unringed gilts exposed to different tests. *Applied Animal Behaviour Science.*

Studnitz, M., Jensen, K.H., Jørgensen, E. and Jensen, K.K. (2003) The effect of nose ringing on explorative behaviour in gilts. *Animal Welfare* 12, 109–118.

Sundrum, A. (1993) Tierschutznormen in der ökologischen Nutztierhaltung und Möglichkeiten zu ihrer Kontrolle. *Deutsche tierärztliche Wochenschrift* 100, 71–73.

Süss, M. (1973) Beitrag zum Verhalten von Milchkühen in Freilaufställen und herkömmlichen Laufställen. Dissertation, Vienna, Austria.

Taschke, A. (1995) Ethologische, physiologische und histologische Untersuchungen zur Schmerzbelastung der Rinder bei der Enthornung. Dissertation, Institut für Nutztierwissenschaften der ETH Zürich, Switzerland.

Taschke, A. and Fölsch D.W. (1997) Ethologische, physiologische und histologische Untersuchungen zur Schmerzbelastung der Rinder bei der Enthornung. *Tierärztliche Praxis* 25, 19–27.

Tauson, R. and Svensson, S.A. (1980) Influence of plumage condition on the hen's feed requirement. *Swedish Journal of Agricultural Research* 10, 35–39.

Taylor, L. and Friend, T.H. (1986) Open-field test behaviour of growing swine maintained on a concrete floor and a pasture. *Applied Animal Behaviour Science* 16, 143–148.

Tober, O. (1996) Zirkadiane Rhythmik ausgewählter Verhaltensweisen von güsten und tragenden Sauen in ganzjähriger Freilandhaltung. *Tierärtztliche Umschau* 51, 111–116.

Trachsler, G. (1993) Erfahrungen mit behornten Kühen in Laufställen. Tagung der ALB-CH, FAT, KAM.

Vaarst, M., Brock, K., Enevoldsen, K. and Kristensen, E.S. (1991) [Health, disease prevention, disease treatment, and reproduction in organic dairy herds. In Danish]. Sundhed, sygdomsforebyggelse og -behandling samt reproduktion i okologiske malkekvægbesætninger. In: Østergaard, V. (ed) *Studier i Kvægproduktionssystemer.* Beretning 699, Danish Institute of Animal Science, pp. 130–187.

Vaarst, M., Enevoldsen, C. and Kristensen, E.S. (1992) [Health, disease control and reproduction in organic dairy herds. In Danish]. Sundhed, sygdomsbekæmpelse og reproduktion i økologiske malkekvægbesætninger 1991–1992. In: Kristensen, T. and Østergaard, V. (eds) *Studier i kvægproduktionssystemer.* Beretning 714, Danish Institute of Animal Science, pp. 176–215.

Wagnon, K.A., Loy, R.G., Rollins, W.C. and Carroll, F.D. (1966) Social dominance in a herd of Angus, Hereford and Shorthorn cows. *Animal Behaviour* 14, 474–479.

Waiblinger, S. (1996) *Die Mensch-Tier-Beziehung bei der Laufstallhaltung von behornten Milchkühen.* Reihe Tierhaltung 24, Ökologie Ethologie Gesundheit, Universität/Gesamthochschule Kassel, Germany.

Waiblinger, S. and Menke, C. (1999) Influence of herdsize on human–animal relationship in dairy farms. *Anthrozoös* 12, 240–247.

Waiblinger, S. and Menke C. (2002) Hörner als Risiko? Der Einfluss von

Management und Mensch-Tier-Beziehung. In: *9. FREILAND-Tagung. Den Tieren gerecht werden – Neue Qualitäten der Tierhaltung, 26 September 2002.* Freiland-Verband, Vienna, pp. 36–43.

Waiblinger, S., Baars, T. and Menke, C. (2001) Understanding the cows: The central role of human–animal relationship in keeping horned dairy cows in loose housing. In: Hovi, M. and Bouilhol, M. (eds) *Human–Animal Relationship: Stockmanship and Housing in Organic Livestock Systems. Proceedings of the Third NAHWOA Workshop, Clermont-Ferrand, 21–24 October 2000.* University of Reading, Reading, UK, pp. 64–78.

Walther, F. (1966) *Mit Horn und Huf.* Paul Parey Verlag, Munich, Germany.

Washburn, S.P., White, S.L., Green, J.T. and Benson, G.A. (2002) Reproduction, mastitis and body condition of seasonally calved Holstein and Jersey cows in confinement or pasture systems. *Journal of Dairy Science* 85, 105–111.

Wood-Gush, D.G.M. (1983) *Elements of Ethology.* Chapman and Hall, London.

Wood-Gush, D.G.M. and Vestergaard, K.S. (1989) Exploratory behavior and the welfare of intensively kept animals. *Journal of Agricultural Ethics* 2, 161–169.

Wood-Gush, D.G.M., Jensen, P. and Algers, B. (1990) Behvaiour of pigs in a novel semi-natural environment. *Biology of Behaviour* 15, 62–73.

Young, R.J., Carruthers, J. and Lawrence, A.B. (1994) The effect of a foraging device ('the Edinburgh foodball') on the behaviour of pigs. *Applied Animal Behaviour Science* 39, 237–247.

9

Assessing Animal Welfare in Organic Herds

Ute Knierim,[1] Albert Sundrum,[2] Torben Bennedsgaard,[3] Ulla Holma[4] and Pernille Fraas Johnsen[5]

[1]Department of Farm Animal Behaviour and Husbandry, University of Kassel, Nordbahnhofstr. 1a, 37213 Witzenhausen, Germany; [2]Department of Animal Nutrition and Animal Health, University of Kassel, Nordbahnhofstr. 1a, 37213 Witzenhausen, Germany; [3]The Royal Veterinary and Agricultural University, Institute for Animal Health and Animal Production, Grønnegaardsvej 2, 1870 Frederiksberg C, Denmark; [4]University of Helsinki, Mikkeli Institute for Rural Research and Training Lönnrotinkatu 3–5, 50100 Mikkeli, Finland; [5]The Danish Animal Welfare Society, Alhamvej 15, 1826 Frederiksberg C., Denmark

Editors' comments

Given the importance of animal welfare in organic animal husbandry, we need tools for measuring how well that goal is being fulfilled. Such tools will be of help not only in certifying organizations but also to farmers, who can subsequently improve their husbandry systems, as well as to consumers, for whom they can guarantee the ethical value in organic animal products. In several European countries, assessment systems are finding practical use in measuring whether an organic farm provides living conditions that fulfil the animal welfare standards promised in organic husbandry. Of course such systems, two of which the authors emphasize here, will be influenced by how animal welfare is viewed, so that different people will give priority to different areas. Although such assessment systems were not developed specifically for organic animal welfare goals, they still are relevant and useful as we face the challenge of developing truly organic ways of assessing animal welfare.

© CAB International 2004. *Animal Health and Welfare in Organic Agriculture* (eds M. Vaarst, S. Roderick, V. Lund and W. Lockeretz)

Relevance of Animal Welfare to Organic Farming

An important objective of organic husbandry is to fulfil high animal welfare standards for ethical reasons, as well as to promote good health and longevity (Boehncke and Krutzinna, 1996; Boehncke, 1997). Moreover, animal welfare is an important motivation for the considerable number of consumers who are prepared to pay higher prices for organic animal products.

Thus, there are three parties interested in information on the status of animal welfare on the individual organic farm: (i) the farmer, together with any possible advisory service; (ii) the consumer; and (iii) the control authority. The different parties focus on different purposes of an animal welfare assessment: respectively advice, certification and control. For farmers, it is important to detect welfare problems to safeguard or improve the current welfare status of their herd, but also to receive a certification of their achievements in order to obtain an adequate economic return for their investments. For the consumer, the certified welfare level or certain minimum standards may form the basis for their purchasing decisions. For the sake of consumer safety and a functioning market, society has an interest in ensuring that product information is clear and reliable, and consequently that marketing rules may be enforced by control authorities. Finally, on a more general level, on-farm welfare assessment can also contribute to the evaluation of housing systems or diagnosis of general welfare problems, and thereby help to develop better management systems. Everything that actually contributes to improving welfare is also in the interest of a fourth party, the animals.

General Difficulties in the Assessment of Animal Welfare

For the purposes mentioned above, science is expected to provide the basis for valid and reliable welfare assessment. This can be fulfilled to a certain degree, but some difficulties and limitations make further improvements necessary, and some limitations are unavoidable. The problems will be briefly discussed.

Animal welfare describes the state of an animal with regard to three concerns: natural living, biological functioning (i.e. health, growth, reproduction, functioning of physiological systems) and the feelings of the animal (Broom, 1996; Fraser *et al.*, 1997; see also Chapter 5). At least according to the common view of science (for another view, see Wemelsfelder *et al.*, 2000), feelings can only be assessed indirectly using animal-based indicators such as behaviour or health. The interpretation of the respective indicators may not always be clear-cut, partly because basic knowledge may be missing. This leaves us with some uncertainty and room for debate when it comes to conclusions about the actual state of the animal.

Additionally, welfare is multidimensional, in that an animal may be well in one respect but not so well in another (Knierim *et al.*, 2001). For example, a cow may experience positive effects arising from affiliative behaviour in a stable social herd but at the same time suffer from injuries and pain because of colliding with inadequate cubicles.

There is no logically or empirically correct way to combine the different welfare dimensions (Fraser, 1995). The weighting of the different aspects of welfare according to their perceived importance always involves a subjective element. A veterinary surgeon may give higher priority to health aspects than would an ethologist, and vice versa (Fraser, 1995; Waiblinger *et al.*, 2001). Also, society's attention to different aspects of welfare changes over time. Whilst in recent years animal health was accepted as a priority, among other reasons because of its relation to consumer health, public opinion in Europe is now prepared to attach more weight to behavioural freedom, as can be seen in current developments in welfare legislation. Therefore, not only scientific knowledge, but also societal values must be taken into account.

From the viewpoint of a practical assessment procedure, a further problem arises. The recording of the indicators mentioned above usually imposes high demands on time, expertise and research facilities, resources that typically are limited for on-farm assessment (Sundrum *et al.*, 1994; Waiblinger *et al.*, 2001). One alternative is not to directly record animal-related indicators, but instead record parameters of the housing and management conditions (design and management criteria) for which a systematic welfare effect has been proven.

Many of the design and management criteria are reliable and easy to measure (Waiblinger *et al.*, 2001). However, their validity as indirect indicators of the likely state of the animal may be inadequate for two reasons. First, individual differences from genetics and ontogenetic experience can make different animals respond differently to the same external conditions (Sundrum, 1997). Second, animal welfare is influenced by multiple interacting factors. For example, the consequences of the feeding space per animal will depend on the quality and frequency of feeding, group size, the social climate in the group, whether there is access to pasture, and perhaps additional factors. On-farm assessment must be able to handle a wide range of combinations of housing design, feeding, other management aspects, climate and human care. Different factors may reduce or increase certain welfare effects, but quantitative knowledge about systematic interaction effects is very limited (Sandøe *et al.*, 1996, 1997; Waiblinger *et al.*, 2001).

While the difficulties are considerable, it is necessary to find ways to assess welfare on the individual farm that are valid, reliable and feasible, and to work on improvements of the different approaches. This means that compromises will have to be accepted. Several projects throughout Europe have taken up the high challenge to develop scientifically based welfare assessment systems. So far, no single system has been generally accepted. The methods that have been developed so far can be divided into three basic approaches: minimum standards, integrated welfare measures and individual integrated system evaluations. Each is described below.

Minimum Standards

In legislation or codes of practice, design and management criteria play a predominant role, usually in the form of minimum standards. The advantages of defining such minimum standards are threefold. First, the decisions on limits can be based on scientific findings regarding the effects of a single factor on animal welfare. Second, non-experts with just basic training can carry out control using simple checklists. Third, minimum standards are transparent in that they directly describe the husbandry conditions. Therefore, minimum standards are very suitable for certification and related control.

However, as described above, single husbandry factors interact with each other and with the characteristics of the animal, so that the actual welfare outcome is not necessarily in accordance with the results obtained from a controlled experiment. Hence, compliance with minimum standards is no guarantee that the animals are well, and possibly vice versa. Apart from this, such standards usually constitute a compromise among different interests related to aspects other than animal welfare, such as economic performances or working conditions. A critical point is that this approach oversimplifies the welfare assessment, as it suggests a yes/no decision, whereas animal welfare is a state that can vary from very poor to very good (Broom, 2001).

The 'Freedom Food' scheme used by RSPCA in the United Kingdom is a welfare assessment scheme based on minimum standards that serves as a certification trademark. The EC-Regulation No. 1804/1999 on organic livestock production is an example of legal minimum standards relating to the welfare of farm animals, among other things. It contains considerably stricter provisions for the protection of calves, pigs and laying hens than the respective conventional welfare Council Directives (Sundrum, 1999). Moreover, it sets up a considerably higher level of regular control of husbandry conditions. The generally stricter requirements and better control can be expected to improve the welfare of organically kept livestock. However, this must be proven in individual cases using animal-related measures.

Integrated Welfare Measures

The TGI systems

To take better account of the multifactorial nature of animal welfare and to allow for a more graded assessment, systems have been developed that evaluate different aspects of welfare by scores, which then may be combined into a final welfare score. Such systems also allow the farmer more flexibility in exactly how to achieve a certain welfare level. The idea of this approach was initially suggested by Bartussek (1985), and the basic concept has been reworked by others (e.g. Sundrum *et al.*, 1994; Beyer, 1998; Bracke, 2001). Bartussek called his system an index system, *Tiergerechtheitsindex* (TGI) in German. In English, Bartussek refers to it as the Animal Needs Index (ANI; Bartussek, 1999) or Housing Condition Score (HCS, Bartussek, 2000). Today, two index systems are used in practice for the assessment of welfare in cattle, pigs and laying hens. These are the TGI 35L (Bartussek *et al.*, 2000), which since 1995 has been the official welfare assessment system for organic farms in Austria, and the TGI 200 (Sundrum *et al.*, 1994), which is used as an advisory tool on organic farms in Germany (Bioland Verband, 2002).

Choice of criteria

Both TGI systems rely heavily on the application of design and to a lesser degree on management criteria, and use only a few semi-quantitative animal-based indicators. They define areas of influence relating either to the animals' freedom to express different behaviour patterns (TGI 200) or to functional areas (TGI 35L), as well as to hygiene and management factors (TGI 200), or light, air and stockpersonship (TGI 35L). These areas are assessed separately, and their scores then combined into a single overall welfare index (for an example of one area of influence, see Table 9.1). Total index and area scores are expected to be positively correlated with the quality of husbandry conditions.

The criteria included in the index systems were chosen because they were perceived as having a particular influence on animal welfare, and scientific findings were available as a basis for the definition of reference values and weighting of scorings. However, the TGI is only partially scientific, and is mostly a pragmatic system. The evaluation of the different conditions is also based on a consensus among people who were involved in the development and possible application of the system. Other highly important factors, such as the human–animal relationship (Rushen and de Passillé, 1992; Sandøe *et al.*, 1996; Boehncke, 1997; Hemsworth and Coleman, 1998; also the NAHWOA Recommendations called for development of welfare assessment tools that included the human–animal relationship) are not included because they cannot be

Table 9.1. Example of TGI assessment scheme for one area of influence from the TGI 35L for cattle: facilities for locomotion and movement (Bartussek *et al.*, 2000, slightly modified).

Column	a				b	c	d	e	f
Points	Loose housing systems Space allowance ('available floor area') (m^2/500 kg)				Tether systems Lying down, lying and rising	Stall dimensions	Longitudinal and horizontal play of tether (m)	Outdoor areas Total access days/year	Pasture days/year
	Dehorned dairy cows	Horned dairy cows	Suckler herds	Young stock, beef cattle					
3.0	≥ 8	≥ 9	≥ 7.5	≥ 6	Comfortable[a]			≥ 270	
2.5	≥ 7	≥ 8	≥ 6.5	≥ 5				≥ 230	
2.0	≥ 6	≥ 7	≥ 5.5	≥ 4	Medium[a]			≥ 180	
1.5	≥ 5	≥ 6	≥ 4.5	≥ 3				≥ 120	≥ 120
1.0			≥ 4.0	≥ 2.5		Comfortable[a]	≥ 0.6/0.4	≥ 50	≥ 50
0.5					Restricted[a]	Medium[a]	≥ 0.4/0.3		≥ 30
0	< 5	< 6	< 4.0	< 2.5	Very restricted[a]	Restrictive[a]	≥ 0.4/0.3		

[a]Definitions in terms of dimensions and systems descriptions not displayed here.

measured in a short time. The human factor is only partially and indirectly estimated, such as by the condition of the equipment and animals' skin or feet. Further criteria are excluded because they must be recorded by experts, for example diagnoses of disease.

Moreover, in the TGI 200, criteria that vary greatly over time, such as climatic factors or feeding rations, are avoided to ensure good repeatability of results (Sundrum, 1997). The TGI 35L tries to solve this problem by procedural rules such as that the assessment should take place in the least favourable season (late winter for cattle). To address the welfare situation of all individuals within the housing system, another requirement is that the conditions of the 25% worst-affected animals must be evaluated if conditions vary greatly among individuals (Bartussek *et al.*, 2000).

An important requirement is that the persons applying the TGI are well trained, so that person-to-person variations are reduced (Sundrum *et al.*, 1994; Amon *et al.*, 1997), especially because many variables must be subjectively estimated (for example cleanliness). An advantage of the TGI assessment is that it can be performed quickly. An initial assessment on a farm takes 30–90 min (average 44 min) if carried out by an experienced assessor and if all relevant documentation is available, such as buildings plans and health records. The follow-up assessments of the same farm take between 10 and 35 min (Bartussek *et al.*, 2000.) On the other hand, this may be too little time to carry out a proper assessment. The schedule may not allow observation of behavioural disturbances such as stereotypies, damaging or redirected behaviours, fear responses or social disturbances.

The grading of the scores for the assessment of the recorded data differs between both TGI systems. Generally no weighting factors are applied, but sometimes weighting is achieved by not allocating all available scores. Moreover, some criteria, such as space allowance or outdoor access, receive a high weight in the total score because they are considered in several areas of influence. However, no scientific or clear empirical basis is yet available for such weighting of criteria. Moreover, possible interaction effects must be ignored despite the claim of the index systems to take account of the multifactorial nature of welfare. Moreover, it is questionable if it is appropriate to allow bad conditions in one area to be offset by good conditions in another. Acknowledging the inherent limits of such offsets, the TGI systems also apply minimum standards, which may originate from legal or private welfare or marketing standards (such as the EC regulation). If those minimum requirements are not met, the calculated TGI is valid only if the deficiencies are removed within a reasonable time (Bartussek *et al.*, 2000). However, even above those minimum standards, compensation between different aspects of welfare may be possible in some but not all areas.

Limitations of the TGI

The main concern regarding the TGI systems is that they measure husbandry conditions or welfare risk factors, but their relationship to the actual welfare state of the animals in the system is unclear. For the reasons stated above (incomplete list of criteria, unknown interaction effects, individual response differences) this relation may even be poor. Different investigations have demonstrated significant but mostly only low linear correlations between certain animal-based parameters and TGI scores. For example, Roiha and Nieminen (1999) and Roiha (2000), in a research project on 14 Finnish organic dairy farms, calculated the correlation between TGI 35L and parameters from health reports from 4 years (on average 5.3 health reports per farm from a Finnish milk recording association). Correlation coefficients mostly ranged between –0.24 and –0.39, indicating that incidences (percentages of first veterinary treatments per year) of udder disorders, ketosis and various parameters of impaired fertility were weakly and negatively correlated with locomotion, social interaction and total scores. Bennedsgaard and Thamsborg (2000) found a moderate correlation (r=0.41, $P < 0.05$, n=23) between TGI 200 scores and factor scores based on production data and disease incidences in dairy cows indicating that herds with higher TGI 200 scores had higher production and better health. It can be questioned, however, whether it is really the multiplicity of criteria, that contributes to these relationships. In a cohort analysis, Sundrum and Rubelowski (2001) showed that the simultaneous consideration of several design criteria (space allowance, feeding space, pen dimension and floor quality) in a

TGI 40 protocol did not enhance predictive value for early losses of fattening bulls when compared to the single criterion of space allowance. Correlations were negative and significant, but low (r=–0.31; $P < 0.01$, n=50, for both space allowance and TGI 40).

To summarize, the TGI systems are operational tools that take account of a broad range of influencing or risk factors for animal welfare. For advisory purposes, they provide systematic checklists and allow comparisons among farms or of the same farm over time. However, the index may not sufficiently indicate the actual animal welfare state. Although minimum standards have the same problem (see above), the way of calculating the TGI also is debatable. Moreover, a score provides little transparency in terms of actual husbandry conditions fulfilled. Therefore, TGI systems appear less suited for certification purposes.

The NAHWOA Recommendations, while recognizing the TGI as a 'useful starting point', offer several suggestions for improved welfare assessment tools, especially for use on organic farms. Among them (several of which are implicit in the previous discussion) are: to make the tool appropriate for solving problems, not just assessing them; to make clear the reason for including each component; to include zoonotic aspects; to include animal health measures, such as somatic cell count; and to extend the scope to include transportation and slaughter.

Other index systems

There are other developments of the idea of index systems, but they are not routinely applied, such as the integrated assessment concept for horses (Beyer, 1998). Others are still in a developmental stage, such as the decision support system to assess welfare in housing systems for pregnant sows (Bracke, 2001) and the assessment system for dairy cows of Capdeville and Veissier (2001) based on the five freedoms (FAWC, 1992).

The assessment method of Beyer (1998) from Germany may be used for evaluation and welfare certification of a housing system, but it is not intended to serve as an advisory tool for the farmer. It differs from the TGI approach in that it attempts to assess the housing situation in relation to the average condition. Moreover, it keeps three welfare areas separate: (i) the housing system itself; (ii) the management of the housing system; and (iii) the management of the exercise yard. The farm may thus be found to have a better housing system than average, while at the same time suffering from worse than average management of the exercise yard. The assessment can be carried out in 1 h with a scheme consisting of 45 questions, with body condition as the only animal-based variable. Such a method allows an assessment relative to other existing housing systems, but does not address the question of cut-off points when animal welfare may be at risk, or of positive goals for good welfare.

The computer-based decision support system of Bracke (2001) in The Netherlands calculates an overall welfare score based on a combination of animal-based, design and management variables. The assessment is derived from scientifically based data on how specific husbandry factors influence welfare on one hand, and on the other from ideas concerning animal needs, taken from the scientific literature and expert interviews (Bracke *et al.*, 1999). The model provides points of reference from the assessment of seven main housing systems for pregnant sows. It is constructed to be flexible in that it allows new information or insights to be incorporated if necessary. It is claimed that this system may be used as an advisory tool for the farmer, as welfare certification for the consumer, and as a means of control by the control authority. However, the system currently requires an experienced user, and questions remain regarding the interpretation of scores in terms of the actual welfare state of the animals. Thus, further development is necessary to make this system applicable in practice.

Capdeville and Veissier (2001) are developing a multidimensional diagnostic tool in France, with dimensions of welfare being defined in terms of the so-called 'five freedoms': (i) freedom from hunger and thirst; (ii) freedom from discomfort; (iii) freedom from pain and injury; (iv) freedom from fear; and (v) freedom to express normal behaviour (FAWC, 1992). It mainly uses animal-based indicators, and the overall welfare assessment is presented in terms of the five freedoms. For example, welfare in a dairy herd may be interpreted as satisfactory with respect to expression of normal behaviour, but insufficient with respect to injuries.

Individual Integrated System Evaluations

The Danish research project 'Ethical accounting in livestock production' developed a method that tries to combine information on animal welfare, production, environmental impact, energy efficiency and economic performance on pig and dairy farms, together with the farmers own ethical values (ascertained in dialogue meetings). The aim of the assessment is to provide farmers with better knowledge of the situation on their farms in relation to their own values and to plan their actions accordingly (Sørensen *et al.*, 1998). Moreover, consequences are assessed for the four main affected parties: the animals, the farmer, the consumers and future generations. The interest of future generations is evaluated by considering environmental aspects such as soil quality and biodiversity (Halberg, 1999).

The assessment of welfare is based on a description of the housing system, the system's application (i.e. management), animal behaviour and animal disease. In dairy herds, the evaluation of the housing system focuses on the feeding facilities, access to water, and space for movement and lying down. Management-related factors such as cleanliness of the lying area and water supplies are recorded several times during the year. The behavioural evaluation relates to recordings of lying behaviour, fear of a known person, and fear reactions during milking (Sandøe *et al.*, 1997; Johanneson *et al.*, 1997). Health data are obtained on several visits to the farm and include factors such as extremely low or high body condition scores, lameness, scab and ringworm infections, diagnoses from claw trimming, reasons for early culling or death, and registration of all disease treatments by veterinarians.

Once a year, the result of the welfare assessment are presented to the farmer and are compared with reference data from other herds as well as to previous results. The welfare assessment is presented as part of the ethical account, which also include sections regarding the year's economic results, the values and expectations of the farmer, annual production, product quality, resource use, nutrient balances, and environmental aspects. The section on animal welfare is typically 10–14 pages long. The farmer receives a written evaluation for the individual variables and a summary of the welfare assessment with general conclusions. The evaluation is written by an animal welfare expert during the project period.

The integration of animal welfare with the environmental impacts of production raises new challenges to the discussion of animal welfare. One example of a conflict of interests is the positive effect on animal welfare of an outdoor production system for pigs and poultry compared to the negative effects on losses of ammonia or control of salmonella and other zoonoses (Sørensen *et al.*, 2001; see also Chapter 11). After 2–3 years for ethical accounting, the farmer and advisers are expected to begin strategic planning. The strategic planning deals with the potential conflicts of interests and outlines the development of the farm for the coming 5–10 years.

The method of welfare assessment used in the ethical accounting requires the observations and their interpretation to be carried out by experts, and further development of the method is needed before it can be applied in practice. In the first version, the welfare assessment required 40–50 h per year for a herd of 80 dairy cows, which was believed to be too expensive for commercial use. The final version of the system is expected to be used by local advisers (animal scientists and veterinarians). After the third year, 68% of the farmers involved in the project reported that the ethical accounting caused them to alter their management concerning animal welfare (Michelsen and el Seady, 1998; Sørensen *et al.*, 2001). In this connection, the NAHWOA Recommendations note that with an appropriately designed assessment, the very fact of participating and keeping records can be beneficial because it increases the farmer's motivation to take action.

Do we need Different Welfare Assessment Approaches on Organic and Conventional Farms?

All these approaches are currently applied on both organic and conventional farms. The current understanding of welfare assessment matches 'organic thinking' closely enough to make this acceptable. However, because organic farming potentially involves special conflicts between animal welfare and other valid goals (e.g. environmental protection, as discussed in Chapter 8), the tools used to assess animal welfare may need to be further adapted for organic systems in particular; we have already mentioned some of the suggestions that the NAHWOA Recommendations give along these lines.

A feature the assessment approaches discussed in this chapter have in common is that they attempt to take multiple interacting factors into account. The animal interacts with many aspects of its housing environment, herdmates or other animals, the feed and water, the stockpeople and possibly other elements. At the same time, actions and reactions of the animal take place on many different levels, including the physiological, physical and behavioural. However, the current state of science does not yet convincingly provide us with appropriate operational tools to determine the result of this complex interaction in terms of the welfare of the animal, at least as long as we strive to know something about the overall welfare of the animal, and not just of single aspects of it.

The attempt of Wemelsfelder *et al.* (2000) to develop a holistic approach for the welfare assessment based on a different theory of science (see Chapter 5) is not yet advanced enough to judge whether it could be used in an on-farm assessment. Even if we have to accept compromises regarding the available assessment tools and bases for interpretation, it nevertheless is important that organic farmers monitor the welfare of their animals for the reasons stated above. Many organic rules are good for the welfare of the animals, but we can also see welfare problems on organic farms. It is in the interest of all parties involved to work on improvements in this regard.

Conclusions

Welfare assessment on farms may serve different aims covering certification and control of welfare on the farm level, diagnosis of welfare problems and provision of an advisory tool for the farmer, and evaluation of housing systems. Consequently, different methods may be appropriate for different goals. If, for example, the goal is to certify that specific husbandry conditions are fulfilled, it may be enough to record the design or management criteria (Johnsen *et al.*, 2001). Consumers will primarily be interested in this kind of information. However, they may also want assurance that the animals are well under these conditions, for which animal-related indicators must be used. For the farmer and adviser, it is important that the assessment reveals not only the welfare state of the animals, but also the connection to the husbandry conditions. For most purposes, therefore, the combined application of design, management and animal-related criteria seems to be most appropriate (Rushen and de Passillé, 1992; Johnsen *et al.*, 2001; Waiblinger *et al.*, 2001). However, available resources and methodological problems limit the use of animal-based criteria in practical welfare assessments. More research is needed for the adequate selection of criteria and recording methods (Sundrum, 1997; Waiblinger *et al.*, 2001).

Fortunately, scientific interest in the operational welfare assessment is steadily increasing, and new projects have started recently at different places in Europe. All assessment systems developed and applied so far have their specific advantages and disadvantages and can be used on both organic and conventional farms. Whilst the approach of integrated welfare assessments rightly emphasizes the multitude of factors influencing animal welfare and is a more flexible assessment tool, we need a better scientific foundation and inclusion of more animal-related criteria before these systems can be regarded as valid. Moreover, the weighting of the different welfare aspects necessarily involves society's consent, not just science. Individual integrated systems tackle some of these problems by considering more animal-related information and evaluating individual welfare aspects separately. However, they are just advisory tools, and are rather time- and resource-consuming. The minimum standards approach has the advantage of transparency regarding the information conveyed and the explicit limitations of the assessment. In general, care should be taken to make explicit what has been assessed – the state of the animal or the husbandry conditions – and the extent to which the assessment is based on sufficient information about the animal itself.

References

Amon, Th., Boxberger, J., Schatz, P. and Kummernecker, C. (1997) Beurteilung der Tiergerechtheit von Milchviehhaltungssystemen anhand einer vergleichenden Modellanalyse TGI 35L/1995 und TGI 200/1994. In: Weber, R. (ed.) *Tiergerechte Haltungssysteme für landwirtschaftliche Nutztiere. Wissenschaftliche Tagung in Zusammenarbeit mit der Internationalen Gesellschaft für Nutztierhaltung (IGN), 23–25 October 1997, Tänikon, Switzerland.* Eidg. Forschungsanstalt für Agrarwirtschaft und Landtechnik, CH-8356 Tänikon TG, pp. 24–35.

Bartussek, H. (1985) Vorschlag für eine Steiermärkische Intensivtierhaltungsverordnung. *Der Österreichische Freiberufstierarzt* 97, 4–15.

Bartussek, H. (1999) A review of the animal needs index (ANI) for the assessment of animals' well-being in the housing systems for Austrian proprietary products and legislation. *Livestock Production Science* 61, 179–192.

Bartussek, H. (2000) How to measure animal welfare? In: Hovi, M. and Trujillo, R.G. (eds.) *Diversity of Livestock Systems and Definition of Animal Welfare. Proceedings of the Second NAHWOA Workshop, 8–11 January 2000 Córdoba, Spain.* University of Reading, Reading, pp. 135–142.

Bartussek, H., Leeb, C. and Held, S. (2000) *Animal needs index for cattle. ANI 35L/2000 – cattle.* Bundesanstalt für alpenländische Landwirtschaft. BAL Gumpenstein, Irdning, Austria.

Bennedsgaard, T. and Thamsborg, S. (2000) Comparison of welfare assessment in organic dairy herds by the TGI200-protocol and a factor model based on clinical examinations and production parameters. In: Hovi, M. and Trujillo, R.G. (eds) *Diversity of Livestock Systems and Definition of Animal Welfare. Proceedings of the Second NAHWOA Workshop, 8–11 January 2000, Córdoba, Spain.* University of Reading, Reading, UK, pp. 143–150.

Beyer, S. (1998) *Konstruktion und Überprüfung eines Bewertungskonzeptes für pferdehaltende Betriebe unter dem Aspekt der Tiergerechtheit.* Vet. Diss., Justus-Liebig-Universität Gießen, Germany

Bioland Verband (2002) *Bioland-Richtlinien.* 26. November 2002, http://www.bioland.de/bioland/richtlinien/erzeuger-richtlinien.pdf, p. 9.

Boehncke, E. (1997) Preventive strategies as a health resource for organic farming. In: Isart, J. and Llerena, J.J (eds) *Resource Use in Organic Farming. Proceedings of the 3rd ENOF Workshop, 5–6 June 1997, Ancona, Italy*, pp. 25–35.

Boehncke, E. and Krutzinna, C. (1996) Animal health. In: Østergaard, T.V. (ed.) *Fundamentals of Organic Agriculture. Proceedings of the 11th IFOAM International Scientific Conference. 11–15 August 1996, Copenhagen, Denmark.* Vol. 1. IFOAM, Werbe Druck, St. Wendel, Germany, pp. 113–124.

Bracke, M.B.M. (2001) Modelling of animal welfare: The development of a decision support system to assess the welfare status of pregnant sows. PhD thesis, Institute of Agricultural and Environmental Engineering (IMAG), Livestock Production and Environmental Engineering Department, Wageningen, The Netherlands.

Bracke, M.B.M., Spruijt, B.M. and Metz, J.H.M. (1999) Overall animal welfare reviewed. Part 3: Welfare assessment based on needs and supported by expert opinion. *Netherlands Journal of Agricultural Science* 47, 307–322.

Broom, D.M. (1996) Animal welfare defined in terms of attempts to cope with the environment. *Acta Agriculturae Scandinavica, Section A, Animal Science Supplement* 27, 22–28.

Broom, D.M. (2001) Coping, stress and welfare. In: Broom, D.M. (ed.) *Coping with Challenge: Welfare in Animals including Humans.* Dahlem University Press, Berlin, pp. 1–9.

Capdeville, J. and Veissier, I. (2001) A method for assessing welfare in loose housed dairy cows at farm level, focusing on animal observations. *Acta Agriculturae Scandinavica, Section A, Animal Science Supplement 30*, 62–68.

Farm Animal Welfare Council (FAWC) (1992) FAWC updates the five freedoms. *Veterinary Record* 131, 357.

Fraser, D. (1995) Science, values and animal welfare: exploring the 'inextricable connection'. *Animal Welfare* 4, 103–117.

Fraser, D., Weary, D.M., Pajor, E.A. and Milligan, B.M. (1997) A scientific conception of animal welfare that reflects ethical concerns. *Animal Welfare* 6, 187–205.

Halberg, N. (1999) Indicators of resource use and environmental impact for use in a decision aid for Danish livestock farmers. *Agriculture, Ecosystems and Environment* 76, 17–30.

Hemsworth, P.H. and Coleman, G.J. (1998) *Human–Livestock Interactions: The Stockperson and the Productivity and Welfare of Intensively Farmed Animals*. CAB International, Wallingford, UK.

Johanneson T., Sørensen, J.T. and Munksgaard, L. (1997) Production environment as component in a welfare assessment system in dairy herds. In: Sørensen J.T. (ed.) *Livestock Farming Systems – More than Food Production. Proceedings of the 4th International Livestock Farming Symposium*. EAAP Publication 89. Wageningen Press, Wageningen, The Netherlands, pp. 251–255.

Johnsen, P.F., Johannesson, T. and Sandøe, P. (2002) Assessment of farm animal welfare at herd level: many goals, many methods. *Acta Agriculturae Scandinavica, Section A, Animal Science Supplement* 30, 26–33.

Knierim, U., Carter, C.S., Fraser, D., Gärtner, K., Lutgendorf, S.K., Mineka, S., Panksepp, J. and Sachser, N. (2001) Good welfare: improving quality of life. In: Broom, D.M. (ed.) *Coping with Challenge: Welfare in Animals including Humans*. Dahlem University Press, Berlin, pp. 79–100.

Michelsen, J. and el Seady, T. (1998) Hvad synes husdyrbrugerne om etisk regnskab for husdyrbrug? In: Sørensen J.T., Sandøe, P. and Halberg, N. (eds) *Etisk regnskab for husdyrbrug*. DSR Forlag, Copenhagen, pp. 113–132.

Roiha, U. (2000) *Luomulehmien hyvinvointi Etelä-Savossa. Luomukotieläintalouden kehittäminen Etelä-Savossa 1996–1999 – hankkeen tutkimusraportti*. Helsingin yliopisto, Maaseudun tutkimus- ja koulutuskeskus, Mikkeli. Julkaisuja 74.

Roiha, U. and Nieminen, T. (1999) Luomunautojen terveys ja hyvinvointi. Helsingin yliopisto, Maaseudun tutkimus- ja koulutuskeskus, Mikkeli. *Julkaisuja* 66, pp. 66–93.

Rushen, J. and de Passillé, A.M.B. (1992) The scientific assessment of the impact of housing on animal welfare: a critical review. *Canadian Journal of Animal Science* 72, 721–743.

Sandøe, P., Giersing, M.H. and Jeppesen, L.L. (1996) Concluding remarks and perspectives. *Acta Agriculturae Scandinavica, Section A, Animal Science Supplement* 27, 109–115.

Sandøe, P., Munksgaard, L., Bådsgaard, N.P. and Jensen, K.H. (1997) How to manage the management factor – assessing animal welfare at the farm level. In: Sørensen, J.T. (ed.) *Livestock Farming Systems – More than Food Production. Proceedings of the 4th International Livestock Farming Symposium*. EAAP Publication 89. Wageningen Press, Wageningen, The Netherlands, pp. 221–230.

Sørensen, J.T., Sandøe, P. and Halberg, N. (eds) (1998) *Etisk regnskab for husdyrbrug*. DSR Forlag, Copenhagen.

Sørensen, J.T., Sandøe, P. and Halberg, N. (2001) Animal welfare as one among several values to be considered at farm level: the idea of an ethical account for livestock farming. *Acta Agriculturae Scandinavica, Section A, Animal Science Supplement* 30, 11–16.

Sundrum, A. (1997) Assessing livestock housing conditions in terms of animal welfare – possibilities and limitations. In: Sørensen, J.T. (ed.) *Livestock Farming Systems – More than Food Production. Proceedings of the 4th International Livestock Farming Symposium*. EAAP Publication 89. Wageningen Press, Wageningen, The Netherlands, pp. 238–246.

Sundrum, A. (1999) EEC-Regulation on organic livestock production and their contribution to the animal welfare issue. In: *Regulation of Animal Production in Europe*. KTBL-Schrift 270. KTBL, Darmstadt, Germany, pp. 93–97.

Sundrum, A. and Rubelowski, I. (2001) The meaningfulness of design criteria in relation to the mortality of fattening bulls. *Acta Agriculturae Scandinavica, Section A, Animal Science Supplement* 30, 48–52.

Sundrum, A., Andersson, R. and Postler, G. (eds.) (1994) *Der Tiergerechtheitsindex-200/1994 – ein Leitfaden zur Beurteilung von Haltungssystemen für Rinder, Kälber, Legehennen und Schweine.* Verlag Köllen, Bonn, Germany, 211 pp.

Waiblinger S., Knierim, U. and Winckler, C. (2001) The development of an epidemiologically based on-farm welfare assessment system for use with dairy cows. *Acta Agriculturae Scandinavica, Section A, Animal Science Supplement* 30, 73–77.

Wemelsfelder, F., Hunter, E.A., Mendl, M.T. and Lawrence, A.B. (2000) The spontaneous qualitative assessment of behavioural expressions in pigs: first explorations of a novel methodology for integrative animal welfare measurement. *Applied Animal Behaviour Science* 67, 193–215.

10

The Role of Humans in the Management of Organic Herds

Mette Vaarst,[1] Francoise Wemelsfelder,[2] Martin Seabrook,[3] Xavier Boivin,[4] and Anita Idel[5]

[1]Department of Animal Health and Welfare, Research Centre Foulum, Danish Institute of Agricultural Sciences, PO Box 50, DK-8830 Tjele, Denmark; [2]Animal Biology Division, Scottish Agricultural College, Bush Estate, Penicuik, Edinburgh EH26 0PH, UK; [3]School of Biosciences, The University of Nottingham, Sutton Bonington Campus, Loughborough LE12 5RD, UK; [4] INRA, Theix, St Genès Champanette, France; [5]League of Pastoral Peoples, 64372 Ober-Ramstadt, Germany

Editors' comments

The people who work with the herd are decisive for the animals' welfare, of course. They make decisions on all levels: they create the framework that aims to meet the goals of organic farming, and they observe and interact with individual animals – to varying degrees, depending on the species and the production system – and intervene in critical situations. All these ways that humans influence the lives of the animals in their households are influenced by their knowledge, experience, ethical values and practical judgement. Some people seem to have a natural gift for immediately perceiving and responding adequately to the different needs signalled by the animals, thus creating optimal conditions for both the individual animal and the herd as a whole. For the rest, this skill needs to be analysed and put it into words, and then practised. In this chapter, the authors discuss some of the complex associations among human choices, animal welfare and organic animal husbandry. They put this field into a broader perspective of knowledge, expectations and different perceptions of daily management

© CAB International 2004. *Animal Health and Welfare in Organic Agriculture* (eds M. Vaarst, S. Roderick, V. Lund and W. Lockeretz)

Introduction

The role of the farmer[1] has been recognized by many (e.g. Hemsworth and Coleman, 1998; Seabrook and Wilkinson, 2000; Lensink *et al.*, 2001) as critical in managing the welfare of farm animals. The farmer is even more important on organic farms, both for the ethical reasons discussed in Chapter 5 and for reasons relating to health and welfare. The multifunctional objectives of organic production may require skills and attitudes that go beyond those normally expected in conventional livestock management. With conventional farming increasingly encompassing environmental goals, organic approaches could offer models to other farmers and animal managers.

Epidemiological studies that seek to identify risk factors for disease often conclude that 'individual herd characteristics' or 'management factors' may explain most of the observed variation. Some studies succeed in identifying specific aspects of management as risk factors for a certain disease, but precise details of the actual management are rarely defined. To address this, it is common to divide the management of animals into elements that either increase or reduce the risk for some disease. Beyond such an analysis, however, it may be relevant to examine the farmer's attitude towards animals and identify different styles of management, and subsequently investigate the effect of these factors on the relations between the farmer, the animal and its surroundings (including other 'stakeholders' such as animal health professionals).

There is a large literature on human–animal relations and on specific management routines affecting animal health and welfare. Here we explore these issues with particular reference to organic farming, starting with the way humans and animals perceive each other. We then consider different sources of farmer knowledge and how these may affect human–animal relations. After this, we discuss the various components of animal management and their potential impact on animal welfare, with most of the examples taken from research on dairy cattle. Finally, looking forward, we propose an ethical framework for the development of organic animal care and management systems.

[1] 'Farmer' is in this chapter defined as the man or woman who manages the herd, takes decision in the herd and takes care of the animals. In some herds, this is the same person as the 'owner', in others not. In some farms, different persons may each take care of or manage a particular group of animals, or they may share the work and responsibilities; here they will still be referred to as 'the farmer'.

Human–Animal Relationships

Animals' perceptions of humans

Animal welfare is influenced not only by the way humans build up a production system, plan strategies for breeding and feeding, and fulfil the animals' needs for natural behaviour and good living conditions. It also is influenced by the direct interaction between humans and animals. This relationship both affects and is affected by how animals perceive humans (Boivin *et al.*, 2001). Estep and Hetts (1992) suggest four main ways in which animals may perceive humans:

- As a danger, which will be due to negative, little, or no contact, so that the animals become fearful. Hemsworth *et al.* (1987) found that inconsistent handling is similar to negative handling, creating fear, harming the human–animal relationship and possibly reducing production.
- As an indifferent object.
- As a provider of food and water.
- As a social partner.

Positive contact between humans and animals is beneficial to animal behaviour: a calm, self-confident and well-balanced person will influence the animal to be non-violent, confident and calm (Seabrook and Bartle, 1992; Seabrook, 1994; Hemsworth and Coleman, 1998). In such cases, humans are perceived not as dangerous threats, but as indifferent objects or social partners. Predictability is important for animals' confidence in humans. Wiepkema (1987) emphasizes the importance of predictability and control over the environment in animals' lives, an argument further underlined by Hemsworth *et al.* (1993) and De Jonge *et al.* (2001).

Humans' perceptions of animals

Just as animals' perceptions of humans clearly affect their welfare, so too, how we view animals undoubtedly also influences how we treat and manage them, and hence also their well-being. Often our relationship with and views of animals have deep historical roots (see Chapter 2) that are strongly linked to broader societal factors. As discussed in Chapter 5, the early scientific literature generally disregarded animals' sentience, strongly tending to objectify them. This tendency may persist today, giving rise to objectified terminologies and models in which animals are 'animal material' when we study them, and 'livestock' rather than 'farm animals' when we farm them. Such an approach subsequently makes it easier to impose industrial and economic values upon individual animals.

The very terms 'management' and 'stockperson' reflect such a perception (Hemsworth and Coleman, 1998).

However, perceptions can change with changing circumstances. This may happen because of consumer demands, a shift in personal experience or in the case of farmers converting to organic production, the adoption of a new approach to farming. Organic farmers must comply with certain legal requirements, which may also change their perception of their animals. For example, in a Danish study, farmers described how they started to reflect upon the keeping of heifers after having been forced to use bedding material (Vaarst, 2000). They realized the benefits to the animals, and regretted that they had not previously questioned keeping heifers on full slatted floors.

This example illustrates that the way we treat animals in our daily lives influences our own perceptions of their needs. When we change the farm to allow the animals to express their basic behavioural needs, our views of animals must be expected to change. The farmer sees the animals interact and behave differently than in a more restricted environment, and consequently learns to understand them as more capable and responsive creatures.

Factors in organic animal production systems that influence the human–animal relationship

On organic farms, animals are perceived as playing an important role in exchanging feed and manure within a closed system. Priority also is given to the well-being and integrity of each animal. Lund *et al.* (2003) suggest that in such farming systems, animals and humans are viewed as partners within a bio-ethical framework. Such principles of organic farming may be expected to influence farmers' own perception of their role in relation to the animals. Although legal requirements can certainly be a force for change, farmers' understanding of organic philosophy is especially likely to bring about such change. Caretakers' sense of moral obligation towards animals is likely to affect their perceptions of their own role in relation to the herd and the individual animal. Verhoog (2000) and Lund *et al.* (2003) discuss various perspectives on our moral obligations toward animals and examine their implications for animal welfare and animal production (see also Chapter 5).

Organic production necessarily entails free-range systems, where minimizing the animals' exposure to humans is viewed as an advantage in that it supports their natural behaviour in an ecologically balanced environment. Boivin *et al.* (1994, 2001) raise the question of whether rearing animals outdoors could increase their fear reaction towards humans, and whether human contact early in life might be necessary to avoid these problems. Seabrook (1998), on the contrary, suggests that when dairy cows can move about and see the 'full person' from different vantage points, that makes them more comfortable in human company. These statements are not contradictory, however; it depends on the situation under which fear reactions occur. Cows may be more comfortable when freely moving around outdoors, but then may become more fearful when confined and treated, such as for artificial insemination, disease treatment or claw trimming.

The actual amount of contact between humans and single animals clearly depends on the species and production system. Whereas dairy cows need to be handled twice a day for milking, sheep living in remote mountain areas do not need such frequent handling. This will undoubtedly affect their perception of human contact during the few times a year when they need to be handled, such as when they are being treated for disease.

Furthermore, Boivin *et al.* (2001) see a potential conflict between the desire to have local and rare breeds in organic farming and the welfare of those animals. These breeds may not have been selected for their responsiveness to humans, but rather for their ability to adapt to local conditions. This applies, for example, to sheep systems in Norway, such as the one described in Chapter 3.

Sources of Farmers' Knowledge

Tacit knowledge of animal management

What kind of knowledge is needed to develop effective interactions with farm animals, and how may the acquisition of such knowledge be fostered by the skills and attributes of a farmer? Skills that develop during interaction with others often are not explicit, and may lead to tacit knowledge (Vaarst, 1995). The daily interaction between farmer and animals is a continuous process of acting, learning, building up of experience, and evaluating this experience in light of the farmer's values and goals.

It is often difficult for a farmer to describe this learning process – it tends to be intangible and related to the immediate context and situation. Through farmers' literature, meetings and advisory services, however, the process can be shared with colleagues, who are fellow 'experts' in the field. Dreyfus and Dreyfus (1986) characterize 'expert action' as skilled behaviour that is carried out intuitively and is based on extensive experience and insight. Such insight emerges from continuous observation, association, analysis, prognosis and weighing of different courses of action; it is not explicitly reflective in a way that enables a farmer to explain what is being done and why.

Tacit knowledge is often handed down directly from previous generations and earlier employers. It is likely that a farmer will have been brought up to think and act in a certain way and not be familiar with alternative arguments. As long as no alternatives are put forward for consideration, in the farmer's mind such actions may be not just the *obvious* way, but the *only* way of doing something. Such tunnel vision certainly does not encourage the evolution of farming methods and attitudes. Reflection and making existing knowledge explicit may be important for the conscious development of organic animal husbandry.

Using tacit knowledge

Recognition of farmers' non-reflective expertise opens up a more general discussion about the nature of knowledge and the importance of experience. As early as the fourth century BC, Aristotle introduced the notion of *techne* to discuss a kind of knowledge that is not in the first place theoretical, but is transferred through practical experience and 'from hand to hand' (Dunne, 1992). The importance of such knowledge is increasingly recognized in the field of human health care, especially in relation to the nursing profession. The role of the nurse is health care, not disease treatment. The subject of interest is the patient, not the disease, and the nurse's actions are aimed at improving the patient's life in hospital. Nurses have often been said to possess an inborn, natural ability for caring, which sometimes has led to an almost mythical belief in their virtue. Taking care of patients is considered something you just pick up through daily routine and practice, something naturally transferred among colleagues.

More recently, however, discussions in some parts of the Western world have suggested that nursing should become more of a profession with explicit standards for what constitutes 'good nursing practice'. Images of Florence Nightingale, whose life was seen as embodying the 'soul' of caring, are now replaced with detailed descriptions of how to make a bed comfortable, how to give injections and how to contact a patient's relatives. Thus nursing becomes explicit and controllable, and open to quality inspection. A nurse who makes the bed properly is a good nurse, because that is how the profession is now defined.

A parallel can easily be drawn with the role of the farmer. As argued above, practical expert knowledge can be viewed as crucial to being a 'good' farmer. However, this knowledge must be made more explicit, and standards must be formulated for what constitutes good practice. Perhaps this is even more applicable in an organic context, where the goals are more complex and the interactions more crucial.

This is not to say that the activities of a 'good' farmer should consist merely of following rules or standards to the letter, because there is no substitute for personal motivation and experience. However, making daily routines explicit may help farmers to better understand and communicate their own motives and insights, and may help to improve the overall management framework for the herd. (For practical examples see Chapter 12.)

The scientific status of expert knowledge

A pressing question in considering the importance of practical expert knowledge is its scientific status. Scientific investigation typically tries to isolate causal elements within a system in order to measure and control them. On the other hand, 'expert knowledge', as described above, is derived from a complex of interdependent factors, many of which are not easily measured and quantified. Scientists often consider this kind of knowledge to be merely 'common sense', which they see as 'subjective' and 'soft' in comparison to controlled experimental knowledge.

However, research by Wemelsfelder *et al.* (2000, 2001) and Wemelsfelder and Lawrence (2001) indicates that such a dismissive judgement of a farmer's expert knowledge may be unjustified. In direct and spontaneous interactions with animals, we perceive and describe their behaviour as expressive of their emotional state (e.g. calm, anxious, content or distressed); the question is how reliable and repeatable such descriptions are (Wemelsfelder, 1997a,b).

To investigate this question, Wemelsfelder and colleagues used specifically designed experimental and statistical procedures to ask observers to assess the behavioural expressions of pigs in their own freely chosen words. This work showed a high level of agreement, repeatability and overall analytical resolution for these assessments (Wemelsfelder *et al.*, 2000, 2001). These studies were subsequently repeated using pig farmers and veterinarians as observers (Wemelsfelder, 2001; Wemelsfelder and Lawrence, 2001; Wemelsfelder *et al.*, 2001). The results indicate that spontaneous and holistic assessments of animal behaviour, although different from traditional kinds of scientific assessments, are not necessarily any less scientific. No one has more experience in observing the expressive repertoire of individual farm animals than the farmer, and this expert knowledge should play a prime role in the management of animal health and welfare. Although practical knowledge is complex and not easily measured, it should be taken seriously when examining factors influencing welfare.

Training to be a 'caretaker' in the organic animal production system

This discussion makes it clear that humans perceive animals and their role on the farm as encompassing more than just their physical and physiological needs. The farmer in an organic system is a caretaker under circumstances where animals are allowed as much access to 'natural behaviour' and 'naturalness' as possible. This gives the caretaker the challenge of behaving towards animals as 'naturally' as possible. This ability, once developed through experience, also must be communicated to others through training. Such training should consist as much as possible of making tacit knowledge explicit, under the guidance of colleagues skilled in 'reading' the behavioural responses of animals and in understanding the needs of animals in a wider perspective.

The Role of the Farmer in Daily Practice

A framework for different management routines

Farmers vary greatly in their attitudes towards animals, and the strategies and daily routines they use to manage their herd are likely to mirror their values and attitudes. Management factors are unlikely to remain static, but rather develop dynamically as the farmer responds to what he or she observes and adjusts the management accordingly (Goodger *et al.*, 1993; Ekman, 1998). Routines may be changed to control a critical situation, or may become more informal and relaxed if no complications occur. For example, Barkema (1998) found that frequent checking of heifer udders was a risk factor for mastitis at first calving. However, it is possible that checking the udders was started because mastitis was a problem. If we can accept that 'local logic' exists in herd management, the challenge is to find the patterns underlying that logic, and to articulate and communicate them as a basis for understanding management routines.

To analyse how different management routines may affect the health and welfare of the herd, we need to examine in more detail the various components of the farmer's role. Seabrook and Bartle (1992) identified three components of that role: managing, operative and empathic. The first refers to management of areas such as effective record keeping, feeding regimes and routine health care. Operative activities refer to the technical ability to perform tasks such as carrying out a specified feeding regime, operating the milking equipment correctly and observing reproduction cycles. Both of these are emphasized in Chapter 12 as important elements in the development of preventive health management. Individual animal-related tasks of a qualitative nature, such as the manner of handling, interaction and contact, are considered aspects of an empathic role, and are not easily quantified.

Examining these aspects of management from an animal welfare perspective, and modifying the categories proposed by Vaarst (1995), it is possible to distinguish three kinds of management routines:

- Those that develop and maintain a relatively fixed physical framework for the whole herd (e.g. those associated with housing).
- Those based on human–animal interaction (e.g. milking).
- Those aimed more specifically at improving animal well-being.

We now examine these routines in more detail in the context of organic farming.

Farm system management routines

The physical living conditions on a farm (e.g. the housing) and associated management strategies form a framework for the whole herd that hardly changes from day to day. The broad goals and priorities for the farm will often be mirrored in the management framework, which is likely to have some built-in flexibility to respond to change. On an organic farm, the framework will be partly structured around the organic standards and is likely to be complex, because the structures will incorporate activities that relate not only to the way that animals live, but also to how they are integrated into the whole farm. For example, grazing management on an organic farm with sheep may be integrated with a cropping programme to make use of grass leys whilst also ensuring parasite management based on clean grazing (Chapter 14). The concept of ethical frameworks that aim to balance animal needs and behaviour with human needs are discussed later in this chapter.

The management structure on any farm will be governed, of course, by factors such as the farmer's perception, motivation and objectives. Noe (1999) and van der Ploeg (1994) offer a model of how a farmer's ideas might be structured and transformed into different farming styles. Figure 10.1 shows that these farming styles can be characterized along two axes, one dealing with financial matters (high versus low input/output), the other with the style of decision-making ('craftsman' vs. 'data-oriented' decision making). Together these axes create four different routes for acting and reacting to the farming environment, and hence dictate how farm structures are developed.

For example, a farmer who is primarily oriented towards high input–output and craftsmanship-like decision making will likely aim for high yields for each animal. This strategy will entail constant fine-tuning of management, handling, feeding, etc., based on a willingness to use a high level of resources in the expectation of high output in return. In contrast, if the farm is based on low input–output and data-oriented decisions, the individual cow will rarely be the focus; rather, the farmer typically will be oriented toward the outside world and market trends. The whole herd may be sold in 1 day if cows do not seem beneficial. All four farming styles have been identified in studies on Danish organic farms (de Snoo, 2002; Vaarst *et al.*, in preparation).

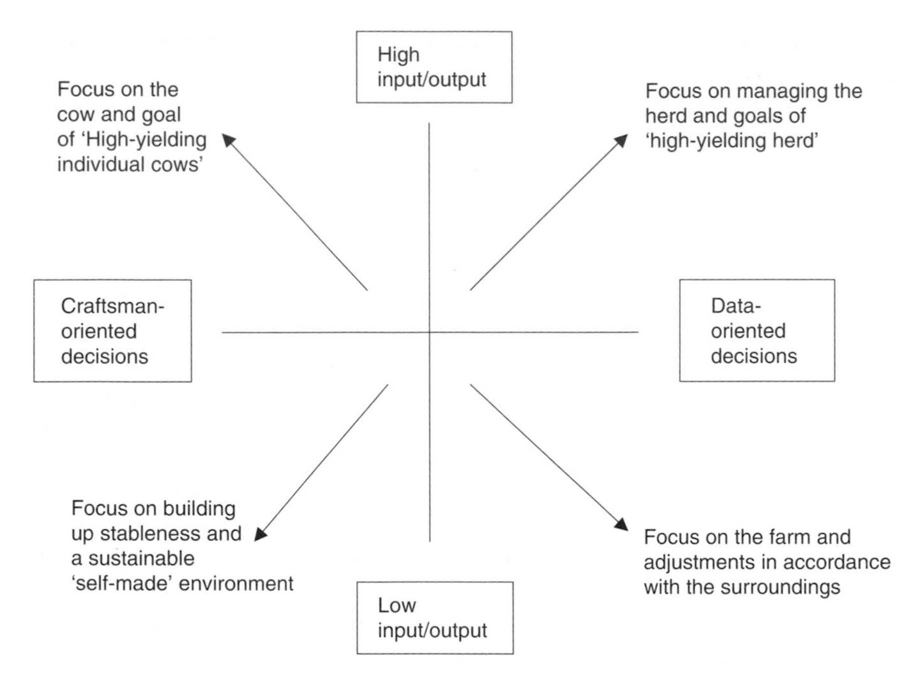

Fig. 10.1. A model for describing different farming styles. Modified after Noe (1999) and van der Ploeg (1993).

Human–animal interactive management routines

We can view interactive management routines as a series of continuous actions and reactions as we observe the animals. Interaction with animals, close observation in particular, may be more critical for organic than conventional farmers, not because of any difference in the way organic farmers view their animals, but because of some practical aspects associated with health management. Chapter 12 discusses the concept of health promotion and recognizes the enhanced role of the farmer in achieving this.

In herds where high production and reproductive efficiency are emphasized, observing animals can enhance their performance, for example through improved heat detection. For an organic farmer, observations of behaviour or early clinical signs of disease also are critical in a preventive approach. Health care strategies that rely not merely on treatment of disease, but also on a more complete understanding of the animal's needs and circumstances, require a close interaction between human and animal. This requires continuous judgements about potentially harmful situations and speedy reaction to them. This is perhaps where the farmer's skills with an organic herd are most clearly recognized. A survey of organic farm advisers, inspectors and veterinarians identified the farmer as crucial in achieving the animal welfare objectives of the organic farming standards (Hovi *et al.*, 2003).

Care-taking routines

One goal of organic animal husbandry is to ensure high levels of animal welfare, not just the absence of suffering. Therefore, care-taking routines play a prominent role. Organic farmers must provide specific living and feeding conditions that enhance the natural behaviour of their herd. In conventional agriculture, the need to increase welfare is often associated with the need to increase human control over an animal's life (Seabrook and Bartle, 1992). In organic agriculture, in contrast, the central concern that animals should lead as natural a life as possible (Alrøe *et al.*, 2001; Lund, 2000; Vaarst *et al.*, 2001) is viewed by some as implying a fundamental respect for an animal's integrity and its natural competence in leading a functional and meaningful life (Wemelsfelder and Birke, 1997; Chapter 5 in this book).

In studies of management routines in Danish organic dairy herds, Vaarst *et al.* (2001) identified a particular type of routine that they called 'care-taking routines'. The presence and intensity of similar routines in conventional herds was not examined in depth, but from treatment records and questionnaires it was clear that the organic farmers used many more care-taking routines (unpublished data). The organic farmers stated that they started to develop these routines mostly after they became organic farmers. Their primary motivation was to keep the cows healthy and to improve their condition when disease occurred, because this was part of 'being organic', i.e. avoiding medication and improving animal welfare. These routines could be described simply as actions to make life more comfortable for the animals. Other management routines, such as improving the hygiene level, may also make life more comfortable for the animals, even though that is not their primary aim. So too, care-taking routines can have benefits beside their primary aim of improving welfare, such as improved production. Seabrook (1994, 2001) provides evidence that care routines such as gently touching or talking to cows during milking may increase milk production.

Organic farming aims to balance production with allowing animals to develop their repertoire of species-specific skills, lead an active social life, and explore and play, and through such activities to develop general competence and well-being. Although natural behaviour of animals generally does not involve contact with humans, some contact is inevitable given the overriding goal of production, which requires regular manipulation of the environment and the animals.

An example of problems that this approach may lead to can be found in a recent study of animal welfare in organic dairy herds (DARCOF,

Askham Bryan College
LIBRARY BOOK

2000). This study identified the management of young calves as a major problem area, caused mainly by keeping calves in groups and by having them stay outdoors at a young age. Farmers found it difficult to manage these situations because of their newness to organic practice and the lack of appropriate regulations.

Vaarst *et al.* (2001) use this case to illustrate how managing the 'natural life' of an organic calf herd demands frequent observation and finely tuned judgement by the farmer. They conclude by suggesting that good animal welfare depends on sensibly combined management of both 'nature' and 'culture': farmers should provide animals with opportunities to fulfil their own needs, but they should engage in daily, unintrusive care-taking to safeguard the calves from unnecessary suffering or death.

Different species place different demands on the humans involved in their management. Actively applying care-taking routines to individual animals will be more relevant for cows than for broilers. This does not mean that the individual counts for less in a broiler flock, because no animal of any species should suffer. Rather, it means that problems and crises might be handled in another way, and that animals in larger flocks will receive less single-animal care. The suffering of animals in larger flocks will be solved by other means, such as when a disease outbreak is handled by treating the whole flock, by moving all diseased animals to a new and isolated pen, or by euthanasia, which stops suffering in a more dramatic way.

Caring for the Animals' 'Natural Lives' on the Organic Farm

Risks to health and welfare posed by natural lives

The interaction between 'natural living' and 'care-taking' goes through the farmer, who is responsible for creating a framework for 'natural living', and whose job is to see to it that the animals are observed and managed properly in such a system and to intervene in case of problems. One example of such a system, described in Chapter 3, is a Scottish farm that enables cows and calves to lead a natural life by being outside all the time, but at the same time seeks to reduce the welfare risks caused by continuous exposure to extreme weather.

The view that natural living is a central feature of animal welfare means that the understanding of an animal's positive and negative experiences must be adjusted in light of what constitutes 'natural living'. What an animal experiences as positive or negative is part of a dynamic learning process. An experience that at first sight may seem negative can become valuable if the animal learns how to behave in a particular situation. The less restrictive a given environment, the more the animal may have to learn before it can move around freely and be comfortable.

This gives the farmer the responsibility of ensuring that this can happen without turning valuable experiences into distinctly negative ones that are accompanied by suffering or other forms of chronic stress.

This is certainly true with regard to the farmer's responsibility to care for diseased animals. The degree of interaction between farmer and animals becomes even more critical, and the concept of 'loving care' for the sick animal plays an important role in the farmer's motivation. The motivation to care should be considered a crucial factor in the general strategy of creating a positive environment that enables an animal to defend itself against disease agents; however, a carefully constructed health plan aimed at promoting positive health is essential to ensure that animals do not suffer unnecessarily from leading natural lives.

Boivin *et al.* (2001) argue that human attitudes towards animals are potentially important in minimizing the use of antibiotics. A good and close human–animal relationship makes it possible for the farmer to note early behavioural deviations associated with disease. When applying so-called 'alternative' treatment methods, such as homoeopathy, the close relationship between farmer and animal also is of great importance. Homoeopathy is common in organic farming, and is encouraged under the EU standards (see Chapter 4). The fundamental principles are described in Chapter 13, which emphasizes the need to consider disease handling and health promotion together, and to view disease as not merely a 'mechanical dysfunction' that can be repaired by animals even though living in an unfavourable environment. This approach suggests a greater knowledge of the animals' circumstances, which can perhaps be really achieved only through a higher level of interaction than is normally associated with managing farm animals. The choice of the correct homoeopathic remedy depends on identifying the 'character type' (constitution) of the animal to be treated. The observational skills of the person administering the treatment (veterinarian or farmer), the communication between the two, and the extent to which the farmer knows the animal are all very important for this treatment method to be effective.

Such a view requires that care-taking be directed towards the individual animal, and ensures that the experience of the individual animal is good. In any herd, animals will be handled, and this must entail minimal stress. In pig herds, animals will be moved around in the systems, and in dairy herds, individual animals are handled during milking, claw trimming, inseminating and treating disease. In Chapter 3, two examples were offered showing that contact between caretakers and animals is important for reducing stress on the animals. The Dutch layer flocks, where the stress of the animals was reduced if the animals were accustomed to having contact with the caretakers, is a good example of farmers iteratively learning and acting. The example of the Scottish cattle herd showed how important it is for the caretaker to walk amongst the animals to reduce the stress on them and hence improve their welfare.

The caretaker's need to compromise between 'natural living' and production goals

Organic animal husbandry is beset by several dilemmas, as was discussed in Chapter 8; moreover, as was vividly illustrated in Chapter 3, the wide diversity of production systems requires the practical role of the caretaker to be approached with respect for local conditions. This includes the ways the caretaker can approach the goal of 'natural living', but it also requires a compromise between meeting the demands for a natural life and one's own judgement of whether a situation is good or bad, for example for the calves to stay outdoors or be with the cow 24 h after calving. One can discuss how to create a life that smacks a little of a 'natural life' in a farm situation. An example of trying to meet animals' needs in weaning is given in Box 10.1.

Problems arising during conversion

Conversion from conventional to organic farming may entail a substantial change in farming style. Vaarst (2000) investigated how such a change might be experienced from the farmers', veterinarians' and cattle health advisers' points of view. She found that during conversion, farmers were preoccupied mostly with changes in the management of fields and crops, whilst regarding changes in housing and feeding regimes of the cow herd as relatively uncomplicated. Some farmers expected an unrealistically rapid decrease in disease incidence after conversion, but this did not prevent them from seeking treatment for their animals when necessary.

On the whole, however, during conversion, farmers did not communicate much with other professionals important to the herd (veterinarians, cattle advisers), who often perceived organic rules regarding disease treatment and other health factors (e.g. prolonged withdrawal time) as illogical. The veterinarians, for their part, frequently expressed scepticism towards organic production in general. They could not see what was different or special about organic agriculture, and did not think it made any difference to animal welfare: the herds, housing systems and grazing routines seemed completely ordinary to them. Consequently, the basis for a fruitful communication may have been quite weak.

The combination of the farmer's reticence in communicating and the veterinarian's scepticism towards conversion seriously affected the health and welfare of several herds. Some farmers said they expected the disease level to decrease just by being organic, without any additional effort. However, this appeared not to be the case. Severe disease problems occurred among small calves staying outdoors in summer, and many of these problems could be traced to farmers' and advisers' unfamiliarity with organic methods (Vaarst *et al.*, 2001). Thus this study shows that it is

Box 10.1. Two examples of gradual weaning, which can be discussed in the light of care-taking in combination with meeting the aim of organic farming to promote a life in 'naturalness'.

Slow weaning of calves (example from Scotland)

A particular management feature of this farm, presented in Chapter 3, is the calf weaning policy, which is intended to minimize stress. It has the following features:

1. Spring-born calves remain with their mothers until early winter.
2. Before weaning in early winter, dams and calves are kept on a sacrifice area of land that will be used for arable production the following year. Calves continue suckling whilst they learn to eat silage. Weaning takes place during late December and early January.
3. Dam and offspring are separated by an electric fence, which allows calves to make physical contact with the dam without being able to suckle.
4. An attempt is made to minimize stress through good stockmanship and a degree of human contact, through quiet handling and walking through groups of calves with slow deliberate steps and a calming voice. In an extensive system such as this, human contact with animals can reduce behavioural problems later in maturity.

Stepwise weaning of piglets (example from Denmark)

1. Farrowing takes place in a hut, and the sow has access to a grassy area of 100 m^2. During the first 10–14 days of the piglets' life they stay inside a fence (about 1 m^2) outside the hut, and in the hut. The sow can easily step over the fence.
2. Then the fence is removed, and the piglets can run on the grassy area and go past the fence between sow areas to form groups.
3. After about 4–5 weeks, the fences in the sow area are removed, and bigger common huts are placed at the area. Piglets in large groups explore the whole area and return to the huts when they need to suckle.
4. Piglets and sows both are fed concentrate and roughage. This makes the drop in weight less dramatic during the last 3–4 weeks of the suckling period.
5. When the piglets are about 7 weeks old, the sows are removed from the area while the piglets stay. The piglets are moved indoors at an age of 8–9 weeks.

important to recognize that there are different farming styles, and that open dialogue is needed among different professionals about the problems that may occur during a change from one style to another.

Vaarst (2001) uses the terms 'technical conversion' and 'mental conversion' to interpret the different difficulties farmers may experience during conversion (Fig. 10.2). 'Technical conversion' concerns the need to

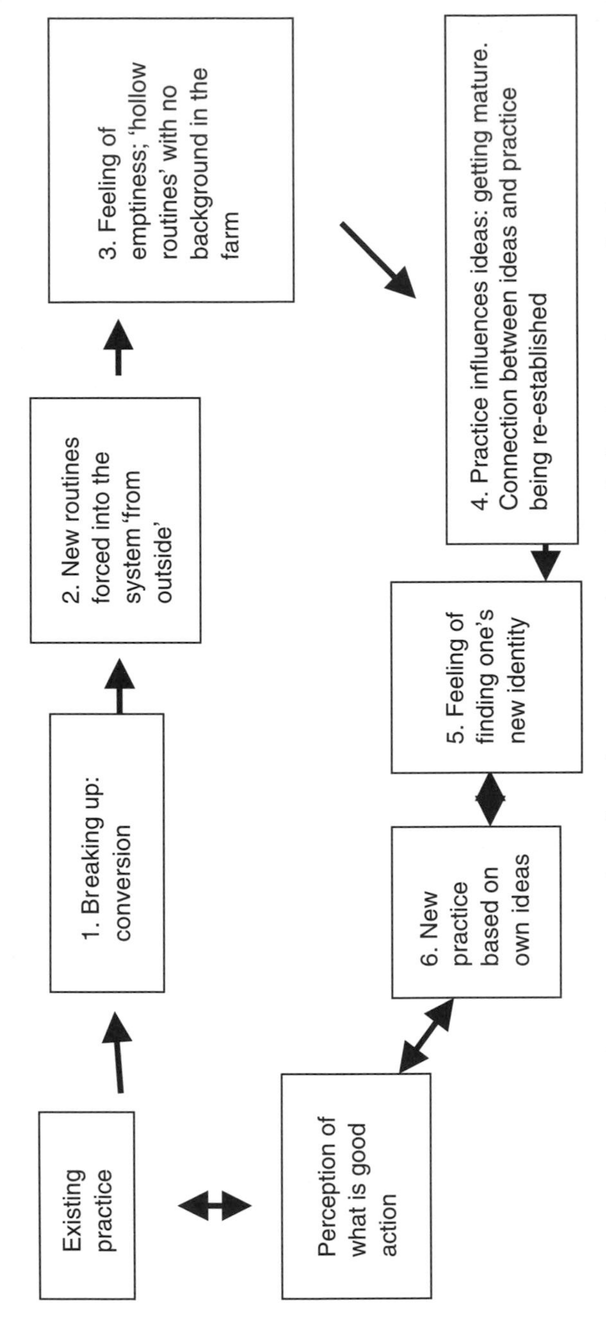

Fig. 10.2. The process of conversion, with 'technical conversion' taking place before 'mental conversion'.

adapt management routines to a new set of rules, while 'mental conversion' indicates the need for a change in the farmer's mind – and the veterinarian's. Perhaps mental conversion occurs only when 'being organic' becomes part of a person's identity, so that one can explain the idea of organic farming from one's own perspective, in one's own terms.

Looking Ahead: Developing an Ethical Framework for Animal Welfare in Organic Farming

Lund (2002) has proposed a bio-ethical framework for considering questions of animal welfare in organic farming. Within this framework, ecosystems, animal species and other relevant natural features are ascribed similar moral status. The focus is the development of the farming ecosystem as a harmonious whole, with care for sub-systems (animals, crops, etc.) given equal consideration. The challenge for the farmer is to combine this ecological approach with an attitude of care for individual animals and their welfare. Lund argues that humans and animals should be viewed as equal and mutually dependent participants within the overall farming system. She speaks of an 'ethical contract' between farmer and animals that morally obliges the farmer to ensure that the animals lead as good and healthy a life as is feasible within the larger system. Thus animals have moral status within the ethical contract as partners with a right to proper care.

Obviously, the contract is asymmetric in that the animals do not have duties towards their human caretakers. However, as Rutgers and Heeger (1995) argue, this asymmetry does not take away from our duty to respect the fundamental integrity of animals. They define animal integrity as 'the wholeness and completeness of the animal and the species-specific balance of the creature, as well as the animal's capacity to maintain itself independently in an environment suitable to the species' (p. 45).

Many challenges lie ahead in gaining a better understanding of the role of the farmer in the organic herd. This role cannot be considered independently of the goals of animal welfare in organic husbandry, and we should therefore seek to study these aspects together, as mutually enhancing parts of human–animal relationships. Each farmer acts in a way that is logical in the context of the particular farm, and it is not always possible for others to follow that logic. However, rather than ignore or deny the existence of 'local logic', we should study it, and make complex individual patterns of decision making explicit and available for communication and discussion.

It may then become possible to develop guidelines for the different levels of herd management that farmers can accept and adopt, while still maintaining their individual preferences, goals and values. Harmony is possible only if the general and the particular, the theoretical and the practical, can be integrated meaningfully in daily practice. Thus the challenge will be to create a foundation for specific organic care-taking that allows adjustments to be made in accordance with the goals for each organic herd. The challenge for organic animal husbandry is to develop management routines that take care not only of the individual farm animal, but also of the organic farming system as a whole.

References

Alrøe, H.F., Vaarst, M. and Kristensen, E.S. (2001) Does organic farming face distinctive livestock welfare issues? A conceptual analysis. *Journal of Agricultural and Environmental Ethics* 14, 275–299.

Barkema, H. (1998) Udder health on dairy farms: a longitudinal study. PhD thesis, Utrecht, The Netherlands.

Boivin, X., Le Neindre, P., Garel, J.P. and Chupin, J.M. (1994) Influence of breed and rearing management on cattle reactions during human handling. *Applied Animal Behaviour Science* 39, 115–122.

Boivin, X., Lensink, B.J. and Veissier, I. (2001) The farmer and the animal: a double mirror. In: Hovi, M. and Bouilhol, M. (eds) *Human–Animal Relationship: Stockmanship and Housing in Organic Livestock Systems. Proceedings of the Third NAHWOA Workshop, Clermont-Ferrand, 21–24 October 2000.* University of Reading, Reading, UK, pp. 7–15.

DARCOF (2000) *Biannual Report, 1996–1998.* Danish Research Centre for Organic Farming, Tjele.

De Jonge, F.H., Aarts, M.N.C., Steuten, C.D.M. and Goewie, E.A. (2001) Strategies to improve animal welfare through 'good' stockmanship. In: Hovi, M. and Bouilhol, M. (eds) *Human–Animal Relationship: Stockmanship and Housing in Organic Livestock Systems. Proceedings of the Third NAHWOA Workshop, Clermont-Ferrand, 21–24 October 2000.* University of Reading, Reading, UK, pp. 40–44.

De Snoo, A. (2002) Farming styles in organic Danish dairy farms and the relation to health strategies with an emphasis on mastitis treatment. Master thesis, Scottish Agricultural Colleage and Danish Institute of Agricultural Sciences.

Dreyfus, H. and Dreyfus, S. (1986) *Intuitive Ekspertise. Den bristede drøm om tænkende maskiner.* Munksgaard Nysyn, Denmark.

Dunne, J. (1992) *Back to the Rough Ground. 'Phronesis' and 'Techne' in Modern Philosophy and in Aristotle.* Notre Dame Press, South Bend, Indiana.

Ekman, T. (1998) A study of dairy herds with constantly low or constantly high bulk milk somatic cell count – with special emphasis on management. PhD thesis, Swedish University of Agricultural Sciences. *Acta Universitatis Agriculturae Suecia Veterinaria* 32, Uppsala, Sweden.

Estep, D.Q. and Hetts, S. (1992) Interactions, relationships and bonds: the conceptual basis for scientist–animal relations. In: Davis, H. and Balfour, D. (eds) *The Inevitable Bond. Examining Scientist–Animal Interactions.* Cambridge University Press, Cambridge, pp. 6–26.

Goodger, W.J., Farver, T., Pelletier, J., Johnson, P., DeSnayer, G. and Galland, J.

(1993) The association of milking management practices with bulk tank somatic cell counts. *Preventive Veterinary Medicine* 15, 235–251.

Hemsworth, P.H. and Coleman, G.J. (1998) *Human–Livestock Interactions: The Stockperson and the Productivity and Welfare of Intensively Farmed Animals*. CAB International, Wallingford, UK.

Hemsworth, P.H., Barnett, J.L. and Hansen, C. (1987) The influence of inconsistent handling by humans on the behaviour, growth and corticosteroids of young pigs. *Applied Animal Behaviour Science* 17, 245–252.

Hemsworth, P.H., Barnett, J.L. and Coleman, G.J. (1993) The human–animal relationship in agriculture and its consequences. *Animal Welfare* 2, 33–51.

Hovi, M., Kossaibati, M., Bennett, R., Edwards, S., Robertson, J. and Roderick, S. (2003) Specialist perceptions of animal health and welfare on organic farms – results of a questionnaire survey (submitted manuscript).

Lensink, B.J., Veissier, I. and Florand, L. (2001) The farmer's influence on calves' behaviour, health and production of a veal unit. *Animal Science* 72, 105–116.

Lund, V. (2000) What is ecological animal husbandry? In: Hermansen, J., Lund, V. and Thuen, E. (eds) *Ecological Animal Husbandry in the Nordic Countries. Proceedings from NJF-Seminar No. 303, Horsens, Denmark, 16th–17th September 1999. Report No. 2/2000.* Danish Research Centre for Organic Farming, Tjele, pp. 9–12.

Lund, V., Anthony, R. and Röcklingsberg, H. (2003) The ethical contract as a tool in organic animal husbandry. *Journal of Agricultural and Environmental Ethics* (in press).

Noe, E. (1999) Værdier, rationalitet og landbrugsproduktion belyst ved en microsociologisk undersøgelse blandt danske økologiske og konventionelle kvægbrugere. PhD thesis, Royal Veterinary and Agricultural University, Copenhagen.

Rutgers, B. and Heeger, R. (1995) Inherent worth and respect for animal integrity. In: Dol, M., Fentener van Vliessingen, M., Kasanmoentalib, S., Visser, T. and Zwart, H. (eds) *Recognising the Intrinsic Values of Animals*. Van Gorcum, Aassen, The Netherlands, pp. 41–51.

Seabrook, M.F. (1994) Psychological interaction between the milker and the dairy cow. In: Bucklin, R. (ed.) *Dairy Systems for the 21st Century*. American Society of Agricultural Engineers, St Joseph, Michigan, pp. 49–58.

Seabrook M.F. (1998) Learning from the animals. In: Cowling, C. and Pritchard, R. (eds) *Towards Better Boarding – Living Together, the Individual and the Community*. Milton Abbey Press, pp. 7–10.

Seabrook, M.F. (2001) The effect of the operational environment and operating protocols on the attitudes and behaviour of employed stockpersons. In: Hovi, M. and Bouilhol, M. (eds) *Human–Animal Relationship: Stockmanship and Housing in Organic Livestock Systems. Proceedings of the Third NAHWOA Workshop, Clermont-Ferrand, 21–24 October 2000.* University of Reading, Reading, UK, pp. 23–32.

Seabrook, M.F. and Bartle, N.C. (1992). Environmental factors influencing the production and welfare of farm animals – human factors. In: Phillips, C.J.C. and Piggins, D. (eds) *Farm Animals and the Environment*. CAB International, Wallingford, UK, pp. 111–130.

Seabrook, M.F. and Wilkinson, J.M. (2000) Stockpersons' attitude to the husbandry of dairy cows. *Veterinary Record* 147, 157–160.

Vaarst, M. (1995) Sundhed og sygdomshåndtering i danske økologiske malkekvægbesætninger. PhD thesis, Clinical Institute, Royal Veterinary and Agricultural University, Copenhagen.

Vaarst, M. (2000) Landmændenes oplevelse af omlægning til økologisk drift and omlægning til økologisk drift set fra dyrlægers og konsulenters synsvinkel. In: Kristensen, E.S and Thamsborg, S.M. (eds) *Vidensyntese om sundhed, velfærd og medicinanvendelse ved omlægning til økologisk mælkeproduktion*. Danish Research Centre for Organic Farming, Tjele. DARCOF-Report no. 6, 15–64.

Vaarst, M., Alban, L., Mogensen, L., Thamsborg, S.M. and Kristensen, E.S. (2001) Health and welfare in Danish dairy cattle in the transition to organic production: problems, priorities and perspectives. *Journal of Environmental and Agricultural Ethics* 14, 367–390.

Vaarst, M., Noe, E., Andersen, H.J., Enevoldsen, C., Thamsborg, S.M., Kristensen, T., Enemark, P., Bennedsgaard, T.W., Pedersen, S.S., Sørensen, C., Nissen, T.B. and Stjernholm, T. (2003) Health advisory service in Danish organic herds: development of three different models based on farmers expectations to advisers (manuscript, in preparation).

van der Ploeg, J. (1993) Rural sociology and the new agrarian question – a perspective from The Netherlands. *Sociologia Ruralis* 2, 240–260.

van der Ploeg, J. (1994) Styles of farming: an introductory note on the concept and methodology. In: van der Ploeg, J. and Long, A. (eds) *Born from Within: Practice and Perspectives of Endogenous Rural Development*. Van Gorcum, Assen, The Netherlands, pp. 7–30.

Verhoog, H. (2000) Defining positive welfare and animal integrity. In: Hovi, M. and Trujillo, R.G. (eds) *Diversity of Livestock Systems and Definition of Animal Welfare. Proceedings of the Second NAHWOA Workshop, Córdoba, 8–11 January 2000*. University of Reading, Reading, UK, pp. 108–119.

Wemelsfelder, F. (1997a) The scientific validity of subjective concepts in models of animal welfare. *Applied Animal Behaviour Science* 53, 75–88.

Wemelsfelder, F. (1997b) Investigating the animal's point of view: an inquiry into a subject based method of measurement in the field of animal welfare. In: Dol, M., Kasanmoentalib, S., Lijmbach, S., Rivas, E. and Van der Bos, R. (eds) *Applied Consciousness and Animal Ethics*. Van Gorcum, Assen, The Netherlands, pp. 73–89.

Wemelsfelder, F. (2001) Qualitative welfare assessment: reading the behavioural expressions of pigs. In: Hovi, M. and Bouilhol, M. (eds) *Human–Animal Relationship: Stockmanship and Housing in Organic Livestock Systems. Proceedings of the Third NAHWOA Workshop, Clermont-Ferrand, 21–24 October 2000*. University of Reading, Reading, UK, pp. 16–22.

Wemelsfelder, F. and Birke, L.I.A. (1997) Environmental challenge. In: Appleby, M.C. and Hughes, B.O. (eds) *Animal Welfare*. CAB International, Wallingford, UK, pp. 35–47.

Wemelsfelder, F. and Lawrence, A.B. (2001) The qualitative assessment of animal behavioural expression as an on-farm welfare-monitoring tool. *Acta Agriculturae Scandinavica* Suppl. 30, 21–25.

Wemelsfelder, F., Hunter, A.T., Mendl, M.T. and Lawrence, A.B. (2000) The spontaneous qualitative assessment of behavioural expressions in pigs: first explorations of a novel methodology for integrative welfare assessment. *Applied Animal Behaviour Science* 67, 193–215.

Wemelsfelder, F., Hunter, E.A., Mendl, M.T. and Lawrence, A.B. (2001) Assessing the 'whole animal': a Free-Choice-Profiling approach. *Animal Behaviour* 62, 209–220.

Wiepkema, P.R. (1987) Behavioural aspects of stress. In: Wiepkema, P.R. and Van Adrichen (eds) *Biology of Stress in Farm Animals: An Integrative Approach*. Martin Nijhoff Publishers, Dordrecht, The Netherlands, pp. 113–133.

11

Animal Health and Diseases in Organic Farming: an Overview

Stig Milan Thamsborg,[1] Stephen Roderick[2] and Albert Sundrum[3]

[1]Danish Centre for Experimental Parasitology, Royal Veterinary and Agricultural University, Dyrkegevej 100, DK-1870 Frederiksberg C, Denmark; [2]Organic Studies Centre, Duchy College, Rosewarne, Camborne, Cornwall TR14 0AB, UK; [3]Department of Animal Health and Animal Health, The University of Kassel, Nordbahnhofstrasse 1a, D-37213 Witzenhausen, Germany

---

Editors' comments

So far, the book has discussed the organic goal of good animal welfare in broad terms. When we turn to describing it more specifically, a major element is freedom from disease, although it involves much more than that, as has been stressed already. Organic livestock systems differ from conventional ones in many ways that can be expected to influence the prevalence of disease, such as feeds, use and non-use of different medicines, housing conditions, and stocking density. Ideally, these differences will achieve the goal of fostering animal health and welfare. But empirically, how has it turned out? What is the disease situation on today's organic farms? In actuality, how well do organic systems manage to control disease without, for example, routine preventive use of antibiotics, as is common in conventional systems, and by striving instead to establish conditions that minimize the risk of disease to begin with? We need to discuss future challenges in the light of what we know – here and now – about the disease status of organic herds: what problems present real challenges to the organic herd today, and how can this shape discussions of future developments? In this chapter, the authors give an overview of this critical topic, based both on available literature and their extensive experience and knowledge.

© CAB International 2004. *Animal Health and Welfare in Organic Agriculture* (eds M. Vaarst, S. Roderick, V. Lund and W. Lockeretz)

Introduction

It is an explicit goal of organic farming to promote and sustain health, and thus animal welfare. Health is often discussed in relation to diseases, but it encompasses other components besides the absence of disease. In humans, this larger concept often is associated with the humans' own perception of the situation and their mental/physiological health status, which is difficult to assess in animals. Welfare in a broad sense may encompass these feelings in animals, as discussed in more detail in Chapter 5. This chapter will present an overview of the disease problems that organic livestock production faces. To understand the situation, it is necessary to discriminate between potential problems and real documented health problems.

This chapter has two objectives. The first is to provide an overview of health problems in organic livestock production. The second is to give some examples that illustrate the background to disease patterns observed in organic production, particularly when compared to the much more thoroughly studied and documented situation in conventional farming. We discuss potential problems arising during the conversion period, as well as the general characteristics of organic production in relation to disease risk. Where pertinent, examples will be taken from similar production scenarios (e.g. free-range poultry) when there is a particular lack of information from the organic sector. Although these examples can provide important indicators and lessons, the insights drawn from them must be viewed in the light of some of the fundamental differences between organic and other methods of farming.

According to the EU regulation on organic livestock production (Chapter 4), animal health management should be based primarily on disease prevention, including breed selection, animal husbandry practice, high-quality feed and free-range conditions allowing regular exercise and appropriate stocking densities (CEC, 1999). Other chapters in this book deal with these issues in more depth. However, specific rules or guidelines for the evaluation of preventive measures against disease or the actual health status in the herd are not part of the EU regulation, nor of most national standards. The animal health regulations tend to be a consequence of the overall objectives of organic farming, and hence there is the question as to whether there are conflicts of aims between overall and specific goals. Furthermore, it is important to consider whether identified health and welfare-related problems are related to the system or caused by inadequate management. Chapter 12 presents a framework on how we may start to incorporate the principles of health care in a more structured manner, whilst encompassing a broader approach beyond merely focusing on disease.

As with all livestock production systems there are specific problems inherent in organic farming, so a discussion of disease patterns in organic farming is highly relevant if high levels of welfare are to be achieved and maintained. Organic farming may result in new manifestations of well-known disease problems (e.g. helminthosis in pigs), diseases may re-emerge (e.g. breast blisters in broilers), or well-known problems may remain prevalent (e.g. mastitis in dairy cattle). This all must be seen in a regional and climatic perspective because of marked differences in animal husbandry practices, different conditions for survival and transmission of infectious agents in the external environment, and different national standards (see Chapter 3).

The background to changes in disease patterns that may be associated with conversion to organic production may relate to:

- Differences in housing conditions and outdoor production.
- Treatment thresholds and restrictions on medicine use, e.g. prohibition on preventive medication.
- A change in farmers' attitudes and perceptions.
- Changes in cost–benefit relationships.
- Implementation of new feeding strategies because of restricted availability of specific nutrients.

Any of these could have both positive and negative effects on the range of diseases and their epidemiology, and these effects will be strongly influenced by the associations among the various factors. It can be assumed that the ban on preventive treatments makes disease prevention more complex, e.g. parasite control (Chapter 14 and Thamsborg *et al.*, 1999). However, the extent of this effect will be closely associated with the conditions under which animals are kept, as well as the attitude of the producer to alternative approaches during and after conversion. There also may be other influential factors, such as the introduction of new breeds (e.g. poultry breeds adapted to free-range conditions) or species mix (balancing sheep:cattle ratio in a strategy to reduce parasite burdens, as discussed in Chapter 14).

Extensive feeding with roughage is a further example of a factor that may influence the level and type of diseases. In cattle, it is likely that a high proportion of roughage will be favourable to the rumen environment and will result in less metabolic disease, as demonstrated in Danish studies (Danish Cattle Advisory Board, 1998; Vaarst *et al.*, 1998a). On the other hand, in some cases the energy requirements may not be sufficient. In pig production, the use of roughage, e.g. pH-lowering silage, may reduce the incidence of gastrointestinal bacterial infections like salmonella, dysentery and lawsonia, but increase the prevalence of nematodes (Petkevicius *et al.*, 1999).

Organic livestock production is not static, but has developed and changed in several ways. This should be kept in mind when organic research results are evaluated. Not only have the IFOAM Basic Standards been revised every 2 years (see Chapter 4), but production techniques as well as infrastructure supporting organic farming (such as advisory service or markets for organic feed) have developed – something that may well affect animal health and welfare (Christensen, 1998; Lund and Algers, 2003; Bennedsgaard *et al.*, 2003). Also, the kind of people becoming organic farmers is likely to have changed over the years, from mostly idealistic pioneers to more entrepreneurial farmers attracted by support schemes and market opportunities (e.g. Hayton, 1999; Lund *et al.*, 2002).

Organic Dairy Production

Generally, organic milk production tends to be similar to what is found in the conventional sector, with similar disease problems, e.g. mastitis, lameness and metabolic disease in adult cattle and internal parasite infections in young animals. The extent of these problems varies considerably among farms.

There are specific practices on organic dairy farms that can present a different perspective on the risks associated with disease and health. Examples include non-use of prophylactic antibiotics, more widespread use of straw bedding, restrictions on slatted floors and tethering, a broader range of breeds, and a strong emphasis on production from forage.

Mastitis

Mastitis has been recognized as the main animal health problem in organic dairy herds. Disease incidence appears to be similar, and occasionally higher compared with conventional production (Augustburger *et al.*, 1988; Offerhaus *et al.*, 1993; Krutzinna *et al.*, 1996; Weller and Cooper, 1996; Spranger, 1998; Hovi and Roderick, 1999; Roiha and Nieminen, 1999; Busato *et al.*, 2000; Weller and Bowling, 2000). In contrast, some studies from Norway (Ebbesvik and Løes, 1994; Hardeng and Edge, 2001), Sweden and Denmark (Vaarst and Enevoldsen, 1994) found a clearly lower incidence of treated cases of clinical mastitis in organic than in conventional dairy herds. The Danish study compared 12 well-established organic farms with a group of conventional farms and found fewer udder infections (lower somatic cell counts, 240,000 vs. 347,000 per ml) and treatments for mastitis (0.12 vs. 0.31 treated lactations per cow per year) (Vaarst and Enevoldsen, 1994). Several of the organic farms very seldom used antibiotics. A more recent Danish survey on 18 recently converted farms and 42 conventional farms with dual purpose breeds revealed little or no difference regarding number of mastitis treatments (0.64 vs. 0.57 treated cases per cow per year) (Danish Cattle Advisory Board, 1998).

The apparent change observed in Danish organic herds over time is not only related to different study designs. More free stalls have been introduced in recent years and mean herd size has increased. The many farmers who converted recently (late 1990s) might have a different attitude compared to the old 'pioneers' in the 1980s and early 1990s regarding threshold for treatment and the approach to diseased animals. This has been clearly demonstrated in Denmark by comparison of treatment data for 2000–2001 in organic farms converted before 1990 (0.35 cases per cow year), in 1995 (0.48 cases) and in 1999–2000 (0.46 cases) (Bennedsgaard *et al.*, 2003). The level in conventional farms was 0.60 cases. Time since conversion may thus be an important cause of variation.

With regard to mastitis in non-lactating cattle, Vaarst (1995) did not find incidence rates to be any higher than in conventional herds in Denmark, whereas Hovi and Roderick (1999) found that 50% of surveyed organic herds in England and Wales had relatively high levels of dry period mastitis compared with conventional farms, which had almost none. The differences between these findings could be associated with the widespread use of antibiotics in dry cow therapy in conventional herds in the UK.

Specific problem areas in mastitis control have been identified under organic management, including the lack of effective alternatives to long-acting antibiotics as a control measure (Hovi and Roderick, 1999). Alternative methods of control such as homoeopathy (Chapter 13), whilst popular among organic farmers, have been poorly researched. Inadequate nutrient supply is another potential risk. The nutrient supply in organic herds should be based primarily on home-grown feed, and the amount of bought-in feed material should be limited (Chapter 15). This may present a higher risk associated with unbalanced diets, especially a shortage of energy combined with an excess of crude protein (Sundrum, 2001). Furthermore, supplementation with vitamins, trace elements and minerals is not a routine practice in most organic dairy farms (Krutzinna *et al.*, 1996). The EU regulation does not allow synthetic vitamins to be administered to ruminants, and the natural alternatives often are much more expensive. Although not specifically an organic farming problem, the requirement to provide bedding material for housed animals presents a potential problem with respect to udder health, because growth of pathogens can increase dramatically when bedding is of poor quality. Despite these apparent risks, there is growing evidence that under organic conditions mastitis can be controlled at levels not very different from those in conventional herds.

Reproductive disorders

Studies of the incidence of reproductive disorders in organic dairy herds are rare and cover only a small number of farms. Studies in different countries have shown that the reproductive life of dairy cows is significantly higher in organic than in conventional farms (Offerhaus *et al.*, 1993; Ebbesvik and Løes, 1994; Krutzinna *et al.*, 1996; Reksen *et al.*, 1999). Perhaps the greatest risk to organically managed animals is associated with the energy supplied to animals genetically selected for high yields. Reksen *et al.* (1999) found that the reproductive efficiency of organic herds in Norway was significantly impaired during winter as a consequence of energy deficiency associated with organic feeding regimes. However, this study was performed in the mid-1990s, when there was no market for organic farmers to buy additional organic feed if their own harvests were not enough.

Metabolic disorders

Several studies found lower incidences of metabolic disorders such as ketosis and hypocalcaemia on organic compared with conventional farms (Ebbesvik and Løes, 1994; Vaarst and Enevoldsen, 1994; Krutzinna *et al.*, 1996; Hardeng and Edge, 2001), although not always significantly lower (Hamilton *et al.*, 2002). The lower incidences probably result from the generally lower production levels in organic livestock farming (Vaarst *et al.*, 1993; Boehncke, 1997). However, the comparisons in the Norwegian study were with conventional farms of similar milk yield (Ebbesvik and Løes, 1994). On average, milk yield per cow per year in organic production clearly is lower. German and Swedish studies have indicated differences of about 20% (Krutzinna *et al.*, 1996; Hamilton *et al.*, 2002), although some of the difference may be explained by breed rather than production effects. Production levels were 6.7% and 10.4% lower for dual-purpose breeds and Jersey, respectively, in 1998 in Denmark (Enemark and Kjeldsen, 1999).

Because of the restriction in supplementation with conventionally produced feeds, genetic performance capacities often are not realized. A study of production and health in Danish organic dairy herds found that daily milk production in 2001 was about 2 kg lower than on conventional farms in herds converting between 1995 and 1999, and about 3.7 kg lower for herds converting before 1990 (Bennedsgaard *et al.*, 2003). Although the relationship between milk yield and predisposition to diseases is not well understood, there are good reasons to assume that high-yielding cows are more susceptible to the adverse effects of inadequate conditions than animals with average performance (Butler and Smith, 1989; Wanner, 1995)

Lameness

Comparative data on lameness and claw lesions are not conclusive. Offerhaus *et al.* (1993) showed that lameness was more prevalent on conventional farms. Vaarst (1995, 1998) found no major differences between systems, whereas a more recent Danish survey showed more recorded claw and leg disorders in recently converted organic farms (Danish Cattle Advisory Board, 1998). However, the studies are not directly comparable because they used different methods of recording and because herd structure and management have changed (e.g. more loose houses are being established and Danish organic herds have become bigger).

Parasitic diseases

With few exceptions young cattle (calves, young replacement stock and bulls/steers for fattening) are most exposed to endoparasitic problems. At present, there is little documentation regarding diseases in these age groups on organic farms. Diarrhoea and ill-thrift amongst calves in deep bedding and in association with turnout have been observed as problems in Danish organic farms (Jensen, 1998; Vaarst *et al.*, 2001). Coccidia infections are considered a likely cause. Some of the problems are related to the turnout of calves at a young age (the Danish national standards require them to be put out to graze at 3 months) without adequate feed supplementation and on pastures used repeatedly year after year for cattle. It may be a transitional problem whilst farmers gain experience in managing young calves on pasture. This can be compared to a Swedish study where organic calves generally were found to be in good condition (Hamilton *et al.*, 2002). Post-turnout diarrhoea in calves was observed in 7% of organic and 5% of conventional herds in a Swedish questionnaire survey of parasite control methods in 162 organic and a similar number of conventional dairy herds (Svensson *et al.*, 2000). The control of coccidiosis and other parasitic diseases is discussed further in Chapter 14.

In the same survey, organic farmers appeared highly knowledgeable about nematode control methods other than chemoprophylaxis, which was used on 58% of the conventional farms, mainly in the form of controlled release devices. The methods used on organic farms included change of pasture between seasons, late turnout, and use of aftermath. Despite this apparent awareness, organic farmers more frequently reported diarrhoea in calves in the first grazing season than did conventional farmers (14% versus 6%) (Svensson *et al.*, 2000). This may indicate more parasitic problems in organic farms, but the observations have not been substantiated by studies involving repeated samplings in organic and conventional farms.

Two studies have reported clinical and parasitological findings among first grazing season heifers in organic farms, without including conventional farms. A Danish survey (Vaarst and Thamsborg, 1994) reported clinical signs of parasitic gastroenteritis in heifers in three of 11 farms during the grazing season. Loss-producing sub-clinical infections, as judged by the level of serum-pepsinogen, were seen in 10–15% of the first season grazing heifers. Infections with both gastrointestinal worms and lungworms persisted from one season to the next. Clinical signs suggestive of lungworm disease resulted in anthelmintic treatments on several farms. Similarly, a more recent Swedish study showed that lungworm infection continues to be a significant problem in organic herds (Höglund *et al.*, 2001). However, in this study, the heifers on the 15 farms examined for two seasons had only moderate to low levels of gastrointestinal parasites (99% of faecal egg counts below 500 eggs per g) and a low incidence of diarrhoea. Prophylactic treatments were not applied, and control relied on combinations of management procedures.

In summary, gastrointestinal parasites may pose a problem in organic dairy production but can most likely be controlled by appropriate management routines. Most parasite problems are observed in recently converted herds (Vaarst *et al.*, 1998a), perhaps because there were no plans for surveillance and pasture rotation in this phase.

Some ectoparasitic or vector-borne infections, like *Paraphilaria bovicola*, which is transmitted by flies and parasitizes the subcutaneous tissues of cattle, cannot at present be controlled without preventive use of parasiticides (Hammarberg, 2001). Similarly, control of liver fluke (*F. hepatica*) in areas where it is endemic may require strategic treatments with an appropriate anthelmintic, since management procedures to control the intermediate snail host often are impractical.

Organic Sheep Production

A survey of organic farmers in the UK identified lameness, mastitis, fly strike, fasciolosis and other helminthosis as the most prevalent health problems afflicting organic sheep (Roderick and Hovi, 1999). Preliminary data from a Swedish survey list haemonchosis, diarrhoea, high lamb mortality (3–36%) and lean ewes as the most commonly registered health problems in 37 organic sheep flocks (Lindqvist, 2001). In brief, the most important health problems in organic flocks were connected with feeding and endoparasites but the problems were apparently not more extensive or different from those in conventional flocks (Lindqvist, 2001).

Diverse production across Europe, discussed in more detail in Chapter 3, mean that certain unique production conditions can present particular local biosecurity risks, parasite control problems, and mineral and trace element deficiencies, e.g. the hill farming areas of the UK (Keatinge and Elliott, 1997). The health and welfare in some hill flocks may also be compromised by the limited opportunity for observation of stock as a result of the very large scale and difficult terrain common on these farms. In contrast, higher stocking rates in lowland areas mean that these are more difficult to convert than the extensive hill farms and that risks from internal parasite infections may be higher (Keatinge, 2001).

The movement of animals between farms, often over long distances, adds additional biosecurity and welfare concerns. An example is Sweden, where animals for slaughter are transported from farms in remote areas to organically certified abattoirs, not the nearest local abattoir (Hammarberg, 2001). Although not exceeding the legal maximum time, this procedure, required by organic legislation, appears to contradict other welfare objectives.

There are specific factors in organic sheep production that may influence the epidemiology of some diseases. As with other species, the restrictions or prohibition of chemotherapy and chemoprophylaxis mean reliance on a broader range of approaches. Low stocking densities, as emphasized in the organic standards, play a positive role in control of some diseases, such as helminthosis, blowfly and footrot. Furthermore, the operation of closed flocks, whereby breeding animals for replacement are reared, not purchased, reduces the risk of introducing infection to flocks. The shorter generation interval with sheep means that breeding strategies are a more feasible option than with cattle (see Chapter 16).

Parasites

Internal parasites, particularly gastrointestinal nematodes, are potentially a serious threat to organic sheep and goat production (Keatinge, 1996). A combination of breed resistance, differences in use of anthelmintics, and early exposure to increasing pasture infectivity may change host–parasite relationships in organically reared sheep, which in turn may result in higher tolerance to parasites as well as a broader diversity of nematodes (Keatinge *et al.*, 2002).

These effects are difficult to assess quantitatively. Flocks vary considerably in faecal worm egg counts as a consequence of variations within and between sheep breeds, in parasite species, in nutritional status, and in levels of immunity. These factors may limit the use of this variable as a tool in control programmes (Gray, 2002). However, a recent comprehensive 3-year survey of 152 organic sheep farms in Sweden showed that a management factor such as a new grazing area at turnout could be linked with low faecal egg counts in lambs in early season (Lindqvist *et al.*, 2001).

The adoption of novel anti-parasitic crops (Niezen *et al.*, 1996), biological control (Larsen, 1999), selection for resistance (Stear and Murray, 1994) and more effective nutrition (Holmes, 1993; Coop and Kyriazakis, 1999) have all been identified as potential control options to avoid clinical and sub-clinical disease and associated severe production penalties. However, practical implementation of several of these options has yet to be realized. Consequently, most organic sheep farmers have to rely on grazing management procedures, such as repeated moves to clean pastures (Thamsborg *et al.*, 1999), and regular (or pre-emptive) use of anthelmintics is still part of the control strategy in several places. Roderick and Hovi (1999) found that most UK organic producers use clean or mixed grazing practices as a means of control, frequently in combination with strategic anthelmintic use. Lindqvist *et al.* (2001) estimated that roughly 20% of organic farmers drench ewes around lambing to control *Haemonchus contortus*.

The main organic factor influencing incidence of fasciolosis, or liver fluke (*F. hepatica*), is limitations on treatment. This may partially explain the observed differences in incidence of fasciolosis in slaughter sheep reported from Sweden: respectively 5% and 17% in lambs and older sheep produced organically versus 4% and 8% in conventional lambs and sheep (Hansson *et al.*, 1999).

Clostridial diseases

The use of vaccination remains an option for organic producers. Clostridial vaccination of sheep is common on many conventional sheep farms in the UK, and 44% of surveyed organic farmers were also reported to be vaccinating routinely (Roderick *et al.*, 1996), with 10% of flocks also being vaccinated against pasteurellosis. Commercially available clostridial vaccines are normally multivalent, and the choice of vaccine should be dictated by the disease risk associated with the age group of sheep to be vaccinated and other known disease risks on the farm. For example, farms with known fasciolosis problems are at risk from Black Disease caused by *Clostridium oedematiens*.

Organic Poultry Production

Mainstream poultry units have created the concept of 'in house' biosecurity, where disease challenge is met by preventive medication, vaccination for viral infections or environmental controls (air quality, etc.) for bacterial problems. Organic animal health standards restrict the use of prophylactic medicines, and as a consequence of being free-range, biosecurity can never be the same. Therefore organic farms are potentially at a greater risk from some diseases; the NAHWOA Recommendations call for the standards to give greater emphasis to biosecurity. However, smaller flock sizes, low stocking rates and access to outdoor environments reduce the potential for other diseases associated with dense populations, high levels of inputs, enclosed environments and restricted movement. Certain free-range farms can be very similar in most respects to some organic ones, and may differ significantly only with respect to health management. Berg (2001) surveyed Swedish organic egg producers and concluded that many of the health problems seen in conventionally managed loose-housed or free-range farms were also found on organic poultry farms.

A survey of small-scale organic layer producers in the UK revealed that they do not generally perceive a problem with the health of their flocks (Roderick and Hovi, 1999). However, countrywide data on large Danish flocks (Kristensen, 1998) show that the mortality of laying hens is markedly higher in organic production: 15–20% compared with 4–5% in conventional battery cages and 9–10% in free-range production (unpublished data provided by the Danish Poultry Council, 1997). The reasons have not been fully investigated, but cannibalism and secondary infections are probably the main ones. Other diseases encountered on surveyed organic farms in Denmark were *Pasteurella multicoda* infection, with mortality up to 50%, egg drop syndrome (avian adenovirus infection) and Newcastle disease (Permin and Nansen, 1996). *Histomonas meleagridis* (black head) was observed in a few flocks. Furthermore, cannibalism with secondary *E. coli* infection was noted as a growing problem, in accordance with the mortality data. Ectoparasites were prevalent in organic flocks, particularly the red mite (*Dermanyssus gallinae*) and lice (*Menacanthus stramineus*) (Permin and Nansen, 1996). However, such high mortality figures are not representative for all organic systems. Recent figures from Denmark show lower levels (Anders Permin, KVL, Copenhagen, 2003, personal communication). Swedish figures indicate that there are no big differences between conventional and organic free-range systems. Berg (2001) found mortality to be 6–9% in Swedish organic flocks, while unpublished data from 2002 show 7.3% mortality (Åsa Odelros, Hedesunda, Sweden, 2003, personal communication).

Lampkin (1997) identified coccidiosis, external parasites, and feather pecking and cannibalism as significant potential problems in organic and other free-range systems. Swarbrick (1986) found a range of diseases and welfare problems in conventional free-range farms in England. There is no evidence to suggest that organic poultry are any less at risk from these.

There is no doubt that there are specific health problems associated with free-range production. Coccidiosis, helminth parasites, histomoniasis and ectoparasites occur more frequently, and the UK Farm Animal Welfare Council (FAWC, 1997) regard the diseases carried by wild birds, such as pasteurellosis, salmonellosis and avian tuberculosis, to be risks to free-ranging birds. Organic and other free-range systems present both positive and negative welfare consequences compared with caged and confined poultry. For example, domination and aggression among some individuals as a consequence of being in a large group may have negative consequences, but this is counterbalanced by the very positive opportunity to express a wide variety of natural behaviours (Sandøe and Simonsen, 1992).

Feather pecking and cannibalism

Feather pecking and cannibalism are proving to be a problem on some large organic poultry farms (Lampkin, 1997). Some degree of feather pecking may be a natural preening process, but under deprived conditions, it results in considerable health and welfare problems. Excessive feather pecking is caused by many factors, including nutritional deficiencies and unsatisfactory housing conditions that result in bullying. Although normally associated with overcrowding, under some conditions excessive and aggressive feather pecking is more common in free-ranging than battery systems. For example, in the latter there may be too few birds to allow hierarchical aggression. Cannibalism may follow feather pecking, but may also arise independently. Keeling *et al.* (1988) observed an outbreak of cannibalism in a free-range flock that occurred primarily in the last 8 weeks of lay. Cannibalism and pecking also have been observed in other free-range flocks in England (Swarbrick, 1986).

Green *et al.* (2000) identified a range of management factors in free-range systems that were associated with an increased risk of feather pecking. Many of these factors could inhibit foraging and dust-bathing and increase competition or frustration. Management and housing factors, particularly in relation to understanding the 'needs' of birds, were also shown to be critical in a study of Dutch organic flocks (Bestman, 2000). An important aspect is that chickens should be reared on the floor (Gunnarsson *et al.*, 1999, 2000).

Breed or strain also is important. The use of cage-adapted birds under organic conditions may cause problems associated with adapting to floor conditions, including feather pecking, when laying conditions are not 'optimal' for the breed (Sørensen, 2001). Chickens are by nature territorial and therefore to meet behavioural needs, due consideration should be given to suitable stocking densities and their innate competition for food and water. Sørensen (1996) gives an account of the genetic effects on behaviour in fowls managed in cages or on the floor (including cannibalism and feather pecking), and discusses the importance of the selection of strains suitable for 'ecological management'.

Feather eating, which can result from a deficiency in essential amino acids, can contribute to feather pecking and then to vent pecking. This is an important difference between ranging and caged hens. Free-range hens spend time foraging and less time pecking other birds. It has been suggested that poor feathering and the high mortality rate from cannibalism could be due to inadequate lysine, methionine and threonine in low-protein diets (Ambrosen and Petersen, 1997). Current EU standards for organic livestock production, which prohibit the use of synthetic amino acids in poultry diets, could present a potential risk for the development of a feather-pecking problem in organic flocks. However, this need not be the case, as it depends very much on the quality of alternative ingredients and the extent to which ranging contributes to intake. This latter depends on breed, housing conditions, climate and many other management factors.

Under organic production conditions, beak trimming is regarded as a mutilation (as discussed in Chapter 8) and therefore cannot be seen as a remedy for feather pecking. Such a practice certainly reduces the bird's ability to pick up food, preen and drink. Vigilance, isolation of victims and perpetrators, scatter feeding, ensuring a balanced diet, encouraging ranging and foraging, keeping birds free of ectoparasites, prevention of boredom and stress, rearing of chicks on the floor, and selection of breeds suitable for outdoor rearing all are necessary to minimize risks of cannibalism within organic livestock production. The avoidance of stress and sudden changes in routine, which is particularly important when mobile houses are used, is another important management factor.

Coccidiosis

Coccidiosis is one of the most important poultry diseases. Although there may be some differences among poultry management systems in the risk from coccidia, it is generally accepted that the disease may be found in most systems, both indoor and outdoor, conventional and organic. Birds reared on litter are always at risk. Conventional rearing of organic laying hens has until recently used coccidiostats in the feed. However, legislation soon will require organic chicks to be reared be under organic conditions, and consequently more problems with coccidia infections may be anticipated.

In Austria, where standards prohibit the use of coccidiostats, considerable problems have been reported (Zollitsch *et al.*, 1995). In a French survey, chickens reared organically (poulets biologiques) were shown to have a higher concentration of the parasite in litter than chickens given medicated feed (poulets labels) (Williams *et al.*, 1996). Lampkin (1997) reported a wide range of mortality rates in UK flocks, from 2.5% to 30%. However, there are clear indications that it is feasible to control this disease in organic flocks. Pedersen *et al.* (2003) reported that only three of 24 flocks of organic table birds in Denmark (approximately 35,000 birds in total) had severe problems of coccidiosis requiring medication. The average mortality was 2–4% in all flocks.

It is impossible to remove coccidial oocysts from a farm environment. Hence, control should focus on producing environmental conditions that substantially reduce the number of contaminating oocysts. A special risk of pasture contamination exists in flocks reared outdoors. Good husbandry, including vigilance and prompt treatment, regular pasture rotation, separation of age groups, and keeping litter dry through good hygiene standards and avoiding water spillage are essential elements in the control of coccidiosis. Resting of pasture is recommended to avoid build-up of contamination in soil. Although the oocysts are temperature sensitive, they can withstand the winter outdoors and have been known to survive for up to 2 years (Sainsbury, 1984). In deep litter systems, the combination of dryness, ammonia and bacterial decomposition will destroy oocysts within 2 weeks of their expulsion in faeces.

It is advisable, with good management, to allow some exposure to coccidia so that the birds develop immunity. Although preventive measures such as isolation and disinfection can be important, these are unlikely to achieve reasonable control. The development of effective vaccines allows organic producers much more flexibility in the control of coccidiosis. Vaccination is permitted under organic regulations, and is viewed by Lampkin (1997) as being more appropriate to organic poultry than anticoccidial drugs. So far, selection of poultry for resistance to coccidiosis has not been considered as a control option. Although genetic resistance to some strains of *Eimeria* has been observed in some strains of chicken (Long, 1968), this resistance is only partial, as all of these *Eimeria* strains can produce lethal levels of oocysts.

Helminthosis

Organic birds are liable to pick up internal parasites, some of which have not recently been major problems in the poultry industry under the strictly controlled, continuously housed conditions of intensive rearing. Some helminths require earthworms as intermediate hosts. A study of Danish organic layers showed a high prevalence of infection by the nematodes *Ascaridia galli*, *Heterakis gallinarum* and *Capillaria* spp. (Permin *et al.*, 1999). The high level of *H. gallinarum* is particularly important because it enables the transmission of *H. meleagridis*. The risk of endoparasitic infections was approximately seven times higher in organic than in conventional flocks (Permin and Nansen, 1996). However, the implications of this finding for production and health is largely unknown. Eggs of *A. galli* are capable of transmitting *Salmonella enterica* to hens (Chadfield *et al.*, 2001). A preliminary experiment has shown that continuous *A. galli* infection can influence the behaviour of the hens towards male character, including more aggression (Roepstorff *et al.*, 1999).

Management practices largely determine the type and severity of internal parasite infestation. Unless an efficient rotation is in place, free-range birds face exposure to contaminated pasture. Organic standards require resting of poultry runs for 1 year between batches.

Organic Pig Production

Organic pig production varies significantly within Europe. In some countries, such as the UK and Spain, there are extensive units, where breeding and in some cases fattening is done outdoors. In others, there is a greater tendency to provide housing with exercise areas. These conditions and their potential disease consequences are discussed in more detail in Chapter 3. Organic pig health and welfare has been identified as a concern in several countries (e.g. Leeb and Baumgartner, 2000; Vaarst *et al.*, 2000). Although the available data are limited, health and welfare problems appear to differ between organic and conventional production.

There has been little work published on pig health in organic production *per se*, but there is a growing body of work concerning outdoor production in northern Europe (Potter, 1998), which is of particular relevance to Nordic and UK organic production. Indoor organic systems, such as those in Germany and The Netherlands, allow housing, some access to outdoor areas, and greater space allowance per pig than conventional production. However, they are likely to suffer problems of limb health similar to those in conventional indoor production, such as injuries caused by poor quality or damaged flooring, or infections from poor hygiene. Good management and attention to floor maintenance are needed to prevent such problems. Straw bedding, which is encouraged in organic farming, reduces leg problems in indoor pigs, although it may lead to horn overgrowth because of too little wear (Davies, 1998).

In a postal questionnaire survey carried out in the UK involving 24 farms with organic pigs, Roderick and Hovi (1999) found that farmers perceived the level of disease as low. External parasites were seen as the biggest problem, followed by infertility, whereas diarrhoea and respiratory diseases were considered minor problems. In agreement with this, small Nordic case studies have shown a low incidence of diarrhoea and respiratory disorders but an increased incidence of joint diseases compared with indoor herds (Olsson *et al.*, 1996, survey of 14 herds; Lindsjö, 1996, 22 herds; Vaarst *et al.*, 1998b, one herd; Kugelberg and Johansson, 2001; two herds). The use of antibiotics was very low in the organic herds. On four pig farms in Denmark with indoor housing and access to pasture, Vaarst *et al.* (2000) found lameness, injuries and sunburn as the most common clinical conditions in sows. Fattening pigs showed signs of respiratory diseases, although the overall impression was that respiratory problems are much less in organic systems. Hansson *et al.* (1999) reported fewer pleuritic lesions at slaughter in organic (outdoor) pigs compared with conventional pigs. The lesions of organic pigs were related to parasitic infections or joint problems. It is likely that the more gradual transition resulting from weaning at 7 weeks explains the lower incidence of diarrhoea around weaning, so that treatment with antibiotics can generally be avoided. The use of vaccines against erysipelas, necrotizing enteritis (*Clostidium perfringens*, type C) and parvovirus is routine in Danish organic herds but not in all Swedish herds. The increased risk of erysipelas in outdoor pigs has been emphasized, and vaccination is strongly recommended (Hansson *et al.*, 1999; Kugelberg and Johansson, 2001).

Winckler *et al.* (2001) described housing and management on 35 organic and 40 'Neuland' (welfare label) farms in Germany. They reported that 85% of the organic farms had slatted floors in the activity areas for pregnant sows, and 42% used farrowing crates. They also reported that callosities at the carpus were lowest on a farm with outdoor housing, and highest on a farm that used little bedding in its indoor housing. The authors noted that major changes in housing would be required to comply with EU regulation 1804/99, which must be fully implemented in this area by 2010.

It is evident from several Danish studies that piglet mortality is high in organic systems (mean of 14.7% in five farms; Lauritsen *et al.*, 2000). However, similar observations have been made in conventional outdoor herds and there is no indication that the high level of mortality is related to organic production as such. Most deaths occurred shortly after birth (0–2 days), which is probably related to hypothermia or trauma inflicted by the sow when she lies down. If the ground is cold or the straw bedding is not thick enough, piglets tend to stay very close to the sow. Later, predators may play a role. Management activities such as tooth clipping and tail docking offer potential sites for entry of infection leading to joint illness or polyarthritis. As these interventions are not permitted routinely on organic farms, this represents a substantially lower infection risk for organically managed animals.

Parasitic diseases

Studies based on data from the early 1990s (reviewed by Nansen and Roepstorff, 1999) have indicated major differences in the pattern of endoparasitic infections among organic herds (breeding units outdoors, fatteners inside or outside), conventional herds with straw bedding, and intensive, specific pathogen-free (SPF) herds. Helminth infections like *Oesophagostomum*, *Trichuris* and *Ascaris* were more prevalent in organic herds.

This is to be expected, because the pigs have outdoor access but also get feed with a high level of insoluble fibre (roughage), which may enhance the establishment and fecundity of *Oesophagostomum* infections (Petkevicius *et al.*, 1999). The build-up of immunity to *Oesophagostomum* is low, so that this infection accumulates with age. *Ascaris* is the only helminth species found regularly in SPF herds, although at low levels. *Ascaris* infections are characteristically found in young weaners in organic herds, which indicates a high level of transmission during the suckling period. A more recent (1999) survey of nine Danish organic pig herds has confirmed the high levels of infections: 33%, 13% and 14% prevalence, respectively, of *Ascaris*, *Trichuris* and *Oesophagostomum* in fatteners (Carstensen *et al.*, 2002). However, the levels of *Ascaris* and *Oesophagostomum* were remarkably lower (by about 50%) than reported earlier (Roepstorff *et al.*, 1992). It was concluded that the general hygienic levels had improved from the first to the second study and that permanent pastures were used less extensively. Despite these findings, a single farm having produced organically for 8 years with a stringent 3-year pasture rotation had up to 90% prevalence of *Ascaris* in both weaners and fatteners. This may indicate that pasture contamination can still build up with time.

How these infection levels affect health is largely unknown. Clinical trichuriosis has been observed in pigs under poor hygienic conditions with access to outdoor yards (Jensen and Svensmark, 1996), and also has been reported from an organic farm (Carstensen *et al.*, 2002). Reduced production levels have been observed in sows and suckling piglets (3% lower body weight at weaning) with mixed infections under experimental conditions (Thamsborg *et al.*, unpublished data), but the situation is by no means comparable to production losses seen in infected ruminants. Coccidia infections of *Eimeria* spp. were more prevalent in organic outdoor breeding units than in conventional intensive units, whereas the opposite was observed for the *Isospora suis*. The latter is the only pathogenic coccidia causing severe diarrhoea in pigs within the first week of life in many intensive indoor herds. These problems are not observed in organic herds (Roepstorff and Nansen, 1994).

Other studies have confirmed some of these observations. In a clinical field survey in Austria, Leeb and Baumgartner (2000) found endo- and ectoparasites to be the main problem on 30 farms that kept pigs mostly indoors but with access to outside pens. Pig scab or mange can best be avoided by buying animals from disease-free herds at establishment. Carstensen *et al.* (2002) found no clinical indication of mange or lice on nine organic Danish farms. Vermeer *et al.* (2000) found endoparasites and post-weaning problems to be of particular relevance on ten Dutch farms where pigs were kept indoors with access to outdoor pens.

Food Safety

Several infectious agents can be transmitted from farm animals to humans through the food chain and cause zoonotic diseases in humans. These diseases often are associated with highly industrialized livestock production, but they also can occur in organic production. In particular, the requirements regarding outdoor production of poultry and pigs may increase the risk for zoonotic infections that have an outdoor reservoir, such as *Campylobacter*. With cattle, differences in animal husbandry are not so dramatic: most conventional farms also provide access to outdoors. Outdoor production may facilitate the spread of diseases from one farm to another. Trichinosis is an example of a zoonotic helminth that may have a reservoir in wild pigs and foxes. The risk that such an infection is introduced into outdoor organic pig herds is substantially higher than in indoor production.

Campylobacter

Campylobacteriosis is becoming an increasingly important food-borne zoonotic infection. It was the most common food-borne zoonosis in Denmark in 1999, with broilers being the main source (Danish Zoonosis Centre, 2000). In conventional production, infection with *Campylobacter* is controlled by avoiding access of wild fauna to housed broilers. This is not possible in organic farming. A recent survey has shown that 100% of organic flocks were infected compared with 49% and 37% among extensive and intensive conventional flocks (Heuer *et al.*, 2001). Because of confounding factors such as age (minimum 81 days at slaughter), breed, feed and access to outdoors, it is impossible to attribute the high prevalence in organic production to any single factor. The consequences of this finding for human health are not known. The problem seems to be restricted to poultry, as similar problems have not been reported in cattle and pig production.

Salmonella

Another important zoonotic infection is salmonella. At present, there is little evidence to suggest any major difference in incidence of this infection between organic and conventional production. With layers, *S. enteritidis* is the most important infection (70–80% of isolates). This infection is spread primarily vertically, i.e. from infected parents in a hatchery to the production flocks by infected layers, so that the production system (organic or conventional) may play only a minor role compared with risks at the hatchery. However, this may change in the future when legislation requires the rearing of organically produced replacement hens and parent stock. The risk of introduction of exotic *Salmonella* serotypes from outdoor fauna is not known. It is possible that feeding roughage in organic flocks may reduce the incidence of the infection in poultry, similar to the observations in pigs described below.

In organic pig production, data from Danish herds so far does not indicate any major difference in *Salmonella* incidence (primarily *Salmonella typhimurium*) between systems (unpublished data from the Danish Zoonosis Centre, 2001). The risk of introducing *Salmonella* from wild fauna, particularly birds and rodents, is probably increased, but this has not been fully assessed (Wegener and Nielsen, 1995). However, several factors may reduce the risk: the feeding regime is different in organic production and the stocking rate is generally lower. Grazing, feeding roughage, e.g. silage with a low pH, or other home grown feed stuffs is likely to reduce the incidence of *Salmonella* infection compared with traditional conventional concentrate-based feeding. Several epidemiological studies on conventional farms have shown that commercial concentrate and pelleted feeds are important risk factors for *Salmonella* (reviewed by Dahl, 1997). The protein level and addition of organic acids are also important.

The level of transmission within the herd is likely to decrease if the stocking rate is lowered, depending on factors like pasture rotation and group management. *Salmonella* may survive outdoors for up to 1 year, and therefore the resting period of pastures may be a significant risk factor. Sows alone or sows and fatteners kept outdoors together may also make a difference.

How infected farms are cleared of infections is largely unknown. Disinfection is difficult but rapid change of pastures may be feasible. At present it is very difficult to predict the incidence of *Salmonella* in organic pig farms.

Other infections

In broad terms, the presence of antibiotic-resistant bacteria in animal products is also regarded as a zoonotic infection. A discussion of the prudent use of antibiotics in organic farming is included in Chapter 13.

Bovine spongiform encephalopathy (BSE) is now regarded as a probable zoonosis. It is most likely to have been spread by feeding cattle and other species with meat and bone meal made from infected sheep or cattle. In the UK, the Soil Association has always prohibited the use of meat and bone meal, but this has not excluded BSE from organic farms, presumably as a consequence of purchasing infected but clinically unaffected animals. The question of whether BSE has been found in cattle born and reared entirely under organic management remains unanswered. Within the EU, control and surveillance measures have been put in place to reduce the incidence of BSE and other transmissible spongiform encephalopathies. These measures apply equally to organic and conventional farms.

Future Challenges

Many challenges remain for farmers, extension workers and researchers to secure a high level of animal health and welfare in organic production. First, it is evident that there are disease problems in organic systems, e.g. mortality in poultry, mortality in outdoor piglets and mastitis in cattle, that are unacceptably high, regardless of how they compare with conventional production. The approach to livestock nutrition is a specific example of organic management that is intended to improve health and welfare, but may actually have the opposite effect when not done effectively. The use of natural forages is intended to promote the effective functioning of the rumen, as well as enabling other ecological objectives of organic farming. However, a failure to meet the nutritional demands of livestock can result not only in energy deficit and production loss, but also in deficiency of specific nutrients and indirectly in reduced immuno-competence.

For several parasitic infections of livestock, chemoprophylaxis has become an integral part of conventional production, and lack of treatment is frequently considered a risk factor. In the longer term, many problems can be solved by changing management or production facilities or perhaps breeding stock. However, we must realize that in a few cases it is difficult to find useful alternatives that do not compromise animal welfare, and that both short-term and long-term strategies are needed.

It is imperative to continue developing organic standards to achieve the highest levels of animal welfare. Some regulations may, without intention, be disadvantageous for animal health. The main objective of the present EU directive (CEC, 1999) is to prevent misuse of the organic label, not to define the quality or level of disease control, for example. Standards and regulations thus provide the framework within which high levels of animal health and welfare must be actively pursued. An important challenge for extension workers and researchers is to increase organic farmers' awareness of the importance of herd health, disease prevention and health promotion. Health plans for conversion and beyond, participation in specific health schemes (concerning zoonotic disease in particular) and active participation of and consultation among farmer, veterinarian and adviser are other important tools in this work.

References

Ambrosen, T. and Petersen, V.E. (1997) The influence of protein level in the diet on cannibalism and quality of plumage of layers. *Poultry Science* 76, 559–563.

Augustburger, F., Zemp, J. and Heuser, H. (1988) Vergleich der Fruchtbarkeit, Gesundheit und Leistung von Milchkühen in biologisch und konventionell bewirtschafteten Betrieben. *Landwirtschaft Schweiz* 1, 427–431.

Bennedsgaard, T.W., Thamsborg, S.M., Vaarst, M. and Enevoldsen, C. (2003)

Eleven years of organic dairy production in Denmark – Herd health and production related to time of conversion and compared to conventional production. *Livestock Production Science* 80, 121–131.

Berg, C. (2001) Health and welfare in organic poultry production. In: *Veterinary Challenges in Organic Farming. Proceedings of the 14th Nordic Committee for Veterinary Scientific Cooperation Symposium. Acta Veterinaria Scandinavica* Supplementum 95, 37–45.

Bestman, M. (2000) The role of management and housing in the prevention of feather pecking in laying hens. In: Hovi, M. and Bouilhol, M. (eds) *Human–Animal Relationship: Stockmanship and Housing in Organic Livestock Systems. Proceedings of the Third NAHWOA Workshop, Clermont-Ferrand, 21–24 October 2000*. University of Reading, Reading, UK, pp. 79–88.

Boehncke, E. (1997) Preventive strategies as a health resource for organic farming. In: Isart, J. and Llerena, J.J. (eds) *Resource Use in Organic Farming. Proceedings of the 3rd ENOF Workshop, Ancona, Italy, 5–6 June 1997*, pp. 25–35.

Busato, A., Trachsel, P., Schällibaum, M. and Blum, J.W. (2000) Udder health and risk factors for sub-clinical mastitis in organic dairy farms in Switzerland. *Preventive Veterinary Medicine* 44, 205–220.

Butler, W.R. and Smith, R.D. (1989) Interrelationships between energy balance and postpartum reproductive function in dairy cattle. *Journal of Dairy Science* 72, 767–783.

Carstensen, L., Vaarst, M. and Roepstorff, A. (2002) Helminth infections in Danish organic swine herds. *Veterinary Parasitology* 106, 243–264.

CEC (1999) Council Regulation No. 1804/1999 supplementing Regulation No 2092/91 on organic production. *Official Journal of the European Communities* 42, L222, 1–28.

Chadfield, M., Permin, A. and Bisgaard, M. (1997) Investigation of the parasitic nematode *Ascaridia galli* as a potential vector for salmonella dissemination in poultry. In: *Salmonella and Salmonellosis. Proceedings of an International Symposium. Ploufragan, France, May 20–22, 1997*, pp. 375–376.

Christensen, J. (1998) *Alternativer – natur – landbrug*. Akademisk forlag, Viborg, Denmark.

Coop, R.L. and Kyriazakis, I. (1999) Nutrition–parasite interaction. *Veterinary Parasitology* 84, 187–204.

Dahl, J. (1997) Foderrelaterede risikofaktorer for subklinisk salmonella-infektion. *Veterinærinformation*, nr. 6.

Danish Cattle Advisory Board (1998) Bulletin No. 178, September 28.

Danish Zoonosis Centre (2000) *Årsrapport fra Zoonosecentret 1999*. Annual Report 1999, SVS, Copenhagen.

Davies, Z.E. (1998) The welfare implications of outdoor porcine breeding systems. Thesis submitted for the degree of PhD, Animal and Microbial Sciences, University of Reading, UK.

Ebbesvik, M. and Løes, A.K. (1994) Organic dairy production in Norway – feeding, health, fodder production nutrient balance and economy – results from the '30-farm-project', 1989–1992. In: Granstedt, A. and Koistinen, R. (eds) *Converting to Organic Agriculture*. Scandinavian Association of Agricultural Scientists, Rapport 93, pp. 35–42.

Enemark, P.S. and Kjeldsen, A. (1999) LK-meddelelse no. 353, Landskontoret for Kvæg. Danish Cattle Advisory Board, Skejby, Denmark.

FAWC (1997) *Report on the Welfare of Laying Hens*. Farm Animal Welfare Council, Tolworth, UK.

Gray, D. (2002) Can parasitic gastro-enteritis be used as an indicator of welfare in organic sheep? In: *Proceedings of UK Organic Research 2002, Colloqium of Organic Researchers. Aberystwyth, 26–28th March, 2002*. University of Wales, Aberystwyth, UK, pp. 347–350.

Green, L.E., Lewis, K., Kimpton, A. and Nicol, C.J. (2000) Cross-sectional study of the prevalence of feather pecking in laying hens in alternative systems and its associations with management and disease. *Veterinary Record* 147, 233–238.

Gunnarsson, S., Keeling, L.J. and Svedberg, J. (1999) Effects of rearing factors on the prevalence of floor eggs, cloacal cannibalism and feather pecking in commercial flocks of loose housed laying hens. *British Poultry Science* 40, 12–18.

Gunnarsson, S., Yngvesson, J., Keeling, L.J. and Forkman, B. (2000) Rearing with early access to perches impairs spatial skills of laying hens. *Applied Animal Behaviour Science* 67, 217–228.

Hamilton, C., Hansson, I., Ekman, T., Emanuelsson, U. and Forslund, K. (2002) Health of cows, calves and young stock on 26 organic dairy herds in Sweden. *Veterinary Record* 150, 503–508.

Hammarberg, K. (2001) Animal welfare in relation to standards in organic farming. Experiences and reflections from a Swedish outlook. In: *Veterinary Challenges in Organic Farming. Proceedings of the 14th Nordic Committee for Veterinary Scientific Cooperation Symposium. Acta Veterinaria Scandinavica* Supplementum 95, 27–31.

Hansson, I., Hamilton, C., Forslund, K. and Ekman, T. (1999) En jämförelse av slaktresultat mellan KRAV-uppfödda och konventionellt uppfödda djur. *Svensk Veterinärtidning* Supplementum 29, 17–24.

Hansson, I., Hamilton, C., Ekman, T. and Forslund K. (2000) Carcass quality in certified organic production compared with conventional livestock production. *Journal of Veterinary Medicine* B 47, 111–120.

Hardeng, F. and Edge, V.L. (2001) Mastitis, ketosis, and milk fever in 31 organic and 93 conventional Norwegian dairy herds. *Journal of Diary Science* 84, 2673–2679.

Hayton, A.J. (1999) Dealing with the organic dairy producer. *Cattle Practice, Journal of the British Cattle Veterinary Association* 7, 383–387.

Heuer, O.E., Pedersen, K., Andersen, J.S. and Madsen, M. (2001) Prevalence and antimicrobial susceptibility of thermophilic *Campylobacter* in organic and conventional broiler flocks. *Letters in Applied Microbiology* 33, 269–274.

Höglund, J., Svensson, C. and Hessle, A. (2001) A field survey on the status of internal parasites in calves on organic dairy farms in southwestern Sweden. *Veterinary Parasitology* 99, 113–128.

Holmes, P.H. (1993) Interactions between parasites and animal nutrition: the veterinary consequences. *Proceedings of the Nutrition Society* 52, 113–120.

Hovi, M. and Roderick, S. (1999) An investigation for the incidence, treatment strategies and financial implications of mastitis in organic and conventionally managed UK dairy herds. An attachment to a Scientific Report to the Ministry of Agriculture, Fisheries and Food, University of Reading, UK.

Jensen, A.M. (1998) Sundhedsmæssige udfordringer ved økologisk kvæghold – opdræt. *Proceedings of the Annual Meeting, Danish Cattle Advisory Board, February*.

Jensen, T.K. and Svensmark, B. (1996) Trichuriasis hos udendørs slagtesvin. *VeterinærInformation* 2, 3–7.
Keatinge, R. (1996) Controlling internal parasites without anthelmintics. A review prepared for Conservation and Woodland Policy Division, Ministry of Agriculture, Fisheries and Food, UK. ADAS, Newcastle upon Tyne, UK, 63 pp.
Keatinge, R. (2001) Organic sheep meat production. In: Younie, D. and Wilkinson, J.M. (eds) *Organic Livestock Production*. Chalcombe Publications, Lincoln, UK, pp. 145–158.
Keatinge, R. and Elliott, J. (1997) Organic sheep and beef production in the hills and uplands. Project Report (1997) Fourth Year of Organic Production. ADAS Redesdale, Newcastle upon Tyne, UK.
Keatinge, R., Jackson, F., Kyriazakis, I. and Cork, S. (2002) Developing parasite control strategies in organic systems. In: *Proceedings of UK Organic Research 2002, Colloqium of Organic Researchers. Aberystwyth, 26–28th March, 2002.* University of Wales, Aberystwyth, UK, pp. 341–346.
Keeling, L.J., Hughes, B. O. and Dun, P. (1988) Performance of free-range laying hens in a polythene house and their behaviour on range. *Farm Building Progress* No. 94, 21–28.
Kristensen, I.S. (1998) Økologisk æg-, kød- og planteproduktion 1996/97. In: *Studier i økologiske jordbrugssystemer*. DJF-report 1, 95–166.
Krutzinna, C., Boehncke, E. and Herrmann, H.-J. (1996) Die Milchviehhaltung im Ökologischen Landbau. *Berichte über Landwirtschaft* 74, 461–480.
Kugelberg, C. and Johansson, G. (2001) Infektionssjukdomar och ektoparasiter hos slaktsvin. *Svensk Veterinärtidning* 53, 197–204.
Lampkin, N. (1997) *Organic Poultry Production*. Welsh Institute of Rural Studies, University of Wales, Aberystwyth, UK.
Larsen, M. (1999) Biological control of helminths. *International Journal for Parasitology* 29, 139–146.
Lauritsen, H.B., Sørensen, G.S. and Larsen, V.A. (2000) Organic pig production in Denmark. In: *Ecological Animal Husbandry in the Nordic Countries. Proceedings from NJF-seminar No. 303, 1999. DARCOF-report* No. 2, 113–118.
Leeb, T. and Baumgartner, J. (2000) Husbandry and health of sows and piglets on organic farms in Austria. Animal health and welfare aspects of organic pig production. In: Alföldi, T., Lockeretz, W. and Niggli, U. (eds) *IFOAM 2000: The World Grows Organic. Proceedings of the 13th International IFOAM Scientific Conference, Basel, Switzerland, 28 to 31 August, 2000*, p. 361.
Lindsjö, J. (1996) Grisar ute! En översikt av rutiner och hälsoläge i svenska besättningar med slaktsvinsuppfödning utomhus. Specialarbete 34. Dept. of Animal Hygiene, Swedish University of Agricultural Sciences, Skara.
Lindqvist, A. (2001) Animal health and welfare in organic sheep and goat farming. Experiences and reflections from a Swedish outlook. In: *Veterinary Challenges in Organic Farming. Proceedings of the 14th Nordic Committee for Veterinary Scientific Cooperation Symposium. Acta Veterinaria Scandinavica* Supplementum 95, 27–31.
Lindqvist, A., Ljungström, B.L., Nilsson, O. and Waller, P.J. (2001) The dynamics, prevalence and impact of nematode parasite infections in organically raised sheep in Sweden. *Acta Veterinaria Scandinavica* 42, 377–389.
Long, P.L. (1968) The effect of breed of chickens on resistance to Eimeria infections. *British Poultry Science* 9, 71–78.

Askham Bryan College
LIBRARY BOOK

Lund, V. and Algers, B. (2003) Research on animal health and welfare in organic farming: a literature review. *Livestock Production Science* 80, 55–88.
Lund, V., Hemlin, S. and Lockeretz, W. (2002) Organic livestock production as viewed by Swedish farmers and organic initiators. *Agriculture and Human Values* 19, 255–268.
Nansen, P., and Roepstorff, A. (1999) Parasitic helminths of the pig: factors influencing transmission and infection levels. *International Journal for Parasitology* 29, 877–891.
Niezen, J.H., Charleston, W.A.G., Hodgson, J., Mackay, A.D and Leathwick, D.M (1996) Controlling internal parasites in grazing ruminants without recourse to anthelmintics: approaches, experiences and prospects. *International Journal for Parasitology* 26, 983–992.
Offerhaus, E.J., Baars, T. and Grommers, F.J. (1993) Gezonheid en vruchtbaarheid op biologische bedrijven farms. Louis Bolk Institute, Driebergen, The Netherlands.
Olsson A.C., Svendsen, J. and Sundelof, J.A. (1996) *Ekologisk svinproduktion*. Specialmeddelandem, Institutionen for Jordbrukets Biosystem och Teknologi, Sveriges Lantbruksuniversitet. No. 224.
Pedersen, M.A., Fisker, C., Thamsborg, S.M., Ranvig, H. and Christensen, J.P. (2003) New production systems: evaluation of organic broiler production in Denmark. *Journal of Applied Poultry Research* (in press).
Permin, A. and Nansen, P. (1996) Parasitological problems in organic poultry production. Report No. 729. Danish Institute of Animal Science, Tjele, pp. 91–96.
Permin, A., Bisgaard, M., Frandsen, F., Pearman, M., Kold, J. and Nansen, P. (1999) Prevalence of gastrointestinal helminths in different poultry production systems. *British Poultry Science* 40, 439–443.
Petkevicius, S., Nansen, P., Knudsen, K.E.B. and Skjøth, F. (1999) The effect of increasing levels of insoluble dietary fibre on the establishment and persistence of *Oesophagostomum dentatum* in pigs. *Parasite* 6, 17–26.
Potter, R. (1998) Clinical conditions of pigs in outdoor breeding herds. *In Practice* January, pp. 3–13.
Reksen, O., Tverdal, A., and Ropstad, E. (1999) A comparative study of reproductive performance in organic and conventional dairy husbandry. *Journal of Dairy Science* 82, 2605–2610.
Roderick, S. and Hovi, M. (1999) Animal health and welfare in organic livestock systems: identification of constraints and priorities. A report to the Ministry of Agriculture, Fisheries and Food. University of Reading, Reading.
Roderick, S., Short, N. and Hovi, M. (1996) Organic livestock production – animal health and welfare research priorities. Technical Report, VEERU, University of Reading, Reading, UK.
Roepstorff, A. and Nansen, P. (1994) Epidemiology and control for helminth parasites of pigs under intensive and non-intensive systems. *Veterinary Parasitology* 54, 69–85.
Roepstorff, A., Nansen, P. and Møller, F. (1992) Parasitter i økologiske slagtesvineproduktion. In: *Forskning i økologisk svineproduktion*. FØJO Report 1, pp. 61–63.
Roepstorff, A., Nørgård-Nielsen, G., Permin, A. and Simonsen, H.B. (1999) Male behaviour and male hormones in *Ascaridia galli* infected hens. *Proceedings of the 17th International Conference of the World Association for the Advancement of Veterinary Parasitology, Copenhagen.*

Roiha, U. and Nieminen, T. (1999) Luomunautojen terveys ja hyvinvointi. Helsingin yliopisto. Maaseudun tutkimus- ja koulutuskeskus, Mikkeli. *Julkaisuja* 66, pp. 66–93.

Sainsbury, D. (1984) *Poultry Health and Management*. Granada Technical Books, London, UK.

Sandøe, P. and Simonsen, H.B. (1992) Assessing animal welfare: where does science end and philosophy begin? *Animal Welfare* 1, 257–267.

Sørensen, P. (1996) Avlsmateriale til økologisk fjerkræproduktion. Report No. 729. Danish Institute of Animal Science, Foulum, pp. 81–90.

Sørensen, P. (2001) Breeding strategies in poultry for genetic adaptation to the organic environment. In: Hovi, M. and Baars, T. (eds) *Breeding and Feeding for Animal Health and Welfare in Organic Livestock Systems. Proceedings of the Fourth NAHWOA Workshop, Wageningen, 24–27 March 2001*. University of Reading, Reading, pp. 51–61.

Spranger, J. (1998) Richtliniengemäße Prävention und Therapie in der Tierhaltung des Ökolandbaus am Beispiel der Mastitis der Kuh. *Deutsche tierärztliche Wochenschrift* 105, 321–323.

Stear M.J. and Murray, M. (1994) Genetic resistance to parasitic disease – particularly of resistance in ruminants to gastrointestinal nematodes. *Veterinary Parasitology* 54, 161–176.

Sundrum, A. (2001) Organic livestock farming. A critical review. *Livestock Production Science* 67, 207–215.

Svensson, C., Hessle, A. and Höglund, J. (2000) Parasite control methods in organic and conventional dairy herds in Sweden. *Livestock Production Science* 66, 57–69.

Swarbrick, O. (1986) Clinical problems in 'free range' layers. *Veterinary Record* 118, 363.

Thamsborg, S.M., Roepstorff, A. and Larsen, M. (1999) Integrated and biological control of parasites in conventional and organic farming systems. *Veterinary Parasitology* 84, 169–186.

Vaarst, M. (1995) Sundhed og sygdomshåndtering i danske økologiske mælkekvægbesætninger. PhD thesis, Department of Clinical Studies, Royal Veterinary and Agricultural University, Copenhagen.

Vaarst, M. and Enevoldsen, C. (1994) Disease control and health in Danish organic dairy herds. In: Hiusman, E.A. (ed.) *Biological Basis of Sustainable Animal Production. Proceedings of the 4th Zodiac Symposium*. EAAP Publication 67, 211–217.

Vaarst, M. and Thamsborg, S.M. (1994) Nematode infections in organic dairy cattle herds in Denmark. In: *Proceedings of the Baltic-Scandinavian Symposium on Parasitic Zoonoses and Ecology of Parasites, Vilnius, Lithuania, 7–8 September, 1994. Bulletin of the Scandinavian Society for Parasitology* 5, 54–55.

Vaarst, M., Enevoldsen, C. and Jakobsen, P. (1993) Reports on diseases in 14 organic farms in Denmark. *Acta Veterinaria Scandinavica* Suppl. 89, 143–145.

Vaarst, M., Thamsborg, S.M. and Kristensen, E.S. (eds) (1998a) *Sundhed, velfærd og medicinanvendelse ved omlægning til økologisk mælkeproduktion*. FØJO Report No. 6.

Vaarst, M., Feenstra, A.A., Roepstorff, A., Hoegedal, P., Larsen, V.A., Worm, R., Hermansen, J.E. and Lauritsen, H.B. (1998b) Sundhedsforhold ved økologisk svineproduktion illustreret og diskuteret med udgangspunkt i en besætningscase. In: *Forskning i økologisk svineproduktion*, FØJO Report 1, pp. 33–45.

Vaarst, M., Roepstorff, A., Feenstra, A., Høgedal, P., Larsen, A., Lauridsen, H.B. and Hermansen, J. (2000) Animal health and welfare aspects of organic pig production. In: Alföldi, T., Lockeretz, W. and Niggli, U. (eds) *IFOAM 2000: The World Grows Organic. Proceedings of the 13th International IFOAM Scientific Conference, Basel, Switzerland, 28 to 31 August 2000*, p. 373.

Vaarst, M., Alban, L., Mogensen, L., Thamsborg, S.M. and Kristensen, E.S. (2001) Health and welfare in Danish dairy cattle during conversion to organic production: problems, priorities and perspectives. *Journal of Environmental and Agricultural Ethics* 14, 367–390.

Vermeer, H.M., Altena, H., Bestman, M., Ellinger, L., Cranen, I., Spoolder, H.A.M. and Baars, T. (2000) Monitoring organic pig farms in The Netherlands. In: *Proceedings of the 51st Annual Meeting of the European Association of Animal Production, The Hague, The Netherlands, 21–24 August 2000.*

Wanner, M. (1995) Leistungshöhe und Gesundheit der Milchkuh. *Schriftenreihe Institut für Nuztierwissenschaften*, ETH Zürich, 14, 53–61.

Wegener H.C. and Nielsen, B. (1995) Salmonella. In: Feenstra, A. and Andresen, M. (eds) *Rapport om zoonoserisiko i danske udendørs svinehold*. SVS-DS rapport, Copenhagen, Denmark.

Weller, R.F. and Bowling, P.J. (2000) Health status of dairy herds in organic farming. *Veterinary Record* 146, 80–81.

Weller, R.F. and Cooper, A. (1996) Health status of dairy herds converting from conventional to organic farming. *Veterinary Record* 139, 141–142.

Williams, R.B., Bushell, A.C., Reperent, J.M.., Doy, T.G., Morgan, J.H., Shirley, M.W., Yvore, P., Carr, M.M. and Fremont, Y. (1996) A survey of *Eimeria* species in commercially reared chickens in France during 1994. *Avian Pathology* 25, 113–130.

Williams, R.B. (1999) Three enzymes newly identified from the genus *Eimeria* and two more newly identified from *E. maxima*, leading to the discovery of some aliphatic acids with activity against coccidia of the domesticated fowl. *Veterinary Research Communications* 23, 151–163.

Winckler, C., Bühnemann, A., Seidel, K., Küffmann, K. and Fenneker, A. (2001) Label pig production and organic pig farming – a pilot study on housing and welfare related parameters in sows. In: *Animal Welfare Considerations in Livestock Housing Systems. Proceedings of International Symposium of CIGR 2nd Technical Section, Szklarska Poreba, Czech Republic, October 23–25*, pp. 479–490.

Zollitsch, W., Zehetner, S. and Hess, J. (1995) Aktuelle probleme in der Masthünerhaltung im biologischen Landbau in Österreich. In: Dewes, T. and Schimdt, L. (eds) *Beiträge zur 3. Wissenschaftstagung zum ökologischen Landbau vom 21–23. Februar 1995 an der CAU zu Kiel*. Wissenschlaftlicher Fachverlag, Giessen, pp. 57–60.

12

Promoting Health and Welfare through Planning

Malla Hovi,[1] Douglas Gray,[2] Mette Vaarst,[3] Andreas Striezel,[4] Michael Walkenhorst[5] and Stephen Roderick[6]

[1]Veterinary Epidemiology and Economics Research Unit, University of Reading, PO Box 236, Reading RG6 6AT, UK; [2]Scottish Agricultural College, Veterinary Science Division, Mill of Craibstone, Buckburn, Aberdeen AB21 9TB, UK; [3]Research Centre Foulum, Danish Institute of Agricultural Sciences, PO Box 50, DK-8830 Tjele, Denmark; [4]Atzelsberger Str. 10, D-91094 Bräuningshof, Germany; [5]Forschungsinstitut für biologischen Landbau, FiBL (Research Institute of Organic Agriculture), Ackerstrasse, CH-5070 Frick, Switzerland; [6]Organic Studies Centre, Duchy College, Rosewarne, Camborne, Cornwall TR14 0AB, UK

Editors' comments

Having discussed the philosophical and ethical aspects of the lives of the animals, and the practical implications that organic approaches have for various health and disease problems, we need to discuss how improvements can actually be carried out in the practical daily activities of the people involved in the herd. A health plan is a significant first step in improving the health of an organic herd. Prevention of disease is a key part of such a plan, because prevention is especially important in organic management for several reasons, including: the strong emphasis on animal welfare in organic farming; the aim of minimizing or avoiding the use of many kinds of chemicals, including medicines; and the longer withdrawal time after use of many medicines, which imposes great economic costs. However, a health plan must also set forth what will be done when prevention fails and treatment is necessary. The authors of this chapter, all of whom have experience as organic researchers and advisers, illustrate the general concept of a health plan with many detailed examples of how it has been effectively used under a wide variety of practical situations. They also identify areas for further development of this important tool.

© CAB International 2004. *Animal Health and Welfare in Organic Agriculture* (eds M. Vaarst, S. Roderick, V. Lund and W. Lockeretz)

Introduction

Maintenance of animal health and welfare is a major objective of organic animal husbandry. Organic perceptions of animal welfare have been discussed in depth in Chapter 5 and throughout this book, and the special nature of the concept, with animal integrity and natural behaviour at its centre, has been defined. Similarly, development of health and welfare planning requires the concept of animal health in organic livestock production to be defined. Many organic standards established by private certification bodies refer to 'positive health' as a specific organic concept. The 'positive health' approach differs from conventional disease treatment and prevention strategies by going beyond the attempts to minimize the risk factors that cause disease (Soil Association, 2002). An organic strategy aims at creating farming systems in which the well-being and basic needs of the animal are the main priorities and can override production objectives if a conflict arises. Risk factors for poor health and welfare, such as tethering, early weaning and high stocking densities, are minimized from the outset in an effort to create a production system that balances human and animal needs. The first part of this chapter outlines the theoretical understanding and potential practical implications of health promotion in organic farming.

There is a limited published literature that deals with practical health and welfare planning in organic livestock production systems. However, a growing number of programmes and projects are developing systematic approaches to health planning. Later we draw on examples from the UK, The Netherlands, Denmark, Germany and Switzerland. The relationship between farmers and their health advisers will be discussed. We highlight areas of particular interest in health planning on organic farms and discuss specific issues in the conversion from conventional to organic livestock production.

The Concept of Animal Health in Organic Farming

The definition of health by the World Health Organization, formulated in 1946 and aimed at health promotion in human populations, still stands as a comprehensive and explicit identification of all aspects of positive health (WHO, 1946): 'Health is a state of complete physical, mental and social well-being and not merely the absence of disease or infirmity.' More often, 'health' connotes 'absence of disease'. The classical division between body and mind forms the basis of this definition. This concept allows further breakdown of the body into different areas, each of which can be exposed to certain well-defined diseases.

Holistic approaches to health implied in the 'positive health' concept require a broader perspective of animal health and an understanding that a living individual is an entity. Physical, emotional and mental levels are all covered in this understanding, which is commonly discussed in relation to human health. Wackerhausen (1994) suggests a definition of health that is neither fundamentalist (i.e. health can be defined and measured in a similar way in all individuals) nor fully relativist (i.e. no common definitions can be made for health in different individuals). According to this definition, health is a characteristic of a living individual and can be understood as an expression of harmony or balance in the individual at all levels. Consequently, health is a more coherent concept than 'absence of disease', and disease is defined as 'disturbance of health'. This holistic approach to the living individual, with its emphasis on harmony within the organism and its surroundings, reflects the objectives of organic farming better than does the definition of health as 'absence of disease'.

Concentrating on health rather than on disease or production goals moves the focus of health promotion to the animals and their environment, away from the diagnosis of disease and identification of risk factors for a particular disease or condition. A holistic concept of health implies that any healthy individual will react to changes in its surroundings. Within this concept, disease is considered to be a reaction of a living individual to unsuitable surroundings (e.g. small cubicles for dairy cows or high microbial challenge in housing systems with high stocking densities and poor ventilation). In some situations, the disturbance in the interaction between the organism and its surroundings causes severe imbalance that is difficult, perhaps impossible, for the individual to cope with. This imbalance results in disease or injury. In organic livestock systems, the aim is to implement health planning and health promotion that produces interventions at an earlier stage, before such situations arise. Following on this, health planning in organic farming can be described as an effort to promote the general health of an individual animal or a herd by actively improving its living conditions in two ways:

- By using breeding as a tool to select animals that are well suited to the conditions on an individual farm.
- By giving livestock access to species-specific feed and feeding, housing conditions and freedom to express natural behaviour.

Striezel (unpublished) has suggested a classification of health planning concepts (Table 12.1). In this classification, the ecological concept comes closest to the organic aims, with balance as its primary goal.

This concept of health planning can be contrasted with conventional concepts of disease prevention, which searches for solutions to one or more specific disease problems by identifying and removing risk factors. Here the disease is in focus, whereas in the ecological concept the whole animal and the entire system are in focus. The practical implications and some difficulties arising from this definition of health are discussed below.

Table 12.1. Health planning concepts in different frameworks (Striezel, unpublished).

Framework	Criteria for measuring health	Goals
Ecological	Behaviour Excreta (urine, milk, faeces) Epidemiology	Balance
Classical/medical	Analytical parameters Microbiology	Standard values
Economic	Risk factors Output parameters	Optimized production
Molecular biological/ mechanistic	Marker genes	Optimized selection

Health Planning – What is it?

In this chapter, a health plan is understood as a farm-specific plan to promote livestock health and welfare. In a health plan, the goals of a farm are made explicit, areas and actions for improvement are identified, and a plan is made for systematic follow-up of these actions.

A health plan can have various levels, as suggested by Vaarst *et al.* (in preparation):

- Acute problem solving.
- Goal-orientated efforts to avoid particular diseases.
- Long-term health planning based on the farmer's goals for the herd and the farm.

Acute problem solving

Acute problem solving is often carried out in the face of a disease outbreak (e.g. a new viral disease entering a pig herd) or when it is discovered that the production goals are not being achieved because of a health problem (e.g. poor weight gain in young stock because of a heavy burden of internal parasites). Whilst the other, long-term levels of health planning ideally should prevent acute problems, some problems are inevitable, and a good health plan should provide for them. Here we will refer only briefly to acute problem solving, as Chapter 13 deals in detail with treatment and therapy of organic livestock.

Tactical planning

A goal-orientated, tactical strategy to avoid a particular disease is a common approach on both conventional and organic farms. The approach is usually based on epidemiological understanding of the disease and concentrates on eliminating or minimizing the known risk factors for the disease in the animals' immediate environment. Farm accreditation for freedom from some disease is a typical example of this approach for epidemic and contagious diseases, e.g. accreditation for freedom from infectious bovine rhinotracheitis, allowing value-added sale of breeding stock.

Another well-known example of a goal-orientated, tactical approach to a common production disease is the Five Point Plan for mastitis control in dairy cows (Box 12.1). This plan, which was launched in the UK in 1960s, was based on substantial epidemiological research into the risk factors for mastitis in dairy herds (Neave *et al.*, 1966). The aim was to control a complex health problem by developing a simple approach that would be effective on all farms, independent of individual circumstances.

Box 12.1. The Five Point Plan for control of mastitis in dairy herds as an example of tactical, disease-focused health planning (Neave *et al.*, 1966).

1. Routine post-milking teat dipping.
2. Prompt treatment of clinical mastitis with antibiotics.
3. Blanket antibiotic dry cow therapy for the whole herd.
4. Culling of cows with chronic mastitis.
5. Milking machine maintenance with annual testing.

The plan became the cornerstone of mastitis control on UK dairy farms for decades, and undoubtedly contributed to the marked improvement in the udder health status of the national dairy herd in the 1970s and 1980s (Booth, 1997). It had the advantage of consisting of only five clear action points, all applicable to farms regardless of their housing and milking systems, breeds of dairy cows or types of udder pathogens. The focus of the plan was on the disease, i.e. limiting the spread of the various udder pathogens within the herd and limiting pathogens' access to the udder. Although the plan succeeded in reducing the incidence of clinical and sub-clinical mastitis in the UK, its effect has been limited, particularly in the control of environmental mastitis, and any further advances in mastitis control will require farm-specific fine-tuning of the plan (Bradley and Green, 1998) or substantial strategic changes in dairy systems. Whilst this type of tactical action can be important in health planning on an organic farm, its narrow goal orientation and disease focus do not reflect the organic objective of developing a system that is in balance with its environment rather than in constant battle with it.

Strategic planning

Long-term, strategic health planning uses farm-specific goals as its starting point and includes both strategic and tactical elements. Farms converting from conventional to organic production often need to make significant changes in the whole farming system, e.g. introducing legumes into a permanent grazing system to maintain soil fertility without soluble nitrogen fertilizers. Similar strategic planning is necessary to maintain livestock health where the conventional husbandry system relied on routine medicinal inputs, such as in-feed antibiotics or routine drenching with anthelmintics.

Whilst the initial focus of such strategic planning could be on a particular disease that has been identified as a problem on a farm, the approach to planning is based on the idea that the whole farm contributes to the problem and thus can contribute to the solution. It is therefore important to recognize and to take into consideration the farmer's general goals for the farm, not just the goals directly associated with the disease in question. Box 12.2 presents an example of this kind of long-term strategic planning.

Definitions

There is no clear and commonly agreed-upon definition of animal health planning in organic farming, apart from the general description outlined earlier. Most European and North American organic standards do not require formal health planning. The UK national organic livestock standards of 2001, however, explicitly require an animal health plan that emphasizes reducing veterinary medicinal inputs, and that is defined as follows:

> [Livestock health] must be provided for by a plan drawn up by the farmer, preferably working in partnership with a veterinary surgeon and agreed between them during and after conversion, to develop and operate an organic livestock system, which conforms to these Standards. The plan must ensure the development of a pattern of health building and disease control measures appropriate to the particular circumstances of the individual farm and allow for the evolution of a farming system progressively less dependant on allopathic veterinary medicinal products.
>
> (UKROFS, 2000)

Box 12.2. An example of strategic and tactical planning on a farm focusing on several related goals (Hovi, unpublished).

Background
An autumn-calving dairy herd has a problem with acute, clinical mastitis in early lactation. As part of the organic conversion, the farm is planning to reduce the herd size to ensure adequate forage production; as a consequence, the farmer will give up the full-time herdsman, i.e. there will be a need to reduce the work-load on the farmer.

Objectives for the herd

- To reduce infection pressure in newly calved cows in order to reduce the occurrence of acute, clinical mastitis.
- To allow more natural feeding behaviour in early lactation in accordance with the organic requirements.
- To reduce labour requirements on the farm.

Strategic plan to achieve these objectives
To shift from autumn block calving to a more relaxed spring calving system or a year-round calving pattern that will allow outdoor calving during the grazing season, reduce labour requirements during the calving season and allow a shift towards more grazing during the peak production period.

The tactical plans coupled with strategic decisions

- Maintain a reasonable calving interval of 385 days by improving heat detection with Kamar indicators, without increasing labour requirements.
- Serve heifers at a more mature age over the following 2 years.
- Improve cow tracks on the farm to increase the area that can be grazed by the milking herd.

This definition corresponds with that suggested by Striezel and Plate (1994), who recommended as the organic approach to long-term health planning a gradual reduction of chemotherapeutic agents, combined with lowering of livestock densities and breeding for resistance against the most common endemic disease problems. Further definitions of health planning in organic farming arising from the goals and objectives of the health planning process are discussed below.

Health Planning – Goals and Objectives

In most organic standards, the stated objective is to maintain good animal health and welfare. The standards further specify that this objective should be achieved by using appropriate breeds, species-specific husbandry methods, high-quality feed materials, regular exercise and access to pasturage, and appropriate stocking densities. From the point of view of planning, these methods are strategic rather than tactical or operational. In contrast, the term 'herd health and production management planning' in conventionally managed systems usually refers to tactical and operational planning. Furthermore, in conventional systems, the main objectives are often driven solely by fixed production levels (Enting *et al.*, 2000), whereas in organic systems, the stated objective is to optimize production levels so that high health and welfare standards can be maintained.

The Organic Advisory Service Animal Health Plan in the UK summarizes the objectives of organic health and welfare plans as follows (OAS, 2000):

- To provide a standard animal health plan for all organic livestock producers, so they are compliant with the relevant livestock regulations/standard requirements.
- To ensure compliance with best livestock practices and to promote positive and 'holistic' animal health and welfare.
- To monitor animal health and welfare status of organic livestock and preventive and curative methods used in order to ensure the development of a pattern of health building and disease control measures that allow for evolution of a system that is progressively less dependent on allopathic veterinary medicinal products.
- To provide the farmer, the herdsperson/shepherds and the veterinarian, in whose care the animals are, a useful tool to help them in their efforts to improve animal health and welfare on the farm.
- To provide the organic inspectors a useful tool in the measurement and evaluation of health and welfare status of organic livestock farms.

The NAHWOA Partnership further states that such plans would help organic farmers take a holistic and evidence-based approach to disease control on their farms, lead to reduced disease incidence on organic farms, and support consumer confidence in organic livestock products (NAHWOA, 2001).

Health planning on organic farms can have various goals, depending on one's viewpoint. For consumers, the benefits of health planning can range from improved transparency to safer food to improved confidence in animal welfare conditions on organic farms. Organic certification bodies might consider the health plans useful in the inspection and monitoring of the farms. Hovi and Sundrum (2002) suggest that health plans on organic farms should emphasize their use as management and communication tools rather than in assessing compliance with organic standards. Management tools are often perceived by their users as means to identify and solve problems. Whilst health planning in an organic farming context should be viewed as a way to maintain dynamic and iterative planning and monitoring of high health and welfare status of livestock, solving and identifying problems inevitably will remain important.

Although health planning can have various goals within the entire organic production chain, they still must be set at the farm level by the farmer and the stockperson who needs to incorporate them into daily routines on the farm. In a Danish study of health planning on organic dairy farms, the importance of the farm-specific nature of targets and goals was recognized (Vaarst *et al.*, 2001; Vaarst *et al.*, in preparation). The study concluded that the farmer-set goals and targets that underpin a health plan must be accepted by all partners in the planning process. In particular, they should be set by the farmer, without being questioned by the other partners – apart from making sure that they fit within the organic principles and standards. However, the ways to achieve these goals should be decided together by the farmer and his or her advisory partners. In this process, questioning the farmer's approaches is acceptable.

In a 2001 UK survey of 120 organic farmers, 87% had a formal health plan in place within 6 months of it becoming required (Jenkinson and Hovi, unpublished). When asked about the expected usefulness of the planning process, the majority felt that health plans were a necessary part of positive health promotion (Box 12.3). A large proportion (65%) also felt that the planning process had helped them identify shortcomings on their farms. This was in contrast to an interview study of 15 Danish organic dairy farmers (de Snoo *et al.*, 2003), who generally did not perceive health planning as essential for identifying potential health problems.

Box 12.3. Results of a UK questionnaire of organic farmers' expectations and experience on health planning (Jenkinson and Hovi, unpublished); proportion of organic farmers (*n*=104) who agreed with the following statements regarding their expectations on the health plans implemented on their farms:

Health plan necessary to control disease and improve animal welfare	65%
Veterinary involvement in making health plans	50%
Health plan has helped to identify shortcomings in their existent management	65%
Following health plan will lead to substantial changes in veterinary practices	38%
Following health plan will lead to substantial changes in farm buildings	20%
Following health plan will lead to substantial changes in livestock diets	22%

Health Planning – What Should be Included?

As outlined above, a health plan should also act as a management tool that helps the farmer to monitor and assess the impact of changes implemented. This aspect of health planning should make it useful for the involved partners (e.g. the farmer and adviser, who is often the veterinarian) and to the wider stakeholder community (e.g. certification bodies).

Hovi and Gray (2002) suggest that several strategic and tactical issues should be considered in a health and welfare plan suitable for an organic system (Box 12.4). The list of issues can be modified, depending on the individual needs of the farm and the objectives and the methods used by all the stakeholders in the planning process.

Box 12.2 gives an example of strategic and tactical health planning in a UK dairy herd to achieve several interrelated objectives. It demonstrates the complexity of multiple objectives and highlights the need to consider all aspects of farm development. This example makes it apparent that one strategic decision can support another, and that tactical plans are needed to make follow-up and assessment more transparent.

The plans in both Boxes 12.5 and 12.6 focus on the prevention of a specific disease or welfare problem. They are examples of problem-oriented health planning, where a problem and its possible risk factors are systematically identified through inspection and observation. These examples demonstrate the need to focus on larger concepts beyond the health complaint in question, such as the design of the animal's environment or the organization of husbandry work. They also emphasize the need to focus on long-term solutions that take organic principles into consideration. Boxes 12.7 and 12.8 present a different approach, where a production system is in focus and collaboration among farms is needed to achieve the planning goals.

An example of health planning and monitoring approach from Germany (Table 12.2) emphasizes feeding, suggesting that whilst housing is important to the health of a dairy herd, changes in feeding show immediate effects and can be used to monitor overall health. In this approach, health management always starts with the observation of the animals. Analytic parameters can confirm the observations but are not considered necessary if the herd is in balance, i.e. health problems are minimal. The approach depends heavily on the observer's ability to recognize abnormal feeding and metabolism-related conditions before clinical symptoms of a disease are visible and the system shifts out of balance.

Vaccine use and immunization – a special case?

Amongst the issues that health planning should cover is the use of vaccines and antiparasitic agents (Box 12.4). Even in conventional

Box 12.4. A list of issues used in the UK for practical health planning, starting with the conversion process and continuing with the fully converted herd, changing goals and methods along the way (Gray and Hovi, 2002). For all areas, the way of monitoring sufficiency and whether the goals are reached should be included in the plan.

Stocking policy
Herd size and harmony between animals (numbers, species, breeds, age groups, etc.) and between animals and land use. This should provide background information for e.g. parasite control and feeding policy.

Breeding policy
Current breeds and breeding seasons are suitable. Breeding goals are in accordance with organic livestock production (e.g. disease resistance, feeding practices). Fertility and reproduction targets are set so that health and welfare are the primary goal. Reproduction techniques are checked against standards.

Feeding practices and policies
Growth targets are checked against standards and health and welfare goals. Balancing of diets is checked to avoid nutritional deficiencies and health problems caused by inappropriate diet.

Housing and husbandry
Current housing system is checked against standards. Changes are planned with health and welfare in mind. Alternatives to current mutilation practices (e.g. dehorning) are considered to ensure that no welfare problems occur, if these practises are not used anymore.

Herd/flock security
The plan should aim for a closed system but make provisions for bought-in animals. Biosecurity of the farm should be considered from the perimeter inwards.

Health and welfare recording and monitoring
An explicit plan for monitoring diseases and other important events should be set up (see Chapter 9). Monitoring and recording system should be built up over time and reassessed at each revision of the plan.

Identification of problem conditions and planning for these
This should include plans for the treatment and therapy of such cases (see Chapter 13).

Vaccination and parasite control policy
These are the two areas of health planning that are likely to have system-level implications when controlled with reduced veterinary medicinal inputs, and often warrant a section of their own.

Box 12.5. An example of solving a problem of feather pecking and cannibalism involving the whole farm on several levels (Bestman *et al.*, unpublished): health planning to solve feather pecking and cannibalism in a free-range layer flock in The Netherlands

Problem identified

Despite careful breed selection and a decision to rear the pullets on the farm from 1-day old, feather pecking and cannibalism still were an occasional problem. The farmer recognized that poor use of outdoor areas and consequent high stocking densities in the building contributed to the problem.

Approach to health and welfare planning

1. Improved use of outdoor areas by using maize as a cover crop in outdoor run areas:
   (a) It was decided to use maize as a cover crop in outdoor areas around permanent layer houses. The maize was overwintered standing, and cobs were cut to the ground daily for hens to eat.
   (b) To encourage the birds to go outdoors, the flockmaster spent considerable time daily in the outdoor area, particularly in the rearing phase.
   (c) Grain was scattered daily in the outdoor area.
   (d) A cockbird per every 30 hens in each flock was kept to encourage outdoor access.

2. Improved design and management of housing:
   (a) Natural light came through the windows whenever possible, with additional artificial lighting to provide light up to 14 h/day. Windows were covered at sunset in the winter to block direct sunlight from the setting sun.
   (b) A total of 80% of the floor area was covered with straw. Whole straw bales were placed on the floor regularly to be scratched and distributed over the bed.
   (c) Grain was scattered daily on the shed floor.
   (d) Some automated tasks, such as feeding, were changed to manual tasks, in order to increase human presence and to guarantee early observation of any problems.

3. Prevention of ground egg laying:
   (a) Floor egg laying was identified as a cause of vent-pecking and subsequent cannibalism.
   (b) Birds were trained to lay in the nesting boxes for the first 3 months of laying period (up to 4 h daily).

systems, these issues tend to be approached strategically by farmers and their advisers. In conventional systems both vaccines and parasiticides are often considered as insurance against endemic disease risks, and their use may even be compulsory in certain farm assurance schemes to guarantee good animal welfare. In organic systems, these inputs are

Box 12.6. A plan constructed for an outdoor sow herd, based on information from the literature and experience from a Danish project on organic pig herds (Vaarst *et al.*, 1998).

Problem

Lameness in sows appeared to be seasonal (early winter and spring) and affected lactating sows in particular. Minerals and feeds were analysed as a potential cause and lameness levels were assessed in different sow groups. It became apparent that feed was not the cause and that lameness was at unacceptable levels only in outdoor sows.

Health planning approach

1. Improvement of sow huts and their position: Inspection of legs and claws revealed cracks in the horn of the frontal part of the claws, and the sows were observed hitting the claws against the footstep of their huts when entering or exiting, particularly when they were approached by the herdsman without prior warning and left the hut in a hurry and did not 'mind the step'. A small soft rubber moulding was constructed to cover the metal footstep. The huts were placed so that the sows could see the approach path from the hut and were less 'on guard', leaving the hut more carefully.

2. Feeding and water troughs: Automatic feeding stations and water trough had been built as part of the fence. During wet periods, the surrounding areas became very poached and muddy. In cold weather, the mud became hard and formed areas of broken ground. The feeding system was separated from the fencing and was rotated around the grazing area. Because the water supply depended on the piping following the fence, the number of water troughs was increased and the troughs were used in rotation. A system was put in place to monitor changes in animal behaviour associated with changes in feeding and watering points.

3. Organization of the outdoor area: Driving with heavy tractors to fill up the feeding automats and to move huts caused a severe problem for the area and added to the 'frozen mud syndrome'. It was agreed that if the above measures would not solve the problem, reorganization of the outdoor area would be needed to avoid compaction by the tractor.

sometimes excluded from the restrictions on the use of other conventional veterinary medicines, as in the EU regulations on organic livestock production (European Communities, 1999). Some private certification bodies and the UK national standards (UKROFS, 2000) require an organic system to become less dependent on all veterinary medicinal inputs, including vaccines and antiparasitic agents. The use of anthelmintics for the control of internal parasites in ruminants is covered in Chapter 14; in this chapter, we highlight the role of vaccines in organic health planning.

According to the EU regulation, vaccine use is acceptable in organic systems as long as the livestock in question are at risk for a particular

Box 12.7. Health planning for mastitis control in alpine systems in Switzerland (Walkenhorst *et al.*, 2002).

Problem

In Switzerland, most dairy farms in the alpine region, where organic farming is common, use alpine pastures for grazing during the summer. High somatic cell counts (SCC) are a problem during the summer grazing season. The problem is exacerbated by the fact that summer grazing coincides with the last third of the lactation for most cows and by the practice of communal grazing, whereby cows from several herds are managed together.

Health planning approach

1. Long-term improvement of udder health on the valley (lowland) farms as described in Box 12.5 point 1, 'Optimization of cow environment', and point 2, 'Improved monitoring the udder health status and efficacy of therapy' (Notz *et al.*, 2002)

2. Improvement of milking technology during summer grazing:

(a) Many alpine milking systems are old and poor, suffering from lack of investment because of their short-term use (often less than 100 days per year), with bucket milking very common.

(b) Installation of a new vacuum pump and a new vacuum pipe is recommended as a standard improvement on alpine farms.

3. Training of herdsmen:

(a) Herdsmen working in alpine systems often are not experienced, and a seasonal workforce is frequently used.

(b) Training is organized for the herdsmen that focuses on: (i) demonstrating the relation between somatic cell count, mastitis and milk quality; (ii) minimization of new intramammary infections (milking hygiene and technique); and (iii) reduction of stress on the cows during handling.

4. Implementation of a communication and recording system

(a) There is an evident lack of communication between the alpine stockmen, the valley farmers, and the veterinarians.

(b) One responsible person for communication is identified for each alpine site.

(c) A recording system of udder health data of each individual cow and the whole herd is designed, based on California Mastitis Test results (alpine stockmen), test day results of SCC measurements (valley farmer) and diagnosis and treatment data (veterinarian).

disease. In the UK, some certification bodies require proof of past incidence of a disease that the farmer wants to vaccinate against (Chris Atkinson, managing director, Scottish Organic Producer Association, personal communication). This puts the onus on the producer to show that the livestock are at risk despite all feasible precautionary steps having been taken to reduce the risk. Whilst some endemic and epidemic diseases that can be vaccinated against are only a risk when susceptible

Box 12.8. Minimization of antibiotic use in organic dairy herds in Switzerland (Notz *et al.*, 2002).

Problem

A large proportion of antibiotics used on dairy farms in Switzerland are for treating sub-clinical mastitis during lactation or during the dry period. In accordance with the requirements of organic standards and the demands of consumers, a project was undertaken to reduce this use systematically.

Health planning approach

Farm-specific udder health plans were created for each farm in collaboration with the farmer, the local veterinarian and the project worker. The requirements and skills of the farmer and the veterinarian were taken into consideration when designing the plan.

1. Optimization of cow environment:
  (a) On each farm, cow environment was analysed and environmental risk to udder health were minimized by making and documenting suggestions for improvements on housing, feeding and milking technology and hygiene.

2. Improved monitoring the udder health status and efficacy of therapy:
  (a) Udder health data on each farm is regularly studied by the project worker, who reports back and makes recommendations.
  (b) Test day results of SCC and therapeutic measures are followed up with regular milk samples (quarter level) at drying off, after calving and before and after clinical and sub-clinical mastitis therapy.
  (c) Cows suffering from incurable sub-clinical mastitis are culled.

3. Replacement of antibiotics with complementary therapy:
  (a) Complementary and alternative therapy is introduced after the most obvious environmental risk factors for mastitis have been eliminated.
  (b) The homoeopathic treatment protocol is created for the veterinarians not trained in homoeopathy. In case of complications and therapy failure, the practitioners are supported by a skilled project worker.
  (c) Only a few cows with bacterial infections of major pathogens and increased somatic cell counts at the end of the first lactation are treated with long-acting antibiotics at drying-off.
  (d) A herd-specific homoeopathic remedy is introduced for prophylactic use.

animals are exposed to outside challenges (e.g. brought on to the farm without quarantine and testing for infectious bovine rhinotracheitis), other diseases that are routinely vaccinated against in some countries, e.g. in the UK, have multiple causes (e.g. *E. coli*-mastitis or pasteurellosis in sheep). The risk of these diseases can be reduced substantially by good husbandry, lower stocking rates or appropriate feeding. Other diseases against which there are effective vaccines have unpredictable epidemiology and are ubiquitous, making it difficult to assess risk and leaving unvaccinated stock at an unknown risk (e.g. clostridial diseases in sheep).

Table 12.2. Health planning approached focused on continuous observation with emphasis on feeding and metabolism (Striezel, unpublished).

Goals	Issues	Observation (farmer)	Analytic parameters (veterinarian/laboratory)
Maintenance of health	Feeding	Consistency of faeces (visual, tactile, i.e. 'boot-test') Digestion status of the faeces (visual, tactile) Filling of the rumen (visual) Body-condition score (visual)	Milk constituents (fat, protein, urea, etc.) Ration formulation analysis Feed and forage analysis
	Metabolism	Rumination activity (visual) Quantity of the milk (visual)	Analysis of blood, urine, saliva, rumen fluid, etc.
	Deciding what is 'healthy' and what is 'diseased'		
Disease identification	Fertility	Signs of heat Discharge Signs of inflammation	Progesterone analysis Microbiological analysis
	Foot health	Lameness Signs of inflammation Signs of (sub) clinical laminitis	Microbiological analysis
	Udder health	Milk quality (visual) Duration of milking	Cell count in milk Microbiological analysis Conductivity analysis

To help organic farmers and organic certification bodies decide on vaccine use, a project in the UK is developing a qualitative risk assessment procedure that allows the user to explore husbandry, breeding, feeding and other options in limiting disease risk before deciding on vaccine use. The process is carried out on a web-site, accessible to farmers and their advisers. The programme allows the farmer to print out a report of the decision-making process, making the process transparent and allowing the certification body to decide whether all other avenues were explored before resorting to vaccines (see Box 12.9 for a sample report). Whilst the process is based on classical risk factor analysis and its aim is to reduce particular disease risks on a farm, it pushes the user towards fundamental changes in the system and thinking regarding health promotion.

Box 12.9. A sample report from a decision support tool for cattle and sheep farmers (under development, personal communication with authors: M. Hovi and D. Gray): DESTVAC assessment report for Home Farm

Date: 13 February 2003

Basic information
FARM/COMPANY NAME: Home Farm
SPECIES: Sheep
DISEASE: Enzootic abortion of ewes (EAE)

Risk factors for EAE: *(* indicates risk factors relevant to your farm)*
* Aborted ewes (carrier animals) are not culled after abortion season.
* Current disease status of the farm re. enzootic abortion is not known.
* Flock receives bought in female replacements or foster lambs from infected flocks/flocks with unknown EAE status
* Aborted lambs and placentas are not buried but routinely left for carrion.
* Lambing is managed high density fields.
* Lambing fields are used in successive years.

For herds with no infection: Bought-in animals from infected herds/flocks
For herds with no infection: Direct contact with animals from neighbouring infected herd/flock
Poor hygiene at lambing (failure to remove aborted fetuses/placentae and to disinfect contaminated area)
Infected neighbouring flock without direct contact
Large number of carrion eating predators/wildlife (e.g. foxes and seagulls)
Failure to isolate all aborted ewes from other pregnant ewes
Failure to routinely diagnose causes of abortion
For flocks with no infection: transmission in clothing, vehicles by contractors/advisers/shepherds
Lambing personnel involced in lambing in other flocks (of unknown disease status)

Risk management measures relevant to risk factors on your farm:
(indicates management measure you have indicated you can/will implement)*
* Maintain genuinely closed herd/flock
* Cull ewes that have suffered enzootic abortion
* Remove and destroy aborted fetuses, dead fetuses/lambs and placentas from lambing fields.
* Isolate aborting sheep until culling

Achieve disease free accredited herd/flock, and only buy-in replacements from accredited herds/flocks
Investigate infertility and abortions routinely
Do not foster lambs from flocks of unknown status or lamb banks
Encourage outdoor lambing
Maintain low stocking densities in lambing fields
Rotate lambing fields annually
Clean and disinfect lambing pens between batches of ewes

Other risk management measure/s relevant to the disease:
(indicates management measure you have indicated you implement already)*
* Control fox population.
* Known disease status of neighbouring herds/flocks.
* Provision of dedicated clothing and equipment for lambing.

These risk factors could be addressed BY:
Unknown current disease status of the farm. BY: Investigate infertility and abortions routinely.
Lambing is managed high density fields. BY: Maintain low stocking densities in lambing fields.
Lambing fields are used in successive years. BY: Rotate lambing fields annually.
Warning *(when a HIGH RISK risk factor is addressed with LOW or MEDIUM IMPACT measure only)*: No warnings.

Please see attached information on cost of vaccine and economic impact of EAE.

Health Plans – Continuity and Motivation

Continuity

Because health planning on organic farms is strategic, it can be difficult to make it a continuous process. Tactical health and production planning is generally considered to be a continuous process that regularly evaluates the impact of interventions to justify the costs incurred. An example of such tactical planning would be a decision to introduce a new, more expensive antibiotic for routine mastitis therapy in a dairy herd to improve poor cure rates. The farmer and the veterinarian would both be keen to assess whether the added investment was worth it. Treated cases would be checked 7 days post-treatment, with the results reviewed after 6 months.

When planning becomes strategic, it is more difficult to identify the measurable outcomes of interventions, particularly when they may primarily have long-term effects. For this reason, many fundamental changes implemented on organic farms, for instance during conversion, may not be adequately monitored and evaluated. However, it is important that health and welfare plans be continuously assessed, both to maintain motivation and to benefit from the learning process that is inbuilt in such assessments.

This is not always the case, as was demonstrated by a Danish survey of 15 organic dairy farmers (de Snoo *et al.*, 2003). Seven of the farmers saw health planning as an opportunity to discuss specific issues when the veterinarian was visiting the farm to treat a sick animal, i.e. health planning was not viewed as a continuous process by either partner. Another Danish study found a lack of systematic follow-up in most cases where a health advisory service contract had been implemented on an organic dairy farm (Vaarst *et al.*, in preparation). The study concluded that the lack of continuity was due mainly to the absence of basic communication procedures that would have allowed the partners to make health planning into a systematic process. The study also identified two features of process-orientated health planning on organic farms that would encourage continuity:

- Efficient, systematic and goal-driven use of data from the herd to identify problem areas and potential improvements.
- Good formal communication of common findings and conclusions to make meaningful follow-up possible, e.g. creating 'common memory' by taking notes of each meeting and distributing them to all partners.

Motivation

Motivation regarding health planning is closely and explicitly linked to follow-up. Follow-up of strategic and tactical planning is usually carried out in a partnership among the farmer, the stockperson and outside advisers such as the veterinarian. This part of health promotion is often performed in the comfort of an office or a farm kitchen, whilst the actual day-to-day health promotion is carried out through daily routines of observation, recording, handling of animals and responding to unpredictable changes in the environment. A large part of this effort is not visible to anybody, apart from the farmer or the stockperson performing the daily routines. Systematic follow-up is a useful way of making these efforts more visible and giving feedback on them. Regarding this feedback link, however, it is vital that the plan's goals are genuinely those that the farmer has chosen. This will ensure that the person primarily responsible for the daily implementation of health promotion will have a vested interest in the assessment of these efforts.

Many farmers converting to organic production may initially find it difficult to modify existing conventional models of health planning to the organic requirements and the new farm goals. Planning based on set production targets, formal farm assurance protocols or veterinary-led statutory systems may appear at odds with strategic thinking, and the motivation to carry on with them may be lost. However, a UK survey of recently converted organic livestock farms found that farmers who had formal health plans as part of another farm assurance scheme before conversion were significantly more likely to have implemented an organic health plan and to think that health planning would be a helpful management tool (Jenkinson and Hovi, unpublished). Farmers' motivation arising from existing health plan templates designed for conventional farm assurance suggests that they have a positive view of their experience with a systematic approach to animal health.

In Denmark, conventional livestock producers are motivated to establish health plans because they must have a health advisory service contract with a local veterinarian to have access to prescription medicines. This motivation was difficult to transfer to organic farms, where access to conventional medicine was less important and veterinarians found it difficult to see a role for their services. A project was established to produce guidelines for a new health advisory service contract for organic cattle herds, as shown in Box 12.10 (Vaarst *et al.*, in preparation).

Health Planning and the Conversion Process

Conversion from conventional to organic production needs special attention, as many strategic and structural changes happen on a farm during

Box 12.10. Guidelines for health advisory service contracts for organic livestock farm in Denmark (Vaarst *et al.*, 2001).

Part A: Preliminary conditions

1. Herd-specific issues determine the level of health advice.
2. The agreement between the farm and the veterinarian confirms that both parties are committed to actively integrate organic principles of animal health and welfare into the work.
3. A written summary of each meeting will be made, and the contract will be assessed annually.

Part B: Choice of advisory model

1. The guidelines propose three basic models (Vaarst *et al.*, 2003), from which the partners choose when initiating the contract. The models set guidance to the level of involvement of the different partners, and are:
   (a) Theme-orientated;
   (b) Monitoring-orientated;
   (c) Process and analysis-orientated.

Part C: Goals

1. Long-term strategic and short-term tactical goals will be listed.
2. The way in which the advisory service contract will help the farm to achieve these goals will be outlined.

Part D: Structure of work/meetings, areas of responsibility

1. Who makes the agenda?
2. Who takes responsibility for systematic follow up?
3. Participation in a local network of organic farmers and their advisers.

this period. These changes offer both threats to and opportunities for health and welfare planning.

The conversion process often is focused primarily on establishing crop rotations that will guarantee adequate soil fertility without artificial fertilizers. This process usually is carefully planned, based on production targets and farm system analysis, using maps, soil and forage analysis and quantification of production.

Conversion of livestock to organic production involves a minimum period that the animals have been fed organic rations. This period varies, depending on the certification standards, but is always substantially shorter than the period required for land conversion. This often makes any planning for livestock conversion operational in nature, aiming at a quick implementation of the requirements concerning veterinary medicines or housing. This approach ignores the opportunities that the conversion period and the first year of organic management often offer. During this phase, major changes in production levels and husbandry techniques are likely, and strategic rather than operational planning should be emphasized, taking into consideration system-level changes that affect animal health and welfare (Box 12.4). Health and welfare planning should start at the beginning of the land conversion period to allow more time for real change.

The health promotion and planning strategies related to conversion will vary from farm to farm, but they should always be based on three key elements (Gray and Hovi, 2001):

- The health status of the herd before conversion (e.g. determining existing endemic disease levels, and planning accordingly).
- All the planned changes on the farm (e.g. changes in crop production and their effect on herd size).
- The farmer's own goals and objectives regarding the livestock enterprise, remembering that they are likely to change significantly during conversion and the first years of organic production.

Future Challenges to Health Planning on Organic Farms

In 2001, NAHWOA concluded that the existing evidence did not support consumers' perceptions and the organic standards' objectives of high animal welfare and health status in organic livestock (NAHWOA, 2001). The Network also concluded that both health and welfare on organic livestock farms could be improved by more widespread use of health plans that take into account the local conditions, the farmer's individual goals and motivations, and the organic standards. However, moving towards having all organic livestock farms implement and use formal health promotion plans faces several challenges: lack of sympathetic partners and advisers to help farmers design 'organic' health plans; lack of resources on organic farms to focus on health planning; and the difficulty in deciding on overall targets for quantifiable health and welfare standards.

Lack of sympathetic advisers as health planning partners for organic farmers

The expertise and commitment of the farmer is crucial to the success of the plan, as has been emphasized already. The results of the UK farmer survey quoted above suggest that a significant commitment exists among farmers, but it is less clear how committed the veterinary profession is to embracing organic principles as part of health planning on organic farms. The same survey found that only half the farmers had used the help of their veterinarian when designing their health plan. Whilst the survey does not indicate the reasons for this low level of veterinary involvement, we can assume that the main problems included veterinarians' lack of understanding of organic goals and lack of knowledge regarding organic standards. Both continuing professional training and sound undergraduate training of veterinarians regarding organic livestock production is needed, particularly if health planning is made compulsory for organic farms.

The ideal situation is likely to be local veterinary involvement in farm-level health planning that benefits from existing partnerships and local knowledge. However, lack of know-how and an unwillingness among local veterinarians has led to various different models being suggested to secure veterinary involvement on organic farms. One was adopted by the Scottish Organic Producer Association (SOPA), who employ a veterinarian with good understanding and acceptance of organic goals and standards. The veterinarian guides the farmer through the process of health planning by annually evaluating the written submissions and giving feedback, e.g. assessing disease levels and suggesting appropriate tactical measures to promote health and welfare (Chris Atkinson, Operations Manager, SOPA; personal communication). Another model, designed by Vaarst *et al.* (in preparation), was presented earlier in Box 12.10.

Lack of resources on organic farms

Land-based organic livestock production, with closed nutrient cycles, home-grown livestock feed and high standards of husbandry, places high demands on the farmer. Sundrum (2001) suggests that since organic livestock production is more likely to occur on mixed (i.e. crop and livestock farms) than on specialized farms, the farmer's ability to achieve high levels of stockmanship may be limited by the requirements of running a complex, mixed system where livestock husbandry may not always be the highest priority. A major challenge in the near future is likely to be the introduction of significantly improved animal husbandry methods that improve the health of organic livestock. Whilst planning, monitoring and decision-support systems are likely to play a key role, sufficient price premiums for organic livestock products will be needed to compensate for the additional management inputs.

Setting of targets and benchmarking

A more formal and transparent approach to health planning, as suggested in this chapter, will require targets for disease and treatment levels, medicine use, and the quantity and quality of prophylactic measures on organic farms. Whilst it is important that actual target setting remain specific to the individual farm, there will be pressures to set overall targets, for instance on acceptable levels of mortality or morbidity. A quantifiable approach will also lead to direct comparisons between conventional and organic systems, with no real benefit to either. The challenge for the organic livestock industry is to make sure that such comparisons and targets are guided by farmers and that open, flexible and useful benchmarking systems are accessible to all partners involved.

References

Booth, J.M. (1997) Progress in mastitis control – an evolving problem. In: *Proceedings of British Mastitis Conference 1997: Progress in Mastitis Control, Stoneleigh, October 8, 1997*, pp. 3–9.

Bradley, A.J. and Green, M. (1998) Clinical mastitis in six Somerset dairy herds. In: *Proceedings of British Mastitis Conference 1997: Progress in Mastitis Control, Stoneleigh, October 8, 1997*, p. 80.

De Snoo, A. (2002) Farming styles in organic Danish dairy farms and the relation to health strategies with an emphasis on mastitis treatment. Master Thesis, Scottish Agricultural College and Danish Institute of Agricultural Sciences.

Enting, J., Huirne, R.B.M., Dijkhuizen, A.A. and Tielen, M.J.M. (2000) Zovex, a knowledge-integrated computer system to support health management on pig farms. *Computers and Electronics in Agriculture* 26, 13–35.

European Communities (1999) Council Regulation (EC) No. 1804/1999 of 19 July 1999 supplementing Regulation (EEC) No. 2092/91 on organic production of agricultural products and indications referring thereto on agricultural products and foodstuffs to include livestock production. *Official Journal of the European Communities* 24.98.1999. Brussels. L 222 1–28.

Gray, D. and Hovi, M. (2001) Animal health and welfare in organic livestock production. In: *Organic Livestock Farming*. Chalcombe Publications, Lincoln, UK, pp. 45–74.

Hovi, M. and Gray, D. (2002) Animal health plans for organic farms: the UK experience. In: Hovi, M. and Vaarst, M. (eds) *Positive Health: Preventive Measures and Alternative Strategies. Proceedings of the Fifth NAHWOA Workshop, Rødding, Denmark, 11–13 November 2001*. University of Reading, Reading, UK, pp. 132–143.

Hovi, M. and Sundrum, A. (2002) Discussion report: Health planning and management in organic livestock systems. In: Hovi, M. and Vaarst, M. (eds) *Positive Health: Preventive Measures and Alternative Strategies. Proceedings of the Fifth NAHWOA Workshop, Rødding, Denmark, 11–13 November 2001*. University of Reading, Reading, UK, pp. 152–156.

NAHWOA (2001) *Final Report and Statements*. Network for Animal Health and Welfare in Organic Agriculture. Available at www.veeru.reading.ac.uk/organic/Final.htm

Neave, F.K., Dodd, F.H. and Kingwill, R.G. (1966) A method of controlling udder disease, *The Veterinary Record* 78, 521–523.

Notz, Ch., Klocke, P. and Spranger, J. (2002) Development of an antibiotics-free udder health concept on Swiss organic farms subsequent to farm sanitation. *Proceedings of the XXII World Buiatrics Congress*, No. 672–340.

OAS (2000) *Animal Health and Welfare Plan–Template*. Organic Advisory Service, Elm Farm Research Centre, UK.

Soil Association (2002) *Animal Health – the Prevention of Infectious Diseases*. Written evidence to the Royal Society for the inquiry into infectious livestock diseases. SA Policy Paper, January 2002, Bristol, UK.

Striezel, A. and Plate, P. (1994) *Leitfaden zur Tiergesundheit in ökologisch wirtschftenden Betrieben*, 1. Aufl., Bioland e.V., Göttingen, Germany.

Sundrum, A. (2001) Organic livestock farming – a critical review. *Livestock Production Science* 67, 207–215.

UKROFS (2000) *United Kingdom Register of Organic Food Standards: Standards for Organic Livestock Production*. August.

Vaarst, M., Feenstra, A.A., Roepstorff, A., Høgedal, P., Larsen, V.Aa., Worm, R., Hermansen, J.E. and Lauritzen, H.B. (1998) Sundhedsforhold ved økologisk svineproduction illustreret og diskuteret med udgangspunkt i en besætningscase. In: Hermansen, J.E. (ed.) *Forskning i økologisk svineproduction*. Report 1, Danish Research Centre for Organic Farming, pp. 33–45.

Vaarst, M., Noe, E., Nissen, T.B. Stjernholm, T., Sørensen, C., Enemark, P., Thamsborg, S.M., Bennedsgaard, T.W., Kristensen, T., Andersen, H.J. and Enevoldsen, C. (2001) Development of health advisory service in Danish organic dairy herds – presentation of an action research project. In: Hovi, M. and Vaarst, M. (eds) *Positive Health: Preventive Measures and Alternative Strategies. Proceedings of the Fifth NAHWOA Workshop, Rødding, Denmark, 11–13 November 2001*. University of Reading, Reading, UK, pp. 144–151.

Vaarst, M., Noe, E., Andersen, H.J., Enevoldsen, C., Thamsborg, S.M., Kristensen, T., Enemark, P., Bennedsgaard, T.W., Pedersen, S.S., Sørensen, C., Nissen, T.B. and Stjernholm, T. (in preparation) Health advisory service in Danish organic herds. Development of three different models based on farmers expectations to advisers.

Wackerhausen, S. (1994) *Et åbent sundhedsbegreb. Mellem fundamentalisme og realisme. Sundhedsbegreber, filosofi og praksis.* Aarhus University Press, Aarhus, Denmark.

Walkenhorst, M., Spranger, J., Klocke, P., Jörger, K. and Schären, W. (2002) Evaluation of risk factors contributing to udder health depression during Alpine summer pasturing in Swiss dairy herds. *Proceedings of the 14th IFOAM Organic World Congress, Victoria, Canada, 21–24 August, 2002*, p. 98.

WHO (1946) *Preamble to the Constitution of the World Health Organization, as adopted by the International Health Conference, New York, 19–22 June, 1946. Official Records of the World Health Organization*, No. 2, p. 100.

13

Approaches to the Treatment of Diseased Animals

Mette Vaarst,[1] Andrea Martini,[2] Torben Werner Bennedsgaard[1] and Lisbeth Hektoen[3]

[1]Department of Animal Health and Welfare, Research Centre Foulum, Danish Institute of Agricultural Sciences, PO Box 50, DK-8830 Tjele, Denmark; [2]Dipartimento di Scienze Zootechniche dell'Universitá di Firenze, Via del Cascine 5, I-50144 Florence, Italy; [3]Department of Large Animal Clinical Sciences, The Norwegian School of Veterinary Science, Pb 8146 Dep., N-0033 Oslo, Norway

Editors' comments

Despite all good efforts, organic animals may become sick and need treatment, but to encourage a preventive approach and to avoid the use of chemicals that may be harmful to the environment, some kinds of disease treatments are restricted. However, in Europe (in contrast to the USA) antibiotics are allowed for disease treatments in organic herds. Therefore, it is necessary to discuss how to optimize their use – that is, how to decide when giving an antibiotic is the most responsible treatment or when it is unnecessary or will not benefit the animal. In the organic approach, non-chemical treatments are favoured. Therefore, the authors of this chapter also discuss veterinary homoeopathy, a leading alternative to the mainstream medical approach, one that is controversial but is explicitly mentioned in the EU standards. However, whatever treatment approach is used, a way to have it aim for good animal welfare is to extend it to include 'care-taking', not just administration of some material, for example. This implies a change in thinking on the part of all concerned, including the veterinarian.

Introduction

Treatment and care of the diseased animal are important in all livestock systems, but particularly in organic farming, where high standards of welfare are required and there is a desire to reduce reliance on medical

© CAB International 2004. *Animal Health and Welfare in Organic Agriculture*
(eds M. Vaarst, S. Roderick, V. Lund and W. Lockeretz)

treatments. Care of diseased animals should aim at minimizing the suffering of the animal, promoting a fast recovery, and preventing the spread of similar disease conditions within and between herds.

This is an element of organic farming that often is not sufficiently emphasized when organic health strategies are discussed within the veterinary profession. The basic standards of the International Federation of Organic Agriculture Movements (IFOAM, 2000) state that, with regard to the treatment and care of diseased animals, 'the aim should be to find the cause and to prevent future outbreaks by changing management practices'. This approach reflects the important aim of promoting positive health by preventing disease and offering the animals conditions that support their welfare by creating well-balanced systems. The development and application of this health promotion and planning process is discussed in detail in Chapter 12.

Inevitably, though, even when health promotion is well planned and can be considered successful, disease does occur in livestock, and disease treatment must be seen as an inevitable part of animal husbandry. The health-planning process should therefore also include treatment and care strategies.

An aim of organic farming is to reduce the use of chemical substances. However, there are occasions when the welfare and health of the animal overrides this goal, as when a disease must be treated and the most effective treatment is to use a veterinary medicine of chemical origin. In other areas of organic farming, failure to prevent a problem may mean a drop in production and economic losses, but should always stimulate the development of further strategies as alternatives to chemicals. With organic animals, in contrast, failure to prevent disease can still be 'solved' by treating diseases with whatever treatment is available and effective. The challenge is to ensure that the treatment we offer to the livestock in our care still embodies the organic ideal, whilst sparing the affected animal unnecessary suffering. Is it possible to develop 'organic' treatment strategies to use when an animal is sick and we need to do something? What should such treatment strategies include?

Disease treatment strategies are influenced by several factors (Box 13.1), such as the caretaker's[1] perceptions, goals and interaction with the animals, the overall herd policy, and society's expectations and preferences (imposed, for example, through price premiums and restrictions on production practices). The critical role of the caretaker is also discussed in Chapters 10 and 12; here we focus on the caretaker's role when a crisis needs to be solved. We will emphasize the treatment and care of the diseased animal in order to bring about its recovery, but at times, humane

[1] In this context, a caretaker is the person taking care of the animals, including taking responsibility for their health and well-being.

Box 13.1. Factors influencing the treatment on different levels: the treatment in case of disease, forming treatment strategies and forming a treatment policy.

Animal/human level: treatment of the diseased animal
The caretaker's perception:

- Health, disease, and animal welfare.
- Own ethical role in relation to the animal, environment and society (expectation and living up to expectations from society).

The expression of disease shown by the animal.
Professional knowledge of farmer, veterinarians and other involved persons about disease, different therapeutic possibilities, and epidemiology.
Prognosis of the disease case (based on symptoms, experience and professional knowledge).

Herd and farm level: forming a treatment strategy
Costs of treatments.
Strategies in the herd/farming styles/the role of the animals in the herd.
Experience with different treatment methods and treatment strategies.

Society: guidelines for treatment policy
Consumer expectations (less use of medicine, better animal welfare)
→ (political response, decisions and regulations based on these expressed expectations.
Legislation related to disease control, zoonoses, minimizing antimicrobial resistance, animal welfare, etc.

slaughter also is a valid option to end suffering or prevent the further spread of a disease.

We outline and discuss possible organic principles in the treatment of diseased animals in the context of organic farming objectives and philosophy, and illustrate their application with some specific examples. The examples largely involve mastitis, a particularly important and well-described condition in organic herds. It is also the most common disease in dairy herds and the leading reason for the use of antibiotics, an issue we discuss not only because of the importance of antibiotics as a treatment, but also because we need to develop an organic perspective on their use.

Among the alternatives to conventional veterinary medicines, we highlight the use of classical homoeopathy, as this approach is favoured within the EU standards (Box 13.2 and European Communities, 1999) and by many organic farmers. Homoeopathy presents an approach to health and disease very different from that of conventional biomedicine. Finally, we discuss some other treatment alternatives and supplementary forms of care-taking, and the possible role of the veterinarian. The question of whether 'organic' treatment strategies exist will be left open for future

Box 13.2. EU regulations and recommendations to standard development of disease handling made by the Network for Animal Health and Welfare in Organic Agriculture (NAHWOA).

EU standards influencing disease handling (European Communities, 1999)

- A maximum of 10% of non-organic animals may be bought per year. Conventional animals for milk production may not have calved at the time of purchase.
- Phytotherapy, homoeopathy or other non-allopathic treatments shall be preferred to allopathic treatments (antimicrobials, anthelmintics, etc.) if the effectiveness of the treatment is documented.
- All treatments with antibiotics have to be performed by a veterinarian, except follow-up treatments of calves less than 6 months old.
- The withdrawal period after treatment with allopathic medicine is two times the conventional withdrawal time.
- Preventive use of medicines is prohibited.
- A cow loses its organic status if it is treated with allopathic treatments (anthelmintics not included) in more than three disease cases within a year. The animal can be converted to organic status again.
- Use of insecticides (e.g. for prevention of summer mastitis) is not allowed.
- The housing conditions must be adapted to the biological and ethological needs of the animals and diseases should be prevented through selection of robust animals and breeds suited for the production system.

NAHWOA recommendations on disease treatment
Development of standards/regulations

- The role of biosecurity, largely within a closed flock/herd, in maintaining a high health status should be further emphasized in the standards.
- The following suggestions were made in regard to the use of alternative/complementary medicine:
  - The standards should require and implement similar recording for complementary medicine as for conventional medicine.
  - There should be compulsory training in animal health management and the use of complementary medicine for converting farmers.
  - The use of complementary medicine should be presented more clearly in the context of preventive measures and health planning, in order to ensure that conversion changes in health management are not limited to a shift from conventional medicine to alternative/complementary medicine.
- The limit on the number of conventional treatments administered to an animal should be reconsidered. The emphasis should be on avoidance of suffering and disease. It was suggested that the number be based on a maximum average number of treatments per animal in the herd/flock.
- The acceptability of coccidial water-administered vaccines in organic poultry hatcheries should be clarified. In layer systems, serious coccidiosis is unlikely to be a problem in adult birds that have acquired immuni-

ty, as long as hygiene standards are good, stocking densities are not too high and site rotation is practised.

Development of systems, advice, training and tools

- Facilitate a greater and more rational use of alternative therapies and meet the legislative requirements for use of veterinary medicine in some Member States.
- Develop and implement animal health plans that guarantee maximal public and animal health standards in organic livestock production.
- The following suggestions in regard to the technical aspects of health management plans were made:
  - Target-setting should be farm-specific and should be based on both short- and long-term targets. It was suggested that setting of 'national' or 'organic' target levels (e.g. for disease levels) would be counterproductive.
  - A step-wise establishment of health plans would be needed, particularly on converting farms. Existing data may not always be adequate to assess the existing disease situation on the farm and to allow target-setting. In this context, the need to ensure the dynamic nature of health management plans was emphasized. Regular updating and reviewing of the plans should be compulsory.
  - The health management plan should be utilized to help to introduce alternative/complementary therapies on a farm in a way that allows feedback and assessment of the impact of these therapies.

Research needs

- There is little need for more knowledge on the general epidemiology of endoparasites in ruminants. However, understanding of the local situation at the farm level in regard to parasite dynamics is vital for effective control.
- There is a need to establish a better understanding of internal parasite epidemiology in monogastric animals and how these parasites affect animal welfare under free-range conditions.
- Collaborative European projects on parasite epidemiology and control in pigs and poultry are needed to ensure adequate and representative data.
- There is still a great need to research and identify alternative therapies and systems level strategies for parasite control in all species.
- There is a need for controlled field trials in parasite control to produce data for simulation models that could eventually be used as decision support tools.
- There is a need to further develop health and welfare indices and to assess their usefulness to all stakeholders.
- Animal health planning research should be innovative and include participatory: demonstration farms, study groups, socio-psychological studies (e.g. to establish what makes farmer willing to accept and adopt advice).
- There is a need to further explore the research needs and methodologies in complementary/alternative medicine.

Askham Bryan College
LIBRARY BOOK

debate and development; our main aim here is to provide the reader with background and inspiration to take up this debate and continue to develop the ideas.

Organic Standards and Use of Medicines

To minimize the risk of residues from veterinary medicines, organic standards usually require prolonged withdrawal periods for meat and milk from animals treated with these chemicals. The specified periods usually are extensions of the mandatory withdrawal periods based on the 'precautionary principle', rather than on scientific measurements regarding the elimination of residues. Some organic livestock standards, such as the EU regulation, also limit the number of treatments an individual animal may receive without losing its organic status. Furthermore, in some countries, certain veterinary chemicals are banned in organic production (e.g. organophosphates), and the use of others is restricted (e.g. avermectins). In some countries, e.g. Italy, the use of any chemical substance with a withdrawal time longer than 10 days is forbidden. Medication as prophylaxis also is prohibited, for example in-feed use of antibiotics.

The restrictions originate in part from the desire to reduce chemical inputs in general. They also have the aim of making disease treatment less attractive (i.e. more expensive because of the prolonged withdrawal periods or more labour-consuming by requiring individual rather than mass treatment). This should favour efforts for health promotion and disease prevention over treatment, which is made a much more expensive solution to any problem.

It has been suggested that the restrictions on medicine use may lead to unnecessary suffering if animals are left without appropriate treatment (Hovi, 2001; Hovi and Kossaibati, 2002). This appears to be a general concern within the veterinary profession. It is important, therefore, to point out that avoidance of suffering clearly overrules any limits on the use of medicine. That is why treatment is allowed at all in organic farming: the responsibility for the individual animal clearly has first priority where suffering is an issue. For example, the EU regulation states that the most effective treatment should be offered to avoid suffering even where the use of a particular material would lead to the animal losing its organic status. To act appropriately in relation to each animal and individual case of disease therefore becomes a particularly crucial role of the caretaker. The cost–benefit balance should not be a consideration when deciding whether to treat the animal when otherwise it could suffer. The decision should be about which form of treatment, if any, will best prevent suffering and aid recovery. NAHWOA has offered several recommendations for the development of organic standards for the use of veterinary medicines, summarized in Box 13.2.

Current Knowledge about the Use of Medicines in Organic Herds

Although quantitative data reflecting veterinary treatments on organic farms are scarce (Keatinge *et al.*, 2000), there are sufficient survey data that indicate the widespread use of both therapeutic veterinary drugs and 'alternative' treatments, particularly homoeopathy (Halliday, 1991; Gustafsson-Fahlback, 1996; Krutzinna *et al.*, 1996; Bouilhol, 1997; Sciarra and Guntensperger, 1997; Enemark and Kjeldsen, 1999; Fisker, 1999; Hamilton *et al.*, 1999; Hovi and Roderick, 1999; Roderick and Hovi, 1999; Vaarst and Bennedsgaard, 2001). In these studies, most of which were completed before the latest increase in the number of organic herds, the levels of recorded therapy usage appear to be significantly lower on organic than on conventional farms

The usage of 'conventional' and 'alternative' therapies appears to vary considerably among farms and diseases. Hovi and Roderick (2000) found approximately equal use of antibiotics and homoeopathic treatments for mastitis in organic dairy herds in the UK (Tables 13.1 and 13.2). This contrasted sharply with conventional herds, where homoeopathic treatments were very uncommon. The study also clearly illustrates the difference in mastitis incidence between the two farming systems, with 28.9 dry cow mastitis cases per 100 cow-years in organic herds, where blanket antibiotic prevention is prohibited, versus 9.2 cases in conventional dairy herds.

Table 13.1. Mastitis incidence, reported as number of cases per 100 cow-years in the UK (Hovi and Roderick, 2000).

	Organic herds	Conventional herds
Overall	36.4	48.9
Mastitis during lactation	37.6	54.5
Mastitis during dry period	29.9	9.2

Table 13.2. Farmers' choice of treatment (%) in 2050 cases of mastitis, in 16 organic and seven matched conventional herds in the UK (Hovi and Roderick, 2000).

	Organic herds	Conventional herds
Antibiotics	44	98
Homoeopathy	50	0.4
Other	6	1.6

The Mastitis Treatment Situation in Danish Organic Dairy Herds over the Past Decade

The following serves as an example of how treatment strategies are linked to the development of organic animal husbandry.

Recent studies: no difference between organic and conventional herds

Recent studies of somatic cell counts, mastitis treatments and production results show no significant difference in treatment frequency or choice of veterinary medical products between conventional and recently converted organic herds in Denmark (Vaarst and Bennedsgaard, 2001; Vaarst *et al.*, 2002; Bennedsgaard *et al.*, 2003a,b). This is in contrast to a decade earlier, when the mastitis level and mastitis handling approach in organic demonstration herds were described in great detail. (In connection with this change, below we discuss the importance of knowing the background of different treatment patterns in order to develop treatment strategies for organic herds.)

There were variations among herds within each group in somatic cell counts and treatment frequency. The length of treatment was significantly shorter in the organic herds. This difference is easily explained by the fact that conventional farmers can perform post-treatments of dairy cows themselves, provided they have a herd health agreement with their veterinarian. Organic farmers in Denmark must call a veterinarian for both first treatment and post-treatment of a mastitis case, which means that they often choose not to have their cows treated more than once (an average of 1.5 days at one treatment per day (Bennedsgaard *et al.*, 2003b)).

The early 1990s: a different approach to mastitis handling in organic herds compared with conventional herds and newly converted organic herds in the early 2000s

In contrast to these recent studies, an earlier study showed clear differences between how mastitis was handled on organic and conventional herds (Vaarst and Enevoldsen, 1994). From 1991 to 1994, a Danish on-farm research project was carried out in 15 organic dairy herds in parallel with a project in conventional dairy herds. Udder health variables were compared between the two study groups. With regard to clinical mastitis, the pattern of microorganisms involved in the cases did not differ significantly from what is found in conventional herds (Vaarst and Enevoldsen, 1997). Few coli-infections were cultivated, but 20% of the samples were no-growth cases, and could have been potential coli-cases. As shown in Table 13.3, udder health – expressed several ways – was better in the

Table 13.3. Results from organic and conventional herds participating in on-farm studies at Research Centre Foulum in the early 1990s (Vaarst and Enevoldsen, 1994).

	Organic herds		Conventional herds	
Udder health parameters	Median	10–90% centiles	Median	10–90% centiles
Mastitis treatment, % of lactations	5	0–14	31	7–52
SCC, % cows >500,000 ml^{-1}	14	3–26	19	12–32
Sub-clinical mastitis, % cows	28	11–44	43	20–65
SCC (1000 ml^{-1}) at herd level, based on SCSCC	240	148–452	347	213–613
Bulk milk SCC (1000 ml^{-1})	210	90–350	315	200–550

SCC, somatic cell counts' SCSCC, single cow somatic cell counts.
Only herds having monthly milk yield control on the single cow level are included (12 organic and 20 conventional).

organic herds. Mastitis treatments were significantly less frequent in the organic herds, which was also reflected in lower somatic cell counts in bulk milk samples and monthly milk samples from individual cows.

Understanding the background for the differences: what can we learn?

These herds were followed for several years; daily management routines were explored through both interviews and observation, and the way of handling mastitis and preventing disease in general was well described (Vaarst, 1995). With regard to mastitis, no distinctive organic characteristics could be identified. The fact that they received organic feed, were out during summer and had access to straw bedding did not seem to create any clear differences compared with conventional farms. There were wide variations within both the conventional and organic groups.

Much more than on the conventional farms, disease handling on the organic farms was characterized by routines that supported udder health and by immediate reaction to signs of mastitis. These routines included, for example: milking out by hand between machine milkings very early in a mastitis case; providing extra bedding in critical situations (e.g. after calving); and careful inspection of milk and udders. Many of the interventions on organic herds were very labour demanding, such as udder massage and milking out by hand, as mentioned above. Several other strategies may also have improved the herd-level somatic cell count, including culling, drying off single udder glands, and letting cows with high somatic cell counts stay together with suckling calves for a short or long period. The observed differences therefore are not explained simply by the farms' being organic, but rather by the farmers being good herd managers and having time enough to take care of the cows, and by the preventive and health-promoting efforts undertaken as part of the daily routines.

The basic ideas of organic farming can be seen as stimulating such an effort in combination with the restrictive antibiotic policy. In interviews, many of these farmers described a 'conversion in the herd and heart' that made them feel responsible for their own animals in a way that was different from before, when they considered veterinary treatment a sufficient solution to a mastitis case. This seems not to have happened (yet) in some newly converted organic herds in a very recent study (Vaarst *et al.*, 2003). This could be due to pressures on the organic farmer that require less time to be spent per animal and therefore may not allow such time-consuming efforts on behalf of individual animals, groups of animals or the whole herd. However, this is not consistent with organic farmers' claim that animal welfare should be fostered through non-medical methods. The time and personal resources should be present within the production system. Otherwise, organic dairy farming will have an unacceptable level of disease problems.

Perception of Disease

Before we discuss the development of treatment strategies, we explore the background for these observed differences in handling disease between older and more recently converted organic herds. If we understand the perception of disease and how this perception may influence how we choose to react to disease, this may help us to develop treatment strategies in accordance with both animal welfare concerns and organic farming strategies.

Despite the emphasis on alternative approaches in organic standards, the widespread use of homoeopathy, and the apparent desire to reduce chemotherapy generally, the evidence suggests that the approach and perception of disease on organic farms may not differ considerably from what is prevalent in conventional systems. However, our perception of what 'disease' is influences our perception of how it should be handled, with respect to both the treatment of the condition and the care of the diseased animal. If disease is viewed as an attack by microorganisms, the treatment will be directed mainly against those microorganisms. If it is viewed as a natural reaction of a balanced animal to unbalanced living conditions, the treatment must involve at least a change in those conditions. It therefore is important to reflect on this to understand the organic principles on care and treatment of illness and disease. As discussed above and in Chapters 5, 10 and 12, the perception of animal welfare may lead to an understanding of how disease could be perceived as in an organic herd.

Kleinmann (1980) contrasts the concepts of disease and illness. He suggests that 'disease' is the biomedical or professional's picture of a certain condition or reaction, whereas 'illness' is the patient's own perception of that condition. In other words, the patient feels illness, whereas the doctor observes disease. In some cases, an individual may show 'illness' but it is impossible to determine 'a disease'. Nevertheless, the patient who feels ill should receive care or treatment. In another case, a 'disease' may be ascertained without the patient being affected by it at all, e.g. a sub-clinical bacterial infection in the udder of a cow. Also, infertility in an animal can be a reflection of a condition where 'something is wrong'. The observed infertility could, for example, be the animal's reaction to high milk production. In such a case, 'treating' it with a hormone injection to boost its fertility would be a violation of its attempts to maintain balance.

In Chapter 12 and elsewhere in this book, health is described as a fundamental balance within an individual, covering physical, emotional and mental aspects. Health is influenced by the individual animal's interactions with its surroundings. If stressful environmental conditions provoke a certain reaction, that reaction can be viewed as a healthy individual's natural response to those conditions. If the natural reaction does not work, this 'response' can create a diseased individual. The resulting condition can be a 'disease', such as diarrhoea, that will disappear as soon as the environment is corrected, e.g. when the food ration is properly balanced again. Obviously this can happen only when the disease has not caused serious lesions in the digestive organs. In that case, the individual should be helped further to support digestive tract recovery, in addition to correcting the environment or removing the causal factors. Another example is a case of chronic lameness from a sole ulcer, initially caused by poor floor surface and inadequate housing. Medication and treatment are needed to correct the crisis in the individual, and changes in the housing system will be needed to prevent further occurrence of the condition, or the individual's reaction to its surroundings.

Categories of Disease Treatments

What kinds of disease treatments are appropriate in organic herds, and how do we approach them from a critical viewpoint? We classify treatment methods according to a model described by Kleinmann (1980), who distinguishes between 'professional medical schools', 'biomedicine' and 'folk medicine'. A professional medical school is an approach to treatment that has its own well-developed philosophy and practice. Traditional Chinese medicine and classical homoeopathy (Box 13.3) can both be found within this category – they are 'schools' with a philosophy and a practice.

'Biomedicine' (often synonymous with 'Western medicine', 'school medicine', 'allopathic medicine' or 'conventional medicine') can be seen as one of the most dominant of these schools. In some respects, however, biomedicine cannot be regarded as having one coherent philosophy. Some actions within the biomedical framework are based purely on empirical findings, and some are more systematically developed from concepts of causal relations, and some – such as surgery – are developed from 'handcraft' (not even being practised by a medical doctor) to 'art' and 'science.

In contrast to a 'school', the category 'folk medicine' is often characterized by inconsistencies in philosophy and practice, consisting of many different kinds of advice and actions that are tried out in an unsystematic way. Phytotherapy can belong to all three categories: it can be given based on active substances (reflecting a bio-medical disease model), it can be prescribed according to homoeopathic principles (simile principle, but not used as potentized remedies), and it is part of several folk medical traditions. Likewise, potentized remedies can also be used following principles that differ from the medical school of classical homoeopathy, such as within the framework of anthroposophy or as folk medicine.

Categories of Treatment Methods used in Organic Farming

In European organic farming, the whole spectrum of treatment methods is allowed and is used, although in the ideals and standards, 'non-chemical' methods are favoured. The use of biomedicine is viewed by many as the most efficient treatment method, and in order to ensure animal welfare, European organic farming must allow any treatment that can solve a disease problem in a sick animal. The National Organic Standards in the USA do not allow antibiotics at all (USDA, 2000). This influences farmers' choices on different levels: it encourages them to search for alternatives that may be as relevant and responsible as antibiotics as far as animal welfare, but it also means that animals at times are treated with antibiotics and then 'exported' to a non-organic farm. A product from an animal that has been treated with antibiotics will never be offered to the consumer as 'organic'. It could raise questions as to whether organic farming is made partially dependent on the existence of conventional farms to take over antibiotic-treated animals from organic herds, which may not help the overall long-term development of organic animal husbandry.

The European decision to allow some biomedical treatment in organic farming does not keep us from raising the question of whether these treatments are the only responsible way to deal with cases of disease. Neither are we kept from working towards development of treatment strategies that integrate various treatment approaches much more than happens today. The goal is to create a treatment policy for organic farming in which animal welfare is ensured and the organic principles are respected and serve to guide our efforts. The choice of treatment should be guided by the disease situation we are in, and we should continuously be building up and exchanging experience and knowledge on the various treatment options available.

In the following, we first discuss biomedicine, focusing on antibiotic use, after which we present homoeopathy, a professional school that is a prominent alternative to the biomedical approach. Next we discuss various other alternatives, using mastitis in dairy herds as an example, and conclude with a discussion of care-taking routines for sick animals, a relevant supplement to any disease treatment.

An Organic Approach to Biomedicine: Responsible Use of Antibiotics

As mentioned earlier, the principle of avoiding suffering must not be overridden by any other consideration, including consumers' demand for products from animals that have not been treated with antibiotics or other medicines. Livestock must not suffer, and this justifies the use of 'chemical' medicines when this is deemed the most appropriate and responsible course of action in the case of bacterial infections, as in most mastitis cases.

Antibiotics are probably the most important biomedical treatments on both organic and conventional livestock farms. Increasing risk of the spread of antibiotic resistance from their use in both human and animal health care is well recognized (WHO, 2000). Limiting their use to essential uses only and seeking modes of health care that enable their use to be reduced in both human and animal medicine are two strategies being pursued all over the world (Federation of Veterinarians of Europe, undated; WHO, 2000). In many respects, organic livestock farming principles provides a good framework for such strategies. The current EU regulation on organic livestock production (European Communities, 1999) appears to try to balance the responsible use of conventional veterinary medicines, the need to safeguard animal health and welfare, and the requirement to retain consumer confidence and consumer protection. The effort to minimize chemical medicine should focus not on less treatment, but on reducing the need for treatment by providing living conditions that promote good health.

In many disease cases, farmers and veterinarians will consider antibiotics as the most responsible treatment, and it is the method they are most familiar with. They must be used in a manner that minimizes the risk of resistance developing. Moreover, they should not be used just because they are the most convenient method, and the decision whether to use them should be carefully thought through. Furthermore, the prolonged withdrawal periods required by organic standards should not bias the choice towards antibiotics with short mandatory withdrawal periods, as this could increase the risk of choosing the wrong treatment, exacerbating the negative environmental impacts by spreading substances foreign to nature, and speeding the development of resistant bacteria.

In summary, responsible antibiotic use in organic livestock production should have clear targets. These should include:

- Minimizing suffering by reducing infections and avoiding more invasive measures or euthanasia.
- Preventing increased antibiotic resistance.
- Avoiding residues in livestock products.

These targets can best be met with a clear, farm-specific strategy that is communicated to all stakeholders at the farm level, including the veterinarian. This strategy should have the following basic elements:

- Clearly stated criteria for when antibiotic treatment must be considered.
- Treatment based on confirmation of a bacterial diagnosis or knowledge of prevalent pathogens based on ongoing surveillance of the herd.
- Treatment based on confirmed susceptibility to specific antibiotics.
- Avoidance of a mid-treatment change in antibiotics unless this is dictated by the results of bacterial culture.
- Post-treatment prognosis as a determining factor in whether or not to treat and the choice of treatment (e.g. chronic, recurring mastitis with *Staphylococcus aureus* should not be treated with antibiotics).
- Recording of post-treatment success of therapy in an agreed-upon manner (e.g. clinical or bacteriological assessment 7 days after treatment), with these records analysed regularly to guide further strategy.
- Adherence to manufacturers' recommendations for use, with required withdrawal periods calculated accordingly.

Homoeopathy: a Professional Medical School that Offers an Alternative to Biomedicine

Of the many alternatives to conventional veterinary medicines, perhaps the most common, and the one specifically mentioned in the EU regulation on organic livestock production (European Communities, 1999), is homoeopathy. Homoeopathy and biomedicine illustrate two very different approaches to health, in particular as to what constitutes a disease, which has obvious implications on how diseased animals are treated in each system. Organic farming and homoeopathy have similar views on health and disease (Baars and Ellinger, 1997; Søgaard, 1997). The ideas of homoeopathy have developed over a long history (Box 13.3), and understanding them will help in understanding how its practitioners' perceptions of health and disease compare with that of biomedical practitioners.

Using veterinary homoeopathy

Homoeopathic treatment demands knowledge of the patient (current status, reaction pattern, life history) and the patient's living conditions, because 'disease' is viewed as a reaction of the living individual, with a particular background, constitution and reaction pattern, to certain living conditions and factors in the surroundings. This can pose difficulties for

Box 13.3. Historical development of the veterinary classical homoeopathic treatment method.

Many of the ideas of classical homoeopathy are similar to ancient Greek ideas of health and disease. However, it was not until the late 18th century that a coherent and systematic philosophy and a well-defined clinical practice was developed by a German medical doctor and pharmacist, Dr Christian Friedrich Samuel Hahnemann (1754–1843). He developed the so-called 'simile' principle, the cornerstone of homoeopathy: *Similia similibus curentur*, likes are cured by likes. This is the principle that a substance (of plant, mineral or animal origin) that creates certain symptoms in a normal, healthy individual should also be used to treat a sick patient who may be exhibiting similar symptoms. His philosophy of health, disease and disease treatment is described in *Organon* (1810). This principle is based on treating according to an individual's expression of symptoms rather than focusing on the disease. Another principle conflicting with the biomedical school is the principle of potencies, which is a step-wise systematic dilution of the 'pure substance' (which diluted 1:10 is the so-called 'mother tincture'), where the substance is activated (potentization) through shaking the dilution at each step. In this way, it is hypothesized that the healing information from the substance is transferred to the water in which it is diluted, and it is assumed that the patient as a consequence is treated with 'energetic information'. This is not in accordance with normal physical laws, but is the hypothesis of 'water memory'. Both these principles make research methods difficult to design compared with biomedical studies, as discussed by Hektoen (2001).

Veterinary homoeopathy was discussed by Hahnemann in an unpublished manuscript in the early 19th century. It was a rapidly growing field in the 1830s and 1840s, but decreased in importance when antibiotics were introduced in the 1920s. From the 1970s onwards, veterinary homoeopathy has gained in popularity, particularly in the treatment of pet animals, but also as a result of a rapid growth in organic farming during the 1990s. Veterinary homoeopathy is now an internationally recognized discipline.

animals in large herds. The greater amount of personal contact with dairy cows and their histories make them more 'available' for individualized treatment than other farm animals. Milking the cows twice a day and observing their heath and nutritional status give the herd manager the necessary background information for individualized choice of remedy, treatment and support. This is in contrast to biomedical treatment, where knowledge of each cow is less critical in diagnosis and treatment. Healing in homoeopathy attempts to minimize the environmental and behavioural obstacles to cure. This example highlights the importance of environmental actions in conjunction with homoeopathic therapy and emphasizes why it would not be appropriate to treat a battery-caged hen homoeopathically.

Need for research into veterinary homoeopathy

The possible positive and negative implications of homoeopathic therapies, the justification for their emphasis in organic regulations, and the lack of data on their efficacy point to the need for research into homoeopathy in farm animals. There is a long tradition of using unconventional therapies in animal husbandry (Schoen, 1994; Vaarst, 1996; Smith-Schalkwijk, 1999), but it is based mainly on custom, experience and case descriptions, with very little conventional scientific investigation.

In conventional medical research, quantitative approaches dominate. The randomized, double-blind, controlled clinical trial is considered the best way to evaluate therapeutic effect (Altman, 1999), and there are strict protocols for conducting such trials (FEDESA, 1995). The research tests the effect of a specific treatment on a specific disease. Control groups, double-blinding, strictly defined inclusion and exclusion criteria, and statistical testing of hypothesis are essential elements of conventional efficacy testing. Most controlled clinical trials evaluating unconventional therapies have been conducted on humans. Meta-analysis and reviews of clinical trials in acupuncture and homoeopathy have found positive tendencies in favour of these therapies, but there have been few trials of high scientific quality (Kleijnen *et al.*, 1991; Linde *et al.*, 1997; ter Reit, 1999). This is also true for clinical trials of veterinary homoeopathy (Kowalski, 1989; Vaarst, 1996; Waller *et al.*, 1998).

The basis of alternative therapies, particularly homoeopathy and acupuncture, is fundamentally different from that of conventional medicine. Homoeopathy and acupuncture focus on energetic balance, individual therapy and 'the whole animal', suggesting that evaluation of an outcome or a therapy may be different from that used in conventional medicine (Walker and Anderson, 1999). This raises the potential need for a new approach to efficacy research for these therapies. It has been suggested that one can use conventional research methods, provided particular attention is paid to the specific nature of these therapies in research planning and evaluation (Schütte, 1994; Vaarst, 1996; Walker & Anderson, 1999). More use of qualitative research methods may help promote understanding and foster dialogue among different medical traditions and develop the unconventional therapies further.

Questions about the practical application of veterinary homoeopathy

The homoeopathic approach to disease raises several questions in the context of organic principles and practical approaches to livestock care under farm conditions:

- Is it possible to base disease treatments on individual prescriptions within intensive farming (e.g. 'obstacles to cure' in battery cage systems, time pressures on caretakers that prevent adequate observation, and lack of time for individualized treatment)?
- To what extent can research within the natural sciences answer questions that fall outside existing natural scientific understanding (i.e. the effects of homoeopathy)?
- How can conventional and homoeopathic approaches be combined within a single health strategy for a farmer who wants to use both?
- How can livestock production methods be adapted to meet the needs of effective homoeopathic treatment, and at what economic cost?
- How can we deal with the fact that there are several approaches to the use of homoeopathic remedies, e.g. some only use one remedy at a time, whereas others mix different remedies (sometimes referred to as unicism versus pluracism) or extend the philosophical background, as in so-called homotoxicology, or example.

For this more holistic approach to be effective across the organic sector requires greater collaboration between farmers and their veterinarians, including greater receptiveness amongst the veterinary profession towards complementary treatments. This will demand a big effort to build up and exchange knowledge and to find a common framework under which treatment strategies can be carried out (Vaarst, 1997).

Alternatives to Biomedical and Homoeopathic Treatments: From Drying off Glands to Herbal Medicine

One way to move forward in organic treatment methods could be to prohibit all use of chemical medicine or antibiotics, as in the US standards. This means that antibiotics must be replaced by other methods. With mastitis, for example, these could include herbal medicines, extra milking out by hand, combined fluid therapy (e.g. water with glucose, calcium or salts in order to increase the fluid balance of the animal) and pain-relieving palliative treatment, culling strategies emphasizing udder health, and homoeopathy.

A critical view of antibiotic policy has been stimulated in recent years by the increasing concern over resistance. A large proportion of antibiotic use for farm animals is considered unnecessary (Aarestrup, 2000) or has only a vague and undocumented basis. (See Baadsgaard, 2001, for statistical analyses and discussions of antibiotic use in conventional swine production.) The idea that antibiotics are the only responsible treatment method can be challenged. Often they are not followed by any other efforts to support the animal. It can be questioned whether a treatment is 'responsible' without supportive and health-stimulating treatments, such as those discussed below.

There is no literature documenting or discussing the potential consequences of complete replacement of antibiotic treatment by other means. These alternatives could include not only the actions already mentioned, such as massaging the udder or frequent milking out, but also the kinds of care-taking management routines discussed in Chapter 10, such as offering possibilities for rest, soft lying areas and positive human contact. A few examples of such alternatives to biomedical treatments and their possible consequences are discussed below.

Drying off a single gland

An increasingly common method of handling chronic mastitis cases is to blind the affected teats, either simply by not milking the gland or by cutting the teat, as with a trampled teat (Vaarst *et al.*, 2002). This will decrease production, but the cow will partly compensate through the other glands, so that the total milk loss will be only about 15%. The prevalence of three-teat cows has been used as indicator of poor udder health in some investigations (Andersen, 2000). However, a large proportion of the cases are the farmers' choice, where the alternative to actively blinding a teat might be repeated treatments of the same cow because of recurrent udder infection in the same gland (Vaarst *et al.*, 2002).

Making lactating cow into 'suckler aunts'

In Denmark, one way of handling cows with high somatic cell counts is to keep 'suckler cows', which are lactating cows with a relatively high production level. This is done mostly by having some cows, e.g. cows with high somatic cell counts, with three to four calves each, either in single boxes (e.g. 16 m^2 strawed area) or common 'suckler areas'. The introduction of a calf to an unknown suckler cow (called the 'suckler aunt', since it is not the calf's biological mother) is most successful if the calf was first allowed to suckle its own mother (Vaarst *et al.*, 2001). For several reasons, including udder health, keeping the cow with a calf (after calving or as a suckler cow) should not lead to omitting postpartum udder control. Practical experience indicates that cross-suckling may lead to the calf's not milking one gland (perhaps a mastitic gland from which the milk tastes saltier and is less appealing). This increases the need for daily inspection of the cow and calves.

Herbal medicine

Very little has been published on the use of herbal veterinary remedies, but there is potential for future development in this area. Local use of peppermint oil on the affected udder gland is known to be beneficial because of the hyperaemia and increased blood flow it causes. Increased use of traditional Chinese herbal medicine could be appropriate, but it has not been described, nor is it under current investigation. Much use of herbal remedies, whether for oral or local application, belongs in the category of 'folk medicine'.

Potential consequences of a zero-antibiotic treatment strategy

These alternatives to antibiotic treatments can be used in various combinations in organic herds. Together with health planning and a generally improved health promotion and disease prevention effort, they will put increased demands on the caretaker. More time will be needed for the whole process of improved and more systematic observation of animals, caring for individual animals, and improved non-medical efforts to promote health. Structural changes may be needed with a zero-antibiotics strategy, such as more space and improved living conditions under which animals can carry out their natural behaviour. The culling pattern of the herd will be strongly dominated by consistent culling of potential mastitis patients, particularly chronic cases that would demand more human attention and would decrease the quality of the product. (More generally, the whole issue of product quality will depend much more on how the farmer handles problems than when antibiotics are available to solve problems that are not solved some other way.) The selection of animals – also animal breeds – may be different. The selection of some rustic breed like Maremmana Italian cow has always been directed towards disease resistance and adaptation to the harsh environment where the Maremmana cattle live. In fact, these cattle generally are not treated against parasites or other diseases (Martini *et al.*, 2001).

The animal welfare implications of a zero-antibiotics strategy have not been documented or explored. The animal welfare implications of blinding a teat, for example, may range from none (as with a slowly progressing chronic case where the gland is gradually atrophying) to a traumatic effect (as with cutting the teat or blinding a teat of high-yielding glands because of a high SCC in that gland, causing severe pressure).

How can our Knowledge of Farmers' Attitudes be used in the Development of 'Organic' Treatment Strategies?

Calling a veterinarian to adminster an antibiotic is generally regarded as fulfilling one's responsibility to treat a sick animal. This is seen as looking after animal welfare, disease stress in the herd, and the long-term improvement of the product (e.g. milk). Often the disease treatment is simply left there, with no alternatives or supplementary treatments being discussed.

The question is whether this is sufficient for organic farming. As a starting point, the organic view of animal welfare within organic farming emphasizes supporting natural behaviour, 'balance' and caring for the animals that we have taken into the farm 'household'. Trying to understand this in the context of disease handling may call for increased weight on stimulating the animal to reach balance on its own. This can be done, for example, by allowing it to perform its natural behaviour and seek comfort as it chooses. It may also call for increased weight on care-taking to support the animal's own health-restoring processes.

Farmers' perceptions, actions and decisions obviously have a major influence on treatment patterns, including – perhaps especially – in organic herds. Development of treatment strategies will be relevant for many organic herds. By 'treatment strategy' we mean an explicit description of the principles that guide the choices of treatment in a farm. This could cover dry cow treatment, use of homoeopathy and culling criteria, for example. In cases of disease, individual conditions will play a role and individual judgements will be necessary, but a treatment strategy will make the basis of the decisions explicit to everyone involved in treatment decisions and will provide a guideline for choices in individual situations.

If organic treatment strategies take as their starting point the animal welfare view described above, without a doubt this will require the farmer to devote more time and effort to identifying and handling situations that are critical for both individual animals and the whole herd. The results just presented from Danish organic dairy herds show that to a substantial degree, 'old' organic farmers managed to make these extra efforts, which demanded work, but which also gave results. The farmers' attitudes and will to do this, their perception of their own responsibility to the animals, and what it may mean to them to be *organic* farmers will influence their strategy.

The next step is to discuss the role of their veterinarians in all this. The veterinarians perform many of the treatments, but do they see themselves as important partners in developing strategies? Recent Danish studies (Vaarst *et al.*, 2003) indicate that the farmers may ask for 'alternative' viewpoints and additional advice from veterinarians, but they buy and pay for treatments of individual animals, and the veterinarians are not involved in strategy making. If specific organic treatment strategies are to be developed, it is crucial that all partners involved understand and share the ideas of organic farming. This is an absolute prerequisite for supporting each other in the development of treatment strategies that ensure responsible treatment of the individual animal and strive to fulfil the organic goals for animal welfare.

Supporting the Diseased Animal

The EU regulations do not specify proper relief of pain and suffering or optimal housing of the diseased animal, but in the interests of welfare, these need to be addressed by both veterinarians and farmers. In organic animal husbandry, an important element of animal welfare is expression of natural behaviour and as natural a life as possible (Chapters 5, 6 and 7). Creating a framework for this 'natural life' may make it more difficult for the farmer to observe disease cases, and therefore requires enhanced monitoring and observation. The farmer has a responsibility to create an environment where the animals can be observed and good conditions are created for healing.

In the daily life of the animals, the caretaker should intervene as little as possible, but in a crisis should intervene rapidly and effectively when necessary. Occurrence of disease is a situation in which it is necessary to take immediate action. In these cases, the focus is primarily on the individual diseased animal, which may be given living conditions that particularly stimulate healing and give it sufficient peace and rest to go through the phases of its illness. This is one of the basic roles of a caretaker: to observe – including under 'natural' conditions and in big herds – and intervene as needed. This demands certain resources: there should always be enough time for an extra effort to handle a crisis, and there should also be extra space. 'Hospital pens' seem a necessary part of any production environment.

An example: a Danish case study

A Danish case study from the early 1990s provides an example of disease handling that goes beyond medical treatment of the diseased animals

(Box 13.4). It involved a dairy herd that participated in a project on organic dairy farming and consequently was followed closely by the project technician and veterinarians. In this herd, the level of sub-clinical mastitis cases involving *S. aureus* increased dramatically over one summer, when there was an outbreak of teat wounds (virus skin infection).

During the autumn, a plan involving several control measures was made in collaboration with the local veterinarian and the project group. Some of the measures are preventive against further outbreaks, and are done on the level of the whole herd to help it come back from a crisis to a healthy situation.

Box 13.4. An example of disease handling that both treats the actual disease and reduces the risk of further outbreaks. Modified from Vaarst (1995).

Background: An organic Jersey dairy herd of 186 cows experienced a dramatic increase in the level of sub-clinical mastitis involving *Staphylococcus aureus* in autumn 1991, when 56 of the lactating cows were found positive in bacteriological culturing. This increase followed an outbreak of teat wounds over the summer that were assumed to have housed the bacteria, which is common in skin and wounds.

Strategy: The herd managers (nine owners and two staff members) developed the following strategy, which included care taking of single cow as well as strategic decisions on a whole herd level:

1. Grouping of cows and milking of infected cows last. The following winter, further grouping: first lactating cows, then older cows without infection in any gland, finally infected cows.
2. Systematic culling of cows infected by *S. aureus* in one or more udder glands.
3. Teat dip with skin care introduced.
4. Checking and adjustment of milking equipment two to three times during the year.
5. All dry cows taken inside two to three times every week during the summer to be checked, have teat dip with skin care, and be sprayed on the udder with peppermint oil as a kind of fly repellent. (This strategy is not normally recommended. Its possible side effect of creating more or less permanent hyperaemia in the skin can be critical: when used over shorter periods in mastitis udders, the hyperaemia helps the healing process in the udder. According to the experience of the herd owners it kept insects away.)
6. Wound care.

Self-medication and Self-cure

Few reports of 'self-cure' exist, but care-taking in a case of disease basically aims at giving the animal the optimal conditions to manage its own situation, which means stimulating self-cure (defined as 'restoration of health without medical intervention'). Good care-taking to stimulate self-cure can be done on a general level, such as providing extra bedding material, extra fine hay, grooming the animal or covering it with a horse blanket. As discussed above, it can also be done on a more goal-directed level, such as milking out by hand in mastitis cases, washing a sore area with warm soapy water in the case of bruises or massaging an affected area.

Engel (2002) raises interesting issues regarding the potential role of animals' self-treatment. She points to their natural ability for dietary self-regulation and searching for nutrients, plants or substances (e.g. certain minerals) that they sense they need to maintain or reach a physiological balance. One prerequisite for this, of course, is access to a rich environment with sources of these items. Another prerequisite is to give the animals conditions that are as natural as possible to allow them to recognize their own needs. This is definitely an area for future research and development.

Conversion to Organic Farming: a Situation of Chaos and Changes that Affect Disease Treatment Patterns and Needs

In a recent study involving interviews of veterinarians, agricultural scientists advising on cattle herds and newly converted farmers, it became clear that the conversion period often involved changes such as an increase in herd size or a change in the housing system (Vaarst *et al.*, 2001a). This leads to a further demand for changes in daily management. During conversion the focus generally was on the crop production (managing without artificial fertilizers and pesticides), and the interaction between the herd and fields (grazing management and roughage production). In contrast, little attention was paid to health issues specifically related to the dairy herd. The knowledge about and understanding of organic animal husbandry was limited among many animal health professionals (veterinarians), and since organic herds apparently look like conventional herds, there was very little effort spent searching for solutions fitting an organic context and the goals of organic farming. Often the farmer had to explain to the veterinarian the restrictions and conditions on organic production.

This is unacceptable for a farmer who is seeking and paying for advice from a professional adviser who is supposed to base that advice on knowledge – not only about disease and health, but also about factors that can influence the solutions. The finding that the conversion of the herd received very little attention and that animal health professionals know so little about organic farming – in particular its goals – points to future challenges. In a recent Danish study further exploring treatment patterns of newly converted organic farmers, the interviewed farmers noted that their veterinarians kept to their traditional role of treating disease (Vaarst *et al.*, 2003). To a much lesser extent, or not at all, did they step in as partners in health planning, development of organic animal husbandry by discussing what 'organic' means in relation to specific routines or strategies (e.g. outdoor calves), or active development of alternatives to antibiotic treatments. Partnership in developing health planning and alternative treatment strategies for the organic herd is urgently needed. The first step is to encourage farmers and their advisers (veterinarians, cattle health advisers) to become conscious of this need. Furthermore, they should be encouraged to search for and develop strategies that fit each individual farm and some basic organic principles for animal husbandry in general, which are not yet fully developed (e.g. the perception of animal welfare, as discussed in Chapter 5).

Conclusions

The principles of organic livestock keeping were developed separately from other parts of the organic farming system. Whereas high animal welfare has been an implicit goal of organic farming, very little discussion has been taking place on how animal welfare could be understood and supported in practice in the specific context of organic farming. This gap between thinking organically and acting in relation to an acute situation is particularly evident in the case of disease treatment.

It is likely that conventional veterinary medicine will remain a part of treatment strategies in organic systems. Therefore, we offered the responsible use of antibiotics as an example of a possible 'organic' approach to the use of conventional medicines. Because homoeopathy has emerged as the main alternative to conventional therapy in many countries, classical homoeopathy and its links to organic principles was discussed in more detail. Finally, we discussed the support of the diseased animals and the implications of a complete prohibition of antibiotics.

The way in which we treat animals – medical treatment as well as care-taking strategies – must be better influenced by a broader view of health and disease: that the animal must be helped to manage and master the disease situation. The perception of what a disease is will govern the perception of how it should be handled. Disease treatment is a human action, performed when the caretaker feels motivated and finds treatment necessary, according to his or her perceptions, goals and experience.

An experienced person with insight and practical knowledge of animal health and welfare will make complex judgements and decisions about handling a critical situation such as a disease outbreak in an animal or a group of animals. However, there may be several possibilities for establishing a treatment policy in the herd that reflects a more holistic and 'organic' way of thinking.

In each case of disease, it seems difficult to describe an overall pattern or to follow inflexible guidelines, even when it was the farmer who made them. In most cases the decisions will be based on 'expert knowledge' and 'tacit knowledge', as judged by experienced professionals (see Chapter 10). Therefore, organic regulations should be carefully considered, so as to avoid situations where they might lead to omission of care-taking and appropriate disease treatment, with the resulting risk of reduced animal welfare. Giving and discussing guidelines and strategies for disease treatments at the herd level in most cases will be beneficial. It will contribute to a treatment policy that is more explicit and systematic, and less chaotic for each farmer, and that can be discussed in relation to the strategy of the individual farm and organic farming in general.

Whilst there are several open questions in regard to alternative and complementary medicine and decision making in treatment situations, the following conclusions can be drawn:

- Organic livestock farms need a holistic health promotion plan that includes explicit strategies for handling diseased animals. Such a plan should include the use of various treatment methods and care-taking efforts.
- Antibiotic use should always be based on the principles of responsible use.
- After disease treatment, the case should be followed up to build experience that can improve future decisions.
- A cohesive and high-quality research programme for alternative treatment methods needs to be developed to promote appropriate use under field conditions.
- More weight must be given to care-taking as an integrated part of disease treatment. Disease treatment should be regarded in a more holistic way, extending from early detection and subsequent action to follow-up. The learning process based on reflection over each disease case should be regarded as important, both when the patient recovers completely and in cases where the patient never recovers.
- The dialogue on all treatment policies and possibilities should be kept alive, open and critical. All treatment methods can be questioned, discussed and developed in relation to the context in which they are used.
- Regulations on medicine use should be developed with care to avoid inflexibility, which can have the opposite of the intended effect.

Regulations should allow individual action in accordance with the existing possibilities in each herd, based on the ethical judgement of each caretaker.

- All disease outbreaks in single animals or groups should be responded to with strong preventive efforts in the herd.

References

Aarestrup, F.M. (2000) Occurrence, selection and spread of resistance to antimicrobial agents used for growth promotion for food animals in Denmark. *Acta Pathologica, Microbiologica et Immunologica Scandinavica* Suppl. no. 101, 108, p. 48.

Altman, D.G. (1999) *Practical Statistics for Medical Research*. Chapman & Hall, London.

Baadsgaard, N.P. (2001) Development of clinical monitoring methods in pig health management. PhD thesis, Department of Clinical Studies, Royal Veterinary and Agricultural University, Copenhagen, and Department of Animal Health and Welfare, Research Centre Foulum, Tjele, Denmark.

Baars, T. and Ellinger, L. (1997) The relation between organic husbandry and homoeopathy: the prevention approach. In: *Veterinary Homoeopathy in Organic Herds. Relevance, Practical Applicability and Future Perspective. Proceedings of an International Workshop at Research Centre Foulum, Denmark, April 17th–18th, 1997*. Internal Report No. 90, Danish Institute of Animal Sciences, pp. 5–11.

Bennedsgaard, T.W., Thamsborg, S.M., Vaarst, M. and Enevoldsen, C. (2003a) Eleven years with organic dairy production in Denmark – herd health and production in relation to time of conversion and with comparison to conventional production. *Livestock Production Science* 80, 121–131.

Bennedsgaard, T.W., Thamsborg, S.M., Aarestrup, F.M., Enevoldsen, C., Vaarst, M. and Larsen, P. (2003b) Use of veterinary drugs in organic and conventional dairy herds in Denmark with emphasis on mastitis treatment. (Submitted.)

Bouilhol, M. (1997) Livestock farming systems and parasite risk for sheep in organic farming. In: Isart, J. and Llerena, J.J. (eds) *Proceedings of the Third ENOF Workshop, Ancona, 5–6 June 1997*, pp. 149–157.

Enemark, P.S. and Kjeldsen, A.M. (1999) Produktionsresultater fra økologiske besætninger. LK – meddelelse nr. 353, Landskontoret for Kvaeg, Skejby, Denmark.

Engel, C.R. (2002) Acknowledging the potential role of animal self-medication. In: *Proceedings of UK Organic Research 2002, Colloquium of Organic Researchers. Aberystwyth, 26–28th March, 2002*, University of Wales, Aberystwyth, UK, pp. 355–358.

European Communities (1999) Council Regulation (EC) No 1804/1999 of 19 July 1999 supplementing Regulation (EEC) No 2092/91 on organic production of agricultural products and indications referring thereto on agricultural products and foodstuffs to include livestock production. *Official Journal of the European Communities* 24.98.1999. Brussels. L 222 1–28.

Federation of Veterinarians of Europe (undated) *Antibiotic Resistance and Prudent Use of Antibiotics in Veterinary Medicine*. Brussels.

FEDESA (1995) *Good Clinical Practice for the Conduct of Clinical Trials for Veterinary Medicinal Products – GCPV The EU Note for Guidance*. Brussels.

Fisker, C. (1999) Økologisk slagtekyllingeproduktion. MSc thesis, Department of Animal Science and Animal Health, Royal Veterinary and Agricultural University, Copenhagen.

Gustafsson-Fahlback, M. (1996) Djurmiljö och parasitforekomst i utegrishallning – inventering på 12 gårdar. *Jordbruksinformation* 5, Swedish Board of Agriculture.

Halliday, G., Ramsay, S., Scanlan, S., and Younie, D. (1991) *A Survey of Organic Livestock Health and Treatment*. The Kintail Land Research Foundation.

Hamilton, C., Hansson, I., Forslund, K. and Ekman, T. (1999) Djurhälsan i ekologisk mjölkproduktion. In: *Ekologisk djurproduktion, Svensk veterinärtidning* 2, Suppl. 29. Centraltryckeriet, Borås, Sweden, pp. 25–29.

Hektoen, L. (2001) Controlled clinical trials used in the evaluation of clinical effect of homeopathic treatment in farm animals. In: Hovi, M. and Vaarst, M. (eds) *Positive Health: Preventive Measures and Alternative Strategies. Proceedings of the Fifth NAHWOA Workshop, Rødding, Denmark, 11–13 November 2001*. University of Reading, Reading, UK, pp. 42–49.

Hovi, M. (2001) Alternative therapy use on UK organic farms – constraints and pitfalls. In: Hovi, M. and Vaarst, M. (eds) *Positive Health: Preventive Measures and Alternative Strategies. Proceedings of the Fifth NAHWOA Workshop, Rødding, Denmark, 11–13 November 2001*. University of Reading, Reading, UK, pp. 7–13.

Hovi, M. and Kossaibati, M. (2002) The impact of organic livestock standards on animal welfare – a questionnaire survey of cattle veterinarians. *Cattle Practice* 10, 183–189.

Hovi, M. and Roderick, S. (1999) Mastitis in organic dairy herds – results of a two-year survey. In: *Mastitis. The Organic Perspective. A One-Day Conference on Current Research, Prevention, Treatment and Alternative Solutions for the Dairy Sector. Friday 3rd September 1999*. The Soil Association in association with University of Reading, Reading.

Hovi, M. and Roderick, S. (2000) Mastitis in organic dairy herds in England and Wales. In: Alföldi, T., Lockeretz, W. and Niggli, U. (eds) *IFOAM 2000: The World Grows Organic. Proceedings of the 13th International IFOAM Scientific Conference, Basel, Switzerland, 28 to 31 August, 2000*, p. 342.

IFOAM (2000) *IFOAM Basic Standards*. International Federation of Organic Agriculture Movements, Tholey-Theley, Germany. http://www.ifoam.org/standard/

Keatinge, R., Gray, D., Thamsborg, S.M., Martini, A. and Plate, P. (2000) EU Regulation 1804/1999 – the implications of limiting allopathic treatment. In: Hovi, M. and Trujillo, R.G. (eds) *Diversity of Livestock Systems and Definition of Animal Welfare. Proceedings of the Second NAHWOA Workshop, Córdoba, 8–11 January 2000*. University of Reading, Reading, UK, pp. 92–98.

Kleijnen, J., Knipschield, P. and Riet, G. (1991) Clinical trials of homeopathy. *British Medical Journal* 302, 316–322.

Kleinmann A. (1980) *Patients and Healers in the Context of Culture. An Exploration of the Borderland between Anthropology, Medicine, and Psychiatry*. University of California Press, USA.

Kowalski, M. (1989) Homöopatische Arzneimittelanwendung in der veterinärmedizinischen Litteratur. Innaug. Diss., Freie Universität, Berlin.

Krutzinna, C., Boehncke, E. and Herrmann, H.-J. (1996) Organic milk production in Germany. *Biological Agriculture and Horticulture 13*, 351–358.

Linde, K., Clausius, N. and Ramirez, G. (1997) Are the clinical effects of homeopathy placebo effects? A metaanalysis of placebo-controlled trials. *Lancet* 350, 834–843.

Martini, A., Giorgetti, A., Rondina, D., Sargentini C., Bozzi, R., Moretti, M., Pérez Torrecillas, C., Funghi. R. and Lucifero, M. (2001) The Maremmana, a rustic breed ideal for organic productions – experimental experiences. In: Hovi, M. and Baars, T. (eds) *Breeding and Feeding for Animal Health and Welfare in Organic Livestock Systems. Proceedings of the Fourth NAHWOA Workshop, Wageningen, 24–27 March 2001.* University of Reading, Reading, UK, pp. 211–218.

Roderick, S. and Hovi, M. (1999) *Animal Health and Welfare in Organic Livestock Systems: Identification of Constraints and Priorities. A Report to the Ministry of Agriculture, Fisheries and Food.* University of Reading, Reading, UK.

Schoen, A.M. (1994) *Veterinary Acupuncture. Ancient Art to Modern Medicine.* American Veterinary Publications, Goleta, California.

Schütte, A. (1994) Ist Forschung in der Veterinärhomöopathie gerechtfertigt? *Berliner-Münchener Tierärztliche Wochenschrift* 107, 229–236.

Sciarra, C., and Guntensperger, I. (1997) Research on organic dairy cattle in Switzerland. In: *Proceedings of the Third ENOF Workshop, Ancona, 5–6 June 1997*, pp. 159–166.

Smith-Schalkwijkm, M.J. (1999) Veterinary phytotherapy: an overview. *Canadian Veterinary Journal* 40, 891–892.

Søgaard, A.B. (1997) Organic agriculture and alternative medicine: parallels and paradigms. In: Olesen, S.G., Eikard, B., Gad, P. and Høg, E. (eds) *Studies in Alternative Therapy 4, Lifestyle and Medical Paradigms*, pp. 150–164.

ter Reit, G.K., Kleijnen, J. and Knipschild, P. (1999) Acupuncture and chronic pain: a criteria-based meta-analysis. *Journal of Clinical Epidemiology* 43, 1191–1200.

USDA (2000) National Organic Program, *7 CFR Part 205*. Agricultural Marketing Service, United States Department of Agriculture, Washington, DC. http://www.ams.usda.gov/nop/nop2000/nop2/finalrulepages/finalrulemap.htm

Vaarst, M. (1995) Sundhed og sygdomshåndtering i danske økologiske malkekvægbesætninger. PhD thesis, Clinical Institute, Royal Veterinary and Agricultural University, Copenhagen.

Vaarst, M. (1996) *Veterinær homøopati: Baggrund, principper og anvendelse med speciel fokus på økologiske malkekvægbesætninger – et litteraturreview.* Beretning nr. 731. Statens Husdyrbrugsforsøg, Denmark.

Vaarst, M. (1997) The challenge of implementing homoeopathy into the health advisory system in organic herd. Danish experiences. In: *Veterinary Homoeopathy in Organic Herds. Relevance, Practical Applicability and Future Perspective. Proceedings of an International Workshop at Research Centre Foulum, Denmark, April 17th–18th, 1997.* Internal Report No. 90, Danish Institute of Animal Sciences, Tjele, Denmark, pp. 41–45.

Vaarst, M. and Bennedsgaard, T. W. (2001) Reduced medication in organic farming with emphasis on organic dairy production. *Acta Veterinaria Scandinavica* Supplementum 95, 51–57.

Vaarst, M. and Enevoldsen, C. (1994) Disease control and health in Danish organic dairy herds. In: *Biological Basis of Sustainable Animal Production, EAAP Publication no. 67*, pp. 211–217.

Vaarst, M. and Enevoldsen, C. (1997) Patterns of clinical mastitis manifestations in Danish organic dairy herds. *Journal of Dairy Research* 64, 23–37.

Vaarst, M., Alban, L., Mogensen, L., Thamsborg, S.M. and Kristensen, E.S. (2001a) Health and welfare in Danish dairy cattle in the transition to organic production: problems, priorities and perspectives. *Journal of Agricultural and Environmental Ethics* 14, 367–390.

Vaarst, M., Jensen, M.B. and Sandager, A.-M. (2001b) Behaviour of calves at introduction to nurse cows after the colostrum period. *Applied Animal Behaviour Science* 73, 27–33.

Vaarst, M., Paarup-Laursen, B., Houe, H., Fossing, C. and Andersen, H.J. (2002) Farmers' choice of medical treatment of mastitis in Danish dairy herds based on qualitative research interviews. *Journal of Dairy Science* 85, 992–1001.

Vaarst, M., Thamsborg, S.M., Bennedsgaard, T.W., Houe, H., Enevoldsen, C., Aarestrup, F.M. and De Snoo, A. (2003) Organic dairy farmers decision making in the first two years after conversion in relation to mastitis treatment. *Livestock Production Science* 80, 109–120.

Walker, L.G and Anderson, J. (1999) Testing complementary and alternative therapies within a research protocol. *European Journal of Cancer* 35, 1614–1618.

Waller, K.P. (1998) *Veterinär homeopati til lantbrukets djur – en sammenställning av vetenskapeliga försök*. Swedish Agricultural University, Uppsala, Sweden.

WHO (2000) *WHO Global Principles for the Containment of Antimicrobial Resistance in Animals Intended for Food*. World Health Organization. http://www.who.int/emc/diseases/zoo/who_global_principles.html

14

Grassland Management and Parasite Control

David Younie,[1] Stig Milan Thamsborg,[2] Francesca Ambrosini[3] and Stephen Roderick[4]

[1]*Scottish Agricultural College, Craibstone Estate, Buckburn, Aberdeen AB21 9YA, UK;* [2]*Centre for Experimental Parasitology, Royal Veterinary and Agricultural College, Bülowsvej 13, DK-1879 Frederiksberg C, Denmark;* [3]*Dipartimento di Scienze Zootechniche dell'Universitá di Firenze, Via del Cascine 5, I-50144 Florence, Italy;* [4]*Organic Studies Centre, Duchy College, Rosewarne, Camborne, Cornwall TR14 0AB, UK*

Editors' comments

Living outdoors is part of the organic animals' being, and therefore of its well-being. It offers the animal the most 'natural' alternative under farming conditions, but it also means that the farmer has less control over some factors important to the well-being of the individual animal. To meet this challenge, keeping animals outdoors involves extra planning and oversight. The authors of this chapter focus on parasite control in ruminants as a key element of animal health and welfare in grazing animals. The organic farmer must base parasite control on different tactics from those of conventional farming. Since parasiticides may not be used routinely, the risks instead must be minimized through a more ecological and whole-farm approach involving a knowledgeable choice of cropping sequence, the age and species of grazing animals, and the timing of grazing relative to the life cycle of the parasite.

Introduction

Parasitic infections of livestock under organic grazing management are among the most important challenges to achieving the desired high standards of animal welfare. Although organic standards permit the use of most conventional veterinary medicines for therapeutic treatment of sick

© CAB International 2004. *Animal Health and Welfare in Organic Agriculture* (eds M. Vaarst, S. Roderick, V. Lund and W. Lockeretz)

animals, the routine use of veterinary drugs on healthy animals as part of a preventive treatment strategy is prohibited. Instead, organic principles demand that the farmer should adapt the livestock management system to minimize the occurrence of health problems in the first place, that is, to adopt positive health management strategies, as discussed in Chapter 12.

Whilst parasitic diseases may not always have such spectacular effects as an epidemic of acute viral or bacterial disease, production and economic losses may be even greater. This is because nearly all grazing animals are infected almost all the time with some parasites, and ingestion from pasture is almost continuous in many regions. In contrast, with acute diseases, if the animal survives, recovery is usually rapid. Moreover, sub-clinical (hidden) intestinal parasitosis is probably far more important than the occasional clinical outbreaks that result in high mortality. Unfortunately, such sub-clinical events are very rarely diagnosed. In addition, infections can reduce progress in selection for improved production traits because high mortality reduces the intensity of selection. Although exposure to a limited challenge from parasitic infection may be desirable in situations where the development of resistance is desirable, infection can cause severe discomfort.

This chapter briefly considers the epidemiology of the more common parasitic infections, knowledge of which is fundamental to the control strategies employed. Based on a sound understanding of the patterns of these infections, alternative approaches to control that are appropriate within the framework of organic farming will be discussed, including some methods that are still in their infancy regarding application and efficacy. The focus will be on conditions that are particularly important in organic production (see Chapter 11), in particular gastrointestinal infections, either because the conventional approaches to control are broadly unacceptable or restricted, or because organic farming presents specific epidemiological scenarios.

Gastrointestinal Nematode Infections of Ruminants: Epidemiological Considerations

The group of gastrointestinal (GI) nematodes comprises several species adapted to different sections of the gut (abomasum, small and large intestines). Although they differ in pathogenicity, in the rate of development and in larval bionomics, these nematodes are generally regarded as one group. Except for a few species of *Nematodirus* and *Trichostrongylus*, different species respectively infect cattle and sheep.

GI nematodes have no intermediate hosts, and infection takes place only from pasture. Eggs from adult females in the gut are excreted with the host faeces and develop there through L1 and L2 larval stages. The next stage, the L3 larvae, are consumed by the host. Under northern temperate conditions, larvae (e.g. *Ostertagia* (cattle), *Teladorsagia* (sheep), *Haemonchus, Trichostrongylus*) or eggs (*Nematodirus*) can overwinter in the sward and be available for infection of susceptible stock at the start of the growing season. These can survive on pasture during the winter and to early summer the following year in most regions. With *Ostertagia, Teladorsagia, Haemonchus* and *Trichostrongylus*, if the sward is not grazed in spring and early summer, most overwintered larvae die, and normally it is safe to graze after mid-summer. On the other hand, in contaminated swards grazed in spring, overwintered larvae are ingested by susceptible animals, which then excrete eggs. As the temperature rises, the rate of development increases and eggs deposited in April, May and June reach the infective stage from early or mid-July onwards. Clinical signs of parasitic gastroenteritis (PGE) then occur between July and September. This pattern will be different in a Mediterranean climate with very dry summers and wet winters, and control will have to be adapted to the epidemiology in a specific region. *Nematodirus* spp. of sheep survive very long on pasture, with the eggs still able to hatch after more than a year.

If exposed, lambs establish immunity at 4–9 months of age, depending on the species of nematode (e.g. immunity to *Nematodirus* spp. builds up after 4 months). However, even where pastures are initially clear of infection in spring, they can become contaminated by infected animals (e.g. the peri-parturient rise of eggs of *Teladorsagia, Haemonchus* and *Trichostrongylus* in lambing ewes). Inhibited larvae (larvae that have entered an inhibited stage of development over the winter period, *Ostertagia* in cattle in particular) are an important means of carrying infection from one season to the next (Armour and Coop, 1991).

With *Nematodirus* spp. of sheep, hatching of eggs requires a prolonged cold period followed by a period above 10°C. Hence, normally only one generation per year is possible; because of the climatic requirements this normally occurs in spring or early summer (Armour and Coop, 1991). The extent to which lambs become infected depends largely on whether they are ingesting significant quantities of grazed herbage at the stage when *Nematodirus* larvae are infective. This in turn will depend on lambing date, with lambs born early most at risk, although as lambs get older they develop a degree of immunity to the parasite. Outbreaks of nematodirosis are also related to soil temperature changes during the early spring. This has enabled some countries to develop a forecasting system to warn farmers of the risk of an outbreak.

The sheep nematode species *Haemonchus contortus* represents a special case in relation to control without use of chemotherapeutics. This is mainly a subtropical–tropical parasite, but it can cause problems as far north as Norway (Helle, 1973). *Haemonchus* is very prolific, resulting in very high faecal egg counts (thousands of eggs per g of faeces). Because of fast development on pasture, the infection builds up rapidly (3–5 weeks) if the weather is warm and humid. The infection is highly pathogenic and causes high mortality. Severe blood loss and hypoproteinaemia are characteristic of this infection. Diarrhoea is not a clinical feature, in contrast to *Teladorsagia* and *Trichostrongylus* infections in sheep. *Haemonchus* survives poorly over winter in northern areas, which makes anthelmintic treatments during housing effective in controlling the infection on farms. However, anthelmintic resistance is common, and organic farms, where a routine treatment during housing often is not performed, may risk a build-up of infections on the pastures. Some evidence indicates that this is actually the case (Lindqvist *et al.*, 2001; Thamsborg *et al.*, 2001a).

In Europe, housed dairy calves typically infect themselves after turnout in spring, although infection levels are rarely high enough to cause disease. Eggs are passed in faeces following a prepatent period of 2–3 weeks, and peak level of infection can occur 2 months after exposure in the spring. However, this rate of faecal egg output will depend on initial pasture infectivity, and the peak may occur at the end of the growing season if the pasture infection level is low. For example, Eysker *et al.* (1998b) reported that after contamination of the pasture, it takes at least 4 weeks in The Netherlands before a substantial number of larvae are present on the herbage. On ungrazed pasture, overwintered larvae rapidly diminish in spring, and by summer these pastures would be almost completely clean (Eysker, 2001). The aim of parasite control should be to balance infection and the acquisition of immunity (Eysker, 2001), as not exposing calves to infection in their first grazing season may result in the postponement of infection to the second season (Ploeger *et al.*, 1990; Eysker *et al.*, 2000).

Potential for Control of Gastro-intestinal Nematodes by Grazing Management

Several grassland management strategies can minimize the risk of infection by endoparasites in cattle and sheep (Thamsborg *et al.*, 1999). These can be categorized as shown in Table 14.1.

Evasion of worm challenge

A basic concept in using grazing management to control nematode infections, and an important resource for minimizing parasite problems on organic farms, is that of the 'clean pasture'. This is a pasture with minimal risk of infection reaching pathogenic levels, because the contamination (number of infective larvae (L3) in the herbage) is nil or very low when animals are introduced on the pasture. The maximum number of larvae

Table 14.1. Grassland management strategies for control of gastro-intestinal worms in sheep and cattle.

Mechanism	Practice	Comment
Evasion of worm challenge	Use of clean pastures before the expected rise in pasture infectivity, i.e. late June (sheep) or mid July (cattle) Use of pastures grazed by a different species in spring Use of silage or hay aftermaths Use of new grass reseeds Use of annual forage crops	Worm challenge on susceptible grazing animals is evaded by *moving* animals from *contaminated pastures* to clean pastures (i.e. pastures that have not been grazed in the same season by the same species)
Dilution of worm challenge	Lower stocking rates Mixed species grazing Mixed age groups grazing	Worm challenge on susceptible grazing animals is *diluted* by a lower stocking rate or by the presence of another unsusceptible species or (immune) adult stock of the same species
Prevention of worm challenge *(in uninfected stock)*	Use of new grass reseeds Use of silage or hay aftermaths Use of annual forage crops Late lambing (*Nematodirus*) Yearly alternation between livestock species	Worm challenge is prevented by introducing *uninfected stock* to clean pastures

on a pasture that qualifies as 'clean' will vary with the predominant nematode species and the climatic conditions, and in particular the longevity of L3 larvae.

Clean pasture can be provided by new reseeds, by silage aftermaths or by annual forage crops such as forage rape, kale or turnips (*Brassica napus*). Pastures can also minimize pathogenic challenge if they have not been grazed by susceptible stock of the same species either earlier within the same season, or during the previous season (i.e. a yearly alternating clean grazing system) (Cawthorne, 1986). An example is shown below of a system in which each field is managed according to a 3-year rotation:

Year 1 →	Year 2 →	Year 3 →
Sheep	Silage	Cattle

The system can be refined further, for example by moving weaned lambs in late summer to silage aftermaths, or moving some cattle in late summer to graze together with dry ewes after weaning. This system is only possible where each field is sufficiently flat to be cut for hay or silage. Nevertheless, even where cutting is not possible, e.g. on undulating permanent pasture, a 2-year rotation between cattle and sheep will also help to minimize worm challenge. Although recent studies have indicated that the move to clean pasture can be successful in early season without dosing with anthelmintics, provided *Haemonchus* is not present (Githigia *et al.*, 2001), in sheep flocks this preventive approach is difficult to maintain without intervention. The peri-parturient egg rise in ewes effectively makes them infective animals and leads to contamination of otherwise clean pasture. Likewise, *Nematodirus* problems may occur in sheep/cattle systems, since the parasite can cycle through young susceptible cattle if they graze contaminated pasture in the spring. For these reasons, clean grazing strategies may be difficult on organic farms, where preventive medication is not allowed. Furthermore, this approach may select strongly for nematode strains resistant to anthelmintics because the nematodes that can survive the anthelmintic treatment will make up the population on the new pasture after turnout or a move.

The concept of 'safe pastures' has been used to designate pastures with a slightly higher risk compared with clean pastures (e.g. pastures grazed by treated, non-lactating ewes in the previous year), but such designations should be based on local epidemiological knowledge. Dutch studies have indicated a beneficial effect of mowing a pasture that has otherwise been grazed permanently by cattle, but little data on this approach exist for sheep (M. Eysker, unpublished data).

Eysker (2001) demonstrated geographical differences in practices required to achieve clean grazing for cattle, and the dangers in extrapolating results from region to region. In Denmark, Nansen *et al.* (1988) showed the effectiveness of a single move to aftermath in July, before the mid-summer increase in pasture infectivity. However, this would not be sufficient in The Netherlands without anthelmintic treatment (Eysker *et al.*, 1998a). Late turnout of dairy calves on to mown pasture is well recognized and widely practised in The Netherlands to severely reduce primary infections and pasture contamination (Eysker, 2001). In the colder climate of Sweden, the build-up of pasture infectivity is too slow to cause major production losses in calves after reinfection. However, pasture infectivity the following spring was sufficient to cause economic loss in calves within 1 month of turnout, suggesting that clean grazing should mean avoiding grazing the year after pastures are used by calves (Dimander *et al.*, 2000). The most common procedure to control nematodes in organically managed cattle herds in Sweden was to turn calves out on pastures not grazed by any cattle in the current or previous grazing seasons (Svensson *et al.*, 2000).

The strategy adopted on any one farm will depend on the physical resources of the farm and the farming system employed. The ideal management strategy for worm control in any one group of susceptible animals is to make two to four repeated moves to clean pasture over the season and to alternate grazing between species. However, many farms may be limited by having only one species or may be unsuited for sufficient forage production or cultivation, e.g. the sheep-only systems in the marginal areas of Scotland. This will limit the successful adoption of clean grazing. Under these conditions, the only grassland management option may be to restrict the stocking rate of sheep, coupled with strategic treatment based on regular faecal egg counts to predict when a rise in worm infection is about to occur. A system that relies on frequent treatments is unlikely to be acceptable to the organic certification body.

Diluting the challenge

Manipulation of stocking density is among the most important strategies available to the organic farmer, although stocking density is generally regarded not as a management procedure for worm control but rather as an important risk factor. As stocking density is increased, contamination of the pasture with worm eggs and larvae increases, leading to a subsequent rise in infection of grazing animals as the season progresses (e.g. Nansen *et al.*, 1988). Reducing the general stocking density will result in fewer problems with nematode infections (Thamsborg *et al.*, 1996).

The effect of reducing stocking density on parasite infection and liveweight gain in steers (100–200 kg live weight) is shown in Fig. 14.1.

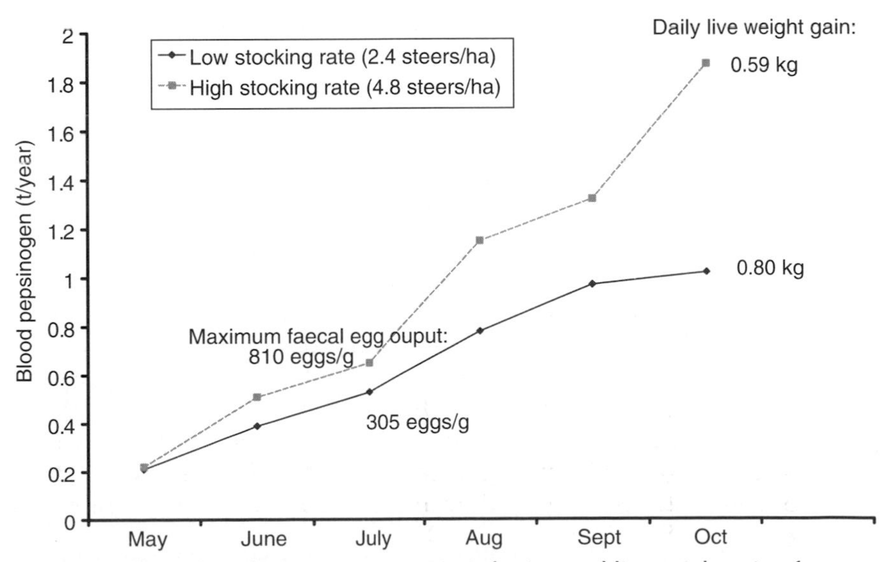

Fig. 14.1. Effect of stocking rate on parasite infection and liveweight gain of Holstein/Friesian steers grazing permanent grassland (Thamsborg, 2000).

Blood pepsinogen is a measure of parasite damage to the gut (abomasal) wall. In this experiment, steers were grazing contaminated pastures, i.e. pastures that had been grazed with susceptible stock in previous years. The adverse effect of the high stocking rate gets worse over time. Stocking rates above or even very close to the carrying capacity of plant production should be avoided.

Grazing of susceptible young animals together with animals of a different species not susceptible to the same parasite, or with older (immune) animals of the same species, in effect reduces the stocking density of the susceptible animals. Susceptible animals will become less infected if the ratio of susceptible to resistant or immune stock is decreased. This is the principle used in mixed grazing, i.e. grazing sheep together with another non-susceptible species on the same pasture, or when weaners graze with older, immune sheep (e.g. dry ewes). In fact, the presence of these immune animals is likely to enhance the beneficial effect of the reduced stocking rate, since by ingesting the herbage these animals will be actively destroying worm larvae and reducing the challenge.

A mixed grazing system can be established in two ways: by grazing both groups of animals simultaneously in the same field (truly mixed grazing), or by dividing the season into two periods and grazing the two groups of animals sequentially (often termed 'alternate grazing'). For example, the grazing area may be grazed by ewes and lambs in the first half of the season (until late June) and thereafter by cattle. Mixed or alternate grazing with sheep is more effective for nematode control in sheep than in cattle (Niezen *et al.*, 1996). In Denmark, promising results have been obtained through both mixed and alternate grazing of cattle and pigs (Roepstorff *et al.*, 2000), although damage to pasture by pigs that are not nose-ringed limits the use of this approach. Also in Denmark, mixed grazing of first- and second-grazing season heifers (Nansen *et al.*, 1990) was shown to be effective in preventing parasitic gastroenteritis. Leader–follower systems, where young cattle are moved ahead of older cattle, is a well-established approach to helminth control (Leaver, 1970).

Nutritional factors

The importance of an adequate protein supply

Nutrition should play a major role in defending grazing livestock from parasitic infection, and appropriate feeding, particularly the supply of adequate protein in late pregnancy and early lactation, will minimize faecal egg output and subsequent contamination of the sward. In organic farming systems, where grazing seasons tend to be later and more prolonged, the supply of home-grown feed is limited. Coupled with the high cost of purchased supplementary feeds, this may be a particular concern in some regions (see Chapter 3 for some examples, and Chapter 15 for a discussion of the potential difficulties of relying on home-grown feeds).

Improved nutrition of the peri-parturient ewe reduces the level of faecal egg output and consequent contamination of the pasture by the grazing ewe. For example, protein supplementation of ewes around lambing may limit the peri-parturient rise in faecal egg counts, depending on the protein level during pregnancy (Houdijk *et al.*, 2001a). The epidemiological importance of this depends on the time of lambing in relation to turnout, if ewes are not outside year-round. In contrast, underfeeding, particularly of protein, reduces the ewe's immunity against worms (Houdijk *et al.*, 2000). Protein supplementation can reduce establishment and alleviate the pathogenic effects of nematode infections (Abbott *et al.*, 1988), particularly if the protein supply was below the animals' requirement (Coop and Kyriazakis, 1999). Faecal egg counts tend to be higher with ewes that have a low condition score during mid-pregnancy (Houdijk *et al.*, 2001a), likewise with yearling compared with older ewes and with multiple-rearing compared with single-rearing ewes (Houdijk *et al.*, 2001b).

It is evident from the above that protein content of the diet is important. For organic farms, this means that a good protein-rich diet will mask the ill-effects of parasitism. Rumen by-pass protein has the best effect, but its availability may be limited in organic farming, particularly since fishmeal and urea are prohibited. Provision of high-quality pasture is imperative. However, supplementation sometimes will be required, depending on the season. This is particularly true for ewes that lamb on pasture or during early grazing season, for lambs that are finished on pasture or during drought, or for outwintered sheep. Supplementation during finishing of lambs is likely to improve production markedly, because nematode infections generally build up in late season and may reduce the lambs' performance.

Whether grazing is supplemented by feeding roughage, concentrates (energy- or protein-rich) or bypass protein will depend on availability and cost–benefit considerations. In young dairy cattle in southern Sweden, Höglund *et al.* (2001) found that a low level of parasitic nematode infections was associated with good management, such as use of parasite-safe pastures and supplementary feeding. Of course, supplementation during the growing season results in lower ingestion of pasture as well as diluting the parasite burden (Jørgensen *et al.*, 1992; Eysker, 2001). Depending on the feedstuff, using supplementation in this manner may be seen as contrary to the organic principle of encouraging natural diets based largely on roughage.

Askham Bryan College
LIBRARY BOOK

Bioactive forages

There is increasing interest in the use of forages containing metabolites that may assist in reducing parasitic infections, the so-called bioactive forages. In contrast to many temperate medicinal plants with a putative anthelmintic activity, the bioactive forages are generally non-toxic, so that correct dosing is not a crucial issue. It should be possible to feed the bioactive plants as a major part of the diet, either cut or grazed; preferably, they should be cultivable in the crop rotation.

Much of the work has been done on sheep, and has initially focused on plants with a high content of condensed tannins (CT). Studies in the southern hemisphere showed that feeding lambs forages with a CT content of 3–8% (dry matter basis) lowered faecal nematode egg counts and worm burdens (Niezen *et al.*, 1993, 1995, 1998). The CT content is near zero in most grasses, around 5% in some temperate fodder legumes, and up to 40% in some tropical browse plants, depending on the method used for determination. The relevant leguminous forages in temperate regions include greater trefoil (*Lotus pedunculatus*), birdsfoot trefoil (*Lotus corniculatus*), sulla (*Hedysarum coronarium*), sainfoin (*Onobrychus viciifolia*), dock (*Rumex obtusifolius*) and *Dorycnium* spp. In other regions of the world, sorghum, cottonseed, quebracho, grape seed and faba bean may be relevant. In most species, it is evident that the CT content varies considerably with cultivar, soil, year, season, etc. Studies in the UK using quebracho as a model CT in the diet at levels of 3–8% have confirmed a 50–60% reduction in establishment of *Trichostrongylus colubriformis* in penned lambs and a 30% reduction in established infections (Athanasiadou *et al.*, 2000a,b).

Quebracho (5% of diet) in a low-protein diet reduced faecal egg counts of *T. colubriformis* by 50% (Butter *et al.*, 2000). *In vitro* studies using the larval development and other assays have shown that purified CT directly reduces preparasitic nematodes. Lambs infected with *Teladorsagia circumcincta* and *Teladorsagia vitrinus* grazing pure stands of sainfoin had 50% lower mean nematode faecal egg counts compared with a reference group on clover-grass (Thamsborg *et al.*, 2001b). Grazing lambs on pastures with *L. corniculatus* in Wales over a whole season also lowered faecal egg counts (C. Marley, Aberystwyth, UK, personal communication, 2000).

Preliminary results with tanniferous forages in ruminants have been promising, although reductions under practical conditions are less than 50%. Much work is needed before they are ready for practical application. In a given region, species suited to the local conditions should be identified. The efficacy is relatively low, and ways must be found of improving this by selection of appropriate cultivars or growing conditions. Use will be governed by the seasonality of the forages and the epidemiology of the parasite infections, if conserved forage cannot be used. The forages need to be integrated into other systems of control, such as repeated moves or biological control. The use of bioactive forages incorporated in the crop rotation for grazing animals or in the diet of housed animals is completely compatible with the principles of organic farming.

Monitoring and intervention

Monitoring of faecal egg counts (FEC) and intervention by anthelmintic treatments when FEC reaches a specified limit (e.g. 400 epg in lambs) is increasingly being advocated. This is particularly true in Denmark, where anthelmintics are available only by prescription on all farms and diagnosis by a veterinarian is required before treatment. Although this approach will avoid clinical disease and will not enhance the development of anthelmintic-resistant nematode strains, it is not proactive. The large production losses from sub-clinical infections will not be avoided if the farmer has no other means of control. In some cases, where disease occurs before faecal egg counts get high, it may be too late for treatment intervention. In the UK, FEC monitoring and treatment is accepted by the certification bodies, but only as part of a wider parasite control policy in which preventive strategies also are adopted.

Another approach involves monitoring of clinical condition, e.g. anaemia in the case of infection caused by *H. contortus* in sheep. In South Africa, the FAMACHA-card has been developed for assessing the colour of eye mucosa (score 1–5) (Van Wyk *et al.*, 1997). It is used monthly, with all animals that score 3 or above getting drenched. Adopting this method can avoid most high faecal egg counts, besides minimizing the risk of anthelmintic resistance. Further, a policy of culling all ewes or lambs treated more than once may increase the level of natural resistance within a flock.

In cattle, monitoring of faecal egg counts or serum pepsinogen (indicative of parasites in the abomasum) in first-season grazing calves 6–7 weeks after turnout may be suggested as a way to determine the overwintering infection on the pasture. In turn this would indicate when it was necessary to move to 'clean' pastures later in the season. However, the approach has not yet been thoroughly tested on-farm.

Lungworm Infections

Epidemiological considerations

Unlike GI nematodes, infective lungworm larvae (*Dictyocaulus viviparus*) reach pasture via the sporangia of the fungus *Pilobolus* spp. within 1 week after faecal deposition (Eysker, 2001). Transmission of the infection from year to year takes places either by overwintering of larvae in the sward or by infected animals carrying larvae or adult worms over from the previous season. Young dairy animals in their first grazing season are most commonly affected, but autumn-born suckled calves also are susceptible. Spring-born suckled calves grazing with their mothers in their first season usually are not affected. The disease tends to be more prevalent on permanent or semi-permanent pastures and in areas with a mild, wet climate. Outbreaks of clinical disease in calves are most common from early July to September, approximately 8 weeks after first exposure as a result of auto-infection of calves. Infection in lactating cows is an increasing problem on both organic and conventional farms. The first sign is a marked drop in milk yields and sporadic coughs, followed by more intense and widespread coughing and loss of appetite. A decline in the use of vaccination in the 1990s, coupled with increased use of anthelmintics for heifers, has been associated with declining exposure, the resulting consequential decrease in acquired immunity, and an increase in disease incidence in adults in conventional farms (Jacobs, 1993; Andrews and de Wolf, 1994; David, 1997). However, this does not explain the increase observed on Danish organic farms without use of anthelmintics and where the vaccine is not available.

Methods of control

Options for controlling lungworm in organic systems are limited. A clean grazing policy will help minimize the exposure of animals to contaminated pasture, although it is not foolproof, because pasture also can be contaminated by aerosol spread by the carrier fungus. In practice, if lungworm is known to have occurred on a farm, the organic farmer will have to rely on vaccination or the strategic use of anthelmintics (Eysker, 2001), although the former is not a desirable long-term strategy in organic farming.

Vaccination does not eliminate contamination of pasture, and vaccinated calves can produce sufficiently high infections to cause disease in susceptible calves grazing with them (Peacock and Poynter, 1980). Because primary infections in susceptible calves usually are low, coming either from low-level excreting carriers or from overwintered larvae, infections need to build up before disease is seen. The lower stocking rates generally used in organic farming may dilute the infection sufficiently to avoid clinical disease, and low stocking rates and repeated moves should be applied if lungworm is known to be a problem. Temporarily housing animals affected by lungworm should be considered, particularly in cows, to avoid lung emphysema due to continued larval challenge on infected pastures.

Biological control options have been investigated, including nematode-trapping fungi (Henriksen *et al.*, 1997) and *Aphodius* beetles (Scarabaeidae) (Gormally, 1993). However, there have been no commercial developments.

Liver Fluke

Options for the control of the large liver fluke (*Fasciola hepatica*) are limited to controlling access to wet pasture areas inhabited by the intermediate host (the mud snail *Lymnaea* spp.), draining these areas to remove the desired habitat, or strategically using specific anthelmintics. Drainage and exclusion clearly are not feasible in many areas. In Holland, lowering the water table has greatly decreased the incidence of fasciolosis over a 40-year period. However, the increasing emphasis on environmental care over production will make such strategies less tenable and may reverse this trend (Eysker, 2001). In fact, increased prevalence has been noted in sheep in Ireland and in cattle in Denmark in the last 2–3 years, perhaps because of more extensive grazing of natural wet areas. Hansson *et al.* (2000) in Sweden were unable to find any significant difference in prevalence between organic and conventional cattle at slaughter. Evasive grazing during the high-risk period in autumn also is an option, depending on the management system and the extent of the problem. Meteorological fluke forecasts have been developed (McIlroy *et al.*, 1990; Goodall *et al.*, 1991), including computerized systems based on historical climatic and epidemiological data (liver condemnation rates), but are not in regular use.

The use of strategic anthelmintics, whilst not suited to an organic approach, offers the only real current alternative where infected pastures are used. Such treatment may be justified since most animals grazing in these areas are likely to be infected. Depending on the type of drug, they can be used strategically to remove immature or mature liver flukes in the animal, thus breaking the life cycle. The severity of the disease can be system- and species-specific. Severe clinical disease occurs in sheep after intake of a large number of larvae (acute disease). There is no build-up in immunity with age; rather, infection accumulates over time. In contrast, yearling cattle, especially if they are housed, probably clear themselves of the infection during the winter or the second season on grass.

There has been some interest in the use of plants with molluscicidal activity, such as *Eucalyptus* spp. (Hammond *et al.*, 1994) and *Euphorbiales* spp. (Singh and Agarwal, 1988). There also is interest in the larvae of the sciomyzid fly *Ilione albiseta* as a potential parasite of the host snail (Gormally, 1987, 1988, 1989). However, these currently are not practical options. Mulcahy *et al.* (1999) report partial success with vaccination using the fluke-derived cathepsin L2 protein, and conclude that vaccination of cattle against liver fluke is technically feasible and is likely to become a commercial reality.

Coccidiosis

Coccidiosis is a disease of the gut, primarily affecting young lambs and calves, caused by single-celled parasites (protozoa) called *Eimeria*. Oocysts containing several *Eimeria* parasites are shed in the faeces of infected animals. They can survive on the ground for more than a year, and if ingested by lambs or calves they break open in the gut and release the parasites, which invade the gut wall (Blewitt and Angus, 1991).

Clinical coccidiosis in sheep results from either residual contamination in the environment or shedding of oocysts by the ewes. The level of environmental infection is probably the most important factor. It follows that the incidence of the disease can be influenced by managing the environment. The risk of infection will be reduced by good indoor hygiene, such as frequent cleaning, dry straw bedding, avoidance of faeces in troughs, and feeding hay in racks and not on a dirty pen floor. Outdoor lambing will generally reduce the risk compared with indoor lambing, depending on grazing management decisions. The best solution is to use a new area for ewes with young lambs or outdoor lambing each year. If a new area is not available, contamination can be reduced through low stocking rates, regular movement of feeding troughs and shelters, and grazing late-born lambs on areas that had not previously been grazed by lambs in the same grazing season. The preventive use of anti-coccidial drugs is not permitted under organic livestock standards, but as with all clinical conditions, clinically affected animals may be treated.

The control of coccidiosis in calves indoors or on pasture follows much the same lines.

Ectoparasites

Ectoparasites pose a serious threat to the health and welfare of livestock, particularly for animals with a thick coat or fleece, such as sheep (e.g. sheep scab), or in certain environments (e.g. ticks). Control is of particular concern under organic conditions, as many of the standard approaches to chemical control either are nor permitted or are restricted to solving short-term problems. Grazing management can minimize the risk of infection from several parasites.

Sheep scab

Sheep scab, caused by the non-burrowing mite *Psoroptes ovis*, can cause severe skin damage, loss of condition and death. It spreads very rapidly, although it cannot survive for more than 1–2 weeks off the host animal. The variable responses of sheep to the mite, and the unpredictable incubation period, course, manifestations and outcome, make this an intriguing and perplexing disease (O'Brien, 1999). Although modifying grassland management to minimize risks often can be a problem, particularly where neighbouring flocks come into easy contact, biosecurity can be enhanced by using secure boundary fencing to prevent potentially infected animals from transferring infection, and by importing animals only from disease-free flocks. (The NAHWOA Recommendantions called for organic standards to give more emphasis to biosecurity and closed flocks.) On larger extensive farms, where exposure to other flocks is most likely, the cost of fencing is likely to be prohibitive. The most practical solutions are first, to ensure good biosecurity by establishing a period of quarantine for any new animals, and second, where there is a known risk from neighbouring flocks, to dip sheep strategically with a synthetic pyrethroid or to inject a macrocyclic lactone endectocide (i.e. ivermectin or milbemycin group). In some countries, the latter treatments may require prior permission from the certifying body.

Several alternatives to the standard chemical control of sheep scab have been researched and have reached varying stages of development. Although little is known about the antigens involved or the mechanism of protection, there is evidence that immunity can be acquired, both naturally and by vaccination using mite proteins (Smith *et al.*, 2001, 2002). The use of entomopathogenic fungi for the control of *Psoroptes* spp. has also been explored (Smith *et al.*, 2000), and promising results have been achieved using neem oil and some insect growth regulators (O'Brien, 1999).

Blow flies

In contrast to the scab mite, the blow flies that cause fly strike spend only a short time on the host sheep, first as eggs and then as larvae. The eggs are laid on wounds or on wet or soiled fleece on various parts of the body, such as the hindquarters following diarrhoea. The larvae then attack the tissues of the host.

Fly strike cannot be controlled by grassland management, except that effective worm control will minimize the occurrence of diarrhoea, and the use of tanniferous forages may affect faecal consistency, reducing the formation of dag (faeces-coated wool) (reviewed by Waghorn *et al.*, 1999). Synthetic pyrethroid insecticides or insect growth regulators can be used for therapeutic or strategic treatment, although, as with other disease treatments, prior permission from the certification body may be required in some countries. Crutching, or the removal of excess wool from the groin area to prevent soiling, also will reduce the risk.

Conclusion

In conventional grassland systems, the main objective is usually to maximize livestock output per hectare through high levels of herbage production and high stocking rates. In contrast, a major priority in managing organic grassland should be the health and welfare of the animals. This remains true even if it means that stocking rates and overall livestock output per hectare are reduced. A more complex grassland system may be required, involving more forward planning, less flexible grazing regimes, short-term grass–clover leys, an additional livestock species to dilute the parasite challenge, increased health monitoring and changes in feeding regimes.

Perhaps the greatest challenge to an organic producer comes during the conversion period. In most cases it is likely that animal health previously was managed largely through the routine use of veterinary drugs. It can be difficult to replace that approach with one based on a sound understanding of the pattern of disease risks, particularly since preventive management must be balanced with the other objectives of organic farming, such as environmental enhancement. Reduced output and a fear of possibly increased health and welfare problems can exacerbate that challenge. The potential threat of increased disease is real, as the part about simply reducing preventive treatments may be pursued more enthusiastically than the proactive establishment of preventive management strategies. A management-based approach requires a higher level of knowledge and skill than that associated with using the safety net of prophylactic treatment programmes.

To some extent this is a question of training and extension, since we already know a great deal about reliable grazing management systems that result in high levels of health and welfare. Nevertheless, research is still required to improve our understanding of soil, plant and animal interactions and to increase the range of realistic alternative parasite control techniques, particularly where land management options are limited. There needs to be further development of grassland management regimes involving, for example, mixed-species grazing, interaction with cutting and nutrition (such as the use of deworming paddocks based on bioactive forage species). Predictive models are needed for developing practical decision-support systems for organic parasite control; the NAHWOA Recommendations called for controlled field trials to provide the data needed for such models. Whilst direct control methods are being studied as alternatives to conventional veterinary drugs (e.g. control of worms by nematophagous fungi), this simply is a replacement of one set of external inputs with another. This may be an appropriate solution under some circumstances, but maintaining the health and welfare of grazing animals in organic systems must still be based on positive, management-based health strategies.

Ultimately, maintaining high standards of health care is always going to be complex on an organic farm, not only because disease and health problems frequently are multifactorial, but also because the production system is attempting to satisfy a range of objectives, not least that of economic efficiency. Hence, there is unlikely to be a single management option for the control of parasitic infections, nor of other kinds. The starting point should be to develop animal health plans that incorporate the multifactorial risks and opportunities and are targeted for the particular farm. This may require incorporating a range of interdependent control options, linked to epidemiological and environmental knowledge.

References

Abbott, E.M., Parkins, J.J. and Holmes, P.H. (1988) Influence of dietary protein on the pathophysiology of haemonchosis in lambs given continuous infections. *Research in Veterinary Science* 45, 41–49.

Andrews, A.H. and Wolf, S. de (1994) Increased prevalence of husk. *Veterinary Record* 134, 152.

Armour, J. and Coop, R.L. (1991) Pathogenesis and control of gastrointestinal helminthiasis. In: Martin, W.B. and Aitken, I.D. (eds) *Diseases of Sheep*, 2nd edn. Blackwell Scientific Publications, Oxford, UK, pp. 122–130.

Athanasiadou, S., Kyriazakis, I., Jackson, F. and Coop, R.L. (2000a) Effects of short-term exposure to condensed tannins on adult *Trichostrongylus colubriformis*. *The Veterinary Record* 146, 713–734.

Athanasiadou, S., Kyriazakis, I., Jackson, F. and Coop, R.L. (2000b) Consequences of long-term feeding with condensed tannins on sheep parasitised with *Trichostrongylus colubriformis*. *International Journal for Parasitology* 30, 1025–1033.

Athanasiadou, S., Arsenos, G. and Kyriazakis, I. (2002) Animal health and welfare issues arising in organic ruminant production systems. In: Kyriazakis, I. and Zervas, G. (eds) *Organic Meat and Milk from Ruminants*. EAAP Publication No. 106. Wageningen Academic Publishers, Wageningen, The Netherlands, pp. 39–56.

Blewitt, D.A. and Angus, K.W. (1991) Cryptosporidiosis and coccidiosis in lambs. In: Martin, W.B. and Aitken, I.D. (eds) *Diseases of Sheep*, 2nd edn. Blackwell Scientific Publications, Oxford, UK, pp. 99–103.

Butter, N.L., Dawson, J.M., Wakelin, D. and Buttery, P.J. (2000) Effect of dietary tannin and protein concentration on nematode infection (*Trichostrongylus colubriformis*) in lambs. *Journal of Agricultural Science, Cambridge* 134, 89–99.

Cawthorne, R.J.G. (1986) Management for the control of parasites. In: Frame, J. (ed.) *Grazing. Occasional Symposium No. 19. Proceedings of a Conference held at Malvern, Worcestershire, UK, 5–7 November 1985*. British Grassland Society, pp. 89–97.

Coop, R.L. and Kyriazakis, I. (1999) Nutrition–parasite interaction. *Veterinary Parasitology* 84, 187–204.

David, G.P. (1997) Survey on lungworm in adult cattle. *The Veterinary Record* 141, 343–344.

Dimander, S.O., Höglund, J., Spörndly, E. and Waller, P.J. (2000) The impact of internal parasites on the productivity of young cattle organically reared on semi-natural pastures in Sweden. *Veterinary Parasitology* 90, 271–284.

Eysker, M. (2001) Strategies for internal parasite control in organic cattle. In: Hovi, M. and Vaarst, M. (eds) *Positive Health: Preventive Measures and Alternative Strategies. Proceedings of the Fifth NAHWOA Workshop, Rødding, Denmark, 11–13 November 2001*. University of Reading, Reading, pp. 59–72.

Eysker, M., Aar, W.M. van der, Boersema, J.H., Dop, P.Y. and Kooyman, F.N.J. (1998a) The efficacy of Michel's dose and move system on gastrointestinal nematode infections in dairy calves. *Veterinary Parasitology* 75, 99–114.

Eysker, M., Aar, W.M. van der, Boersema, J.H., Githiori, J.B. and Kooyman, F.N.J. (1998b) The effect of repeated moves to clean pasture on the build up of gastrointestinal nematode infections in calves. *Veterinary Parasitology* 76, 81–94.

Eysker, M., Boersema, J.H., Kooyman, F.N.J. and Ploeger, H.W. (2000) Resilience of second year grazing cattle to parasitic gastroenteritis following negligible exposure to gastrointestinal nematode infections in their first year. *Veterinary Parasitology* 89, 37–50.

Githigia, S.M., Thamsborg, S.M. and Larsen, M. (2001) Effectiveness of grazing management in controlling gastrointestinal nematodes in weaner lambs on pasture in Denmark. *Veterinary Parasitology* 99, 15–27.

Goodall, E.A., McIlroy, S.G., McCracken, R.M., McLoughlin, E.M. and Taylor, S.M. (1991) A mathematical forecasting model for the annual prevalence of fasciolosis. *Agricultural Systems* 36, 231–240.

Gormally, M.J. (1987) Effect of temperature on the duration of larval and pupal stages of two species of sciomyzid flies, predators of the snail *Lymnaea truncatula*. *Entomologia Experimentalis et Applicata* 43, 95–100.

Gormally, M.J. (1988) Temperature and the biology and predation of *Ilione albiseta* (Diptera: Sciomyzidae) – potential biological control agent of liver fluke. *Hydrobiologia* 166, 239.

Gormally, M.J. (1989) Studies on the oviposition and longevity of *Ilione albiseta* (Dipt.: Sciomyzidae) – potential biological control agent of liver fluke. *Entomophaga* 33, 387–395.

Gormally, M.J. (1993) Laboratory investigations of *Aphodius* beetles (Scarabaeidae) as biological control agents of the cattle lungworm, *Dictyocaulus viviparus* (Nematoda: Dictyocaulidae). *Biocontrol Science and Technology* 3, 499–502.

Hammond, J.A., Fielding, D. and Nuru, H. (1994) Eucalyptus: a sustainable self-delivery molluscicide. *Veterinary Research Communications* 18, 359–365.

Hansson, I., Hamilton, C., Ekman, T. and Forslund, K. (2000) Carcass quality in certified organic production compared with conventional livestock production. *Journal of Veterinary Medicine, series B* 47, 111–120.

Helle, O. (1973) Helminthological problems in Norway. In: Urquhart, G.M. and Armour, J. (eds) *Helminth Disease of Cattle, Sheep and Horses in Europe*. Robert Maclehose and Co., The University Press, p. 59.

Henriksen, S.A., Larsen, M., Gronvold, J., Nansen, P. and Wolstrup, J. (1997) Nematode-trapping fungi in biological control of *Dictyocaulus viviparus*. *Acta Veterinaria Scandinavica* 38, 175–179.

Höglund, J., Svensson, C. and Hessle, A. (2001) A field survey on the status of internal parasites in calves on organic dairy farms in southwestern Sweden. *Veterinary Parasitology* 99, 113–128.

Houdijk, J.G.M., Kyriasakis, I., Jackson, F., Huntley, J.F. and Coop, R.L. (2000) Can an increased intake of metabolizable protein affect the periparturient relaxation of immunity against *Teladorsagia circumcincta* in sheep? *Veterinary Parasitology* 91, 43–62.

Houdijk, J.G.M., Kyriasakis, I., Coop, R.L. and Jackson, F. (2001a) The expression of immunity to *Teladorsagia circumcincta* in ewes and its relationship to protein nutrition depend on body protein reserves. *Parasitology* 122, 661–672.

Houdijk, J.G.M., Kyriasakis, I., Jackson, F. and Coop, R.L. (2001b) The relationship between protein nutrition, reproductive effort and breakdown in immunity to *Teladorsagia circumcincta* in periparturient ewes. *Animal Science* 72, 595–606.

Jacobs, D.E. (1993) Calfhood vaccination for dictyocaulosis. *Veterinary Record* 133, 579.

Jørgensen, R.J., Satrija, F., Monrad, J. and Nansen, P. (1992) Effect of feeding lucerne pellets on trichostrongyle infection in grazing heifers. *The Veterinary Record* 131, 126–127.

Leaver, J.D. (1970) A comparison of grazing systems for dairy herd replacements. *Journal of Agricultural Science, Cambridge* 75, 265–272.

Lindqvist, Å., Ljungstrøm, B.L., Nilsson, O. and Waller, P.J. (2001) The dynamics, prevalence and impact of nematode parasite infections in organically raised sheep in Sweden. *Acta Veterinaria Scandinavica* 42, 377–389.

McIlroy, S.G., Goodall, E.A., Stewart, D.A., Taylor, S.M. and McCracken, R.M. (1990) A computerised system for the accurate forecasting of the annual prevalence of fasciolosis. *Preventive Veterinary Medicine* 9, 27–35.

Mulcahy, G., O'Connor, F., Clery, D., Hogan, S.F., Dowd, A.J., Andrews, S.J. and Dalton, J.P. (1999) Immune responses of cattle to experimental anti-Fasciola hepatica vaccines. *Research in Veterinary Science* 67, 27–33.

Nansen, P., Foldager, J., Hansen, J.W., Henriksen, S.A. and Jorgensen, R.J. (1988) Grazing pressure and acquisition of *O. ostertagi* in calves. *Veterinary Parasitology* 27, 325–335.

Nansen, P., Steffan, P., Monrad, J., Grønvold, J. and Henriksen, S.A. (1990) Effects of separate and mixed grazing on trichostrongylosis in first- and second-season grazing calves. *Veterinary Parasitology* 36, 265–276.

Niezen, J.H., Waghorn, T.S., Waghorn, G.C. and Charleston, W.A.G. (1993) Internal parasites and lamb production – a role for plants containing condensed tannins? *Proceedings of the New Zealand Society of Animal Production* 53, 235–238.

Niezen, J.H., Waghorn, T.S., Charleston, W.A.G. and Waghorn, G.C. (1995) Growth and gastrointestinal nematode parasitism in lambs grazing either lucerne (*Medicago sativa*) or sulla (*Hedysarum coronarium*) which contains condensed tannins. *Journal of Agricultural Science, Cambridge* 125, 281–289.

Niezen, J.H., Charleston, W.A.G., Hodgson, J., Mackay, A.D and Leathwick, D.M. (1996) Controlling internal parasites in grazing ruminants without recourse to anthelmintics: approaches, experiences and prospects. *International Journal for Parasitology* 26, 983–992.

Niezen, J.H., Robertsen, H.A., Waghorn, G.C. and Charleston, W.A.G. (1998) Production, faecal egg counts and worm burdens of ewe lambs which grazed six contrasting forages. *Veterinary Parasitology* 80, 15–27.

O'Brien, D.J. (1999) Treatment of psoroptic mange with reference to epidemiology and history, *Veterinary Parasitology* 83, 177–185.

Peacock, R. and Poynter, D. (1980) Field experience with a bovine lungworm vaccine. *Symposia of the British Society for Parasitology* 18, 141–148.

Ploeger, H.W., Kloosterman, A., Borgsteede, F.H.M. and Eysker, M. (1990) Effect of naturally occurring nematode infections in the first and second grazing season on the growth performance of second-year cattle. *Veterinary Parasitology* 36, 57–70.

Roepstorff, A., Monrad, J., Sehested, J. and Nansen, P. (2000) Mixed grazing with sows and heifers – parasitological aspects. In: Hermansen, J.E., Lund, V. and Thuen, E. (eds) *Ecological Animal Husbandry in the Nordic Countries, DARCOF Report no. 2/2000*, pp. 41–44.

Singh, A. and Agarwal, R.A. (1988) Possibility of using latex of euphorbiales for snail control. *Science of the Total Environment* 77, 231–236.

Smith, K.E., Wall, R. and French, N.P. (2000) The use of entomopathogenic fungi for the control of parasitic mites, *Psoroptes* spp. *Veterinary Parasitology* 92, 97–105.

Smith, W.D., van den Broek, A., Huntley, J., Pettit, D., Machell, J., Miller, H.R.P., Bates, P. and Taylor, M. (2001) Approaches to vaccines for *Psoroptes ovis* (sheep scab). *Research in Veterinary Science* 70, 87–91.

Smith, W.D., Bates, P., Pettit, D.M., van den Broek, A. and Taylor, M.A. (2002) Attempts to immunize sheep against the scab mite, *Psoroptes ovis*. *Parasite Immunology* 24, 303–310.

Svensson, C., Hessle, A. and Höglund, J. (2000) Parasite control methods in organic and conventional herds in Sweden. *Livestock Production Science* 66, 57–59.

Thamsborg, S.M. (2000) Uptake of nematode infections in dairy breed steers reared at different stocking rates. In: Soegaard, K., Ohlsson, C., Sehested, J., Hutchings, N.J. and Kristensen, T. (eds) *Grassland Farming – Balancing Environmental and Economic Demands*. Vol. 5, *Grassland Science in Europe*. European Grassland Federation, pp. 561–563.

Thamsborg, S.M., Jørgensen, R.J., Waller, P.J. and Nansen, P. (1996) The influence of stocking rate on gastrointestinal nematode infections of sheep over a two-year grazing period. *Veterinary Parasitology* 67, 207–224.

Thamsborg, S.M., Roepstorff, A. and Larsen, M. (1999) Integrated and biological control of parasites in organic and conventional production systems. *Veterinary Parasitology* 84, 169–186.

Thamsborg, S.M., Søland, T.M. and Vigh-Larsen, F. (2001a) Klinisk hæmonchose hos får. *Dansk Veterinærtidsskrift* 84, 6–9.

Thamsborg, S.M., Mejer, H. and Roepstorff, A. (2001b) Sainfoin reduces the establishment of nematode infections in grazing lambs. *Proceedings of the 18th International Conference of the WAAVP, Stresa, Italy*, p. 117.

Van Wyk, J.A., Malan, F.S. and Bath, G.F. (1997) Rampant anthelmintic resistance in sheep in South Africa – what are the options? *Proceedings of the 16th WAAVP, August 1997, Sun City, South Africa.*

Waghorn, G.C., Gregory, N.G., Todd, S.E. and Wesselink, R. (1999) Dags in sheep; a look at faeces and reasons for dag formation. *Proceedings of the New Zealand Grassland Association* 61, 43–49.

15

Feeding for Health and Welfare: the Challenge of Formulating Well-balanced Rations in Organic Livestock Production

Werner Zollitsch,[1] Troels Kristensen,[2] Christian Krutzinna,[3] Finnain MacNaeihde[4] and David Younie[5]

[1]Department of Animal Science, University of Natural Resources and Applied Life Sciences Vienna, Gregor Mendel Strasse 33, A-1180 Vienna, Austria; [2]Department of Agricultural Systems, Danish Institute of Agricultural Sciences, Research Centre Foulum, PO Box 50, DK-8830 Tjele, Denmark; [3]Department of Animal Nutrition and Animal Health, The University of Kassel, Nordbahnhofstrasse 1a, D-37213 Witzenhausen, Germany; [4]Teagasc, Johnstown Castle Research and Development Centre, Wexford, Ireland; [5]Scottish Agricultural College, Craibstone Estate, Buckburn, Aberdeen AB21 9YA, UK

Editors' comments

Feeding is a major element of exchange and interaction between the two main components of the farm's agroecosystem: the fields and the animals. Thus, it is an area where the organic principles about harmony on all levels must be carefully thought through, because the interest of the individual animal may conflict with the interests of the whole farm. For example, the feeds that are best for the animals might not be the ones that grow best under the farm's particular conditions. The challenge of reconciling these possibly conflicting interests is different for different climatic regions of Europe. Yet feeding is central to the welfare of the animal. Therefore, the production of an appropriate and well-balanced feed ration is a core area of animal husbandry, one which the authors discuss from a broad perspective. They also consider one particularly controversial aspect of organic feeding: the use of amino acids for monogastric animals.

© CAB International 2004. *Animal Health and Welfare in Organic Agriculture* (eds M. Vaarst, S. Roderick, V. Lund and W. Lockeretz)

Introduction

A basic principle of organic agriculture is that livestock are kept as a part of the farming system, and that their nutrition therefore should be based on home-grown feeds (European Communities, 1999). This principle is intended to guarantee optimum feed quality. At the same time, the nutrient cycles will be kept intact on the levels of the farm, the region and even globally, because the amount of nutrients transferred between these levels will be strictly limited. Therefore, both the sustainability and the productivity of the farming system depend on the internal flow of nutrients as represented by feed and manure (Fig. 15.1). Although this approach implies that the health and welfare of organic livestock cannot be isolated from the whole system, various simplifications are made when discussing the impacts of nutrition on animal health and welfare. Despite these simplifications, a systematic approach is essential if animal health and welfare is to be understood and maintained in organic farming.

Ideally, one principle of organic systems is the maintenance of an optimal degree of diversity and a sufficient buffering capacity against potentially harmful external influences (Boehncke, 1997). In the past few decades, however, economic pressures and consumer demands have made it necessary even for organic farmers to select livestock with very high genetic potential for specific performance traits. Because of technical problems such as the shortcomings of selection within small populations and the complexity of breeding programmes, organic farmers often use the 'conventional genetic potential' of modern breeds, an issue discussed at length in Chapter 16.

In these animals, the genetic progress achieved in certain performance traits has not been accompanied by a similar increase in feed intake. This has resulted in a demand for feeds with a high concentration of the limiting nutrients. Considering, too, that feed can account for up to 50–60% of total production costs, the goal of providing livestock with

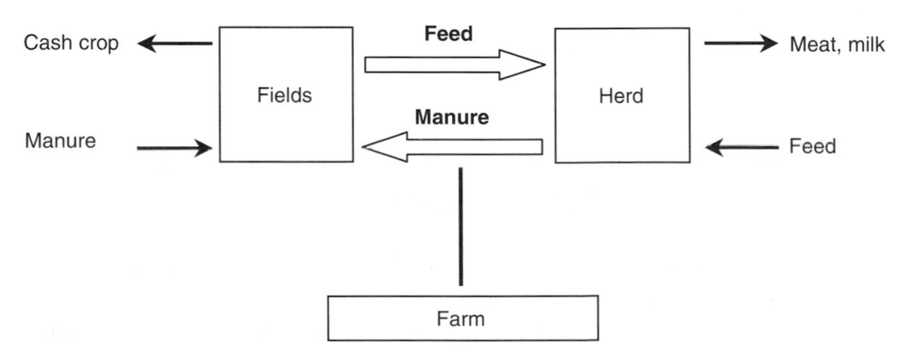

Fig. 15.1. Maintaining the sustainability of organic farming: interaction between crop production and herd feeding.

only organic feeds must be met in a manner that allows the animals' needs to be met without jeopardizing sustainability while also being economically feasible for the farmer. These key elements will define the framework for any strategy that needs to be developed to balance the flow of nutrients around organic animal husbandry.

Feeding Strategies for Organic Livestock

From the viewpoint of animal welfare as well as that of the farmer, organic feeding strategies should include:

- Providing the animal with a quantity of nutrients that is appropriate for a given level of production.
- At the same time, formulating diets using feedstuffs that allow the animals' physiology to function in an optimal way, i.e. to take into account the physiological peculiarities that are an outcome of evolutionary processes.
- Allowing appropriate feeding behaviour by offering naturally structured feedstuffs, employing animal-friendly feeding techniques, and providing a proper environment for feed intake.
- Achieving the goals and values intrinsic to organic farming, such as environmentally sound production methods, intact nutrient cycles, and low use of critical, non-renewable resources.
- Serving the farmers' economic interests.

Obviously, some of the objectives stated earlier contradict these principles somewhat. Given the availability of animals with a high genetic potential for certain performance traits, the animals' high requirements for essential nutrients often cannot be fulfilled with the strict use of farm-grown feeds of the types that the animals adapted to during their evolution. The high limits on the use of concentrates in ruminants' diets given in the EU regulation (European Communities, 1999), which clearly give priority to the goal of sufficient nutrient intake over the second and third objectives described above regarding the animals' physiology and behaviour, may serve as an example.

The Concept of Balancing Nutrient Supply and Nutrient Requirement

If the ultimate task of feeding is to maintain the animal's optimal health and welfare, then the concept of suggested nutrient requirements and animal response needs to be discussed. Usually, nutrient requirements are defined for a certain level of production, using factorial approaches (i.e. calculating requirements as a function of body weight, physiological status and level of performance, using established factors for use of nutrients and energy; ARC, 1988; NRC, 1994, 1998), or even for a certain economic return for a given livestock unit ('econometric feeding', Ahmad *et al.*, 1997). There are some suggestions that certain mineral and vitamin requirements for maintaining health status may be higher (e.g. Corah, 1996; Weiss, 1998), but the concept of 'feeding for animal welfare' has hardly been incorporated into the perspective of conventional animal nutrition. However, some nutrition-related welfare aspects are present in the EU regulation for organic livestock, such as the minimum period for which young mammals have to be fed with natural milk, the requirement of the maximum use of pasturage for herbivores, and the mandatory inclusion of forage in pig and poultry diets (European Communities, 1999; see also Chapter 4).

Even if organic farmers do not intend to feed animals to their full genetic potential, the adaptability of the animal to a variation in nutrient intake is not fully understood. It is obvious that severe undernutrition impairs animal welfare and health, but the question remains whether animal welfare may also be harmed if high yields are to be reached by a maximum nutrient supply and the use of feedstuffs that the animals are not adapted to. If the concept of a desired feed intake (Emmans, 1981) is valid, then in organic farming, which imposes limitations on certain feedstuffs, the use of animals with high genetic merit may pose a problem from the point of animal welfare. In contrast to this theory, moderate underfeeding accompanied by a reduction in yield may help the animal tolerate certain nutritional imbalances better. Furthermore, the range of dietary contents of specified nutrients that allow an animal to maintain health and welfare is influenced by several factors, such as the species' regulative capacity, age, duration of suboptimal supply and intake of other nutrients.

Mineral Nutrition

The availability–demand balance of minerals

These interactions may be demonstrated using minerals as an example. In general, all essential minerals must be present in balanced amounts in animal rations. (Essential minerals include the major elements P, K, Mg, Ca, S and Na, and the trace elements Fe, Mn, Cu, Zn, Mo, Se, Co, I and B.) However, as one would expect, growing animals – calves, lambs, piglets and chicks – have a greater demand for minerals and are more likely to suffer deficiency than mature animals (Rogers *et al.*, 1989; Rogers and Gately, 1992).

Besides the local variations that obviously exist for the mineral content of feedstuffs, the following factors will influence the balance between the amount of minerals required by an animal and the amount supplied by the ration:

- Availability of dietary minerals for the organism. For example, monogastric livestock may consume high amounts of P, but if it is present as phytate, only about 20–40% of it can be absorbed from the small intestine. Furthermore, under such conditions, Ca, Mg, Fe and Zn may be bound as phytates and therefore not be available for the animal.
- Adaptive capacity of the animals, which is based on homeostatic mechanisms of resorption and excretion. For example, the rate of resorption of Fe is significantly increased in cases of Fe deficiency, whereas a varying intake of F and I is balanced by opposite changes in the rate of excretion.
- Existence of body stores from which minerals can be mobilized to a certain extent during phases of sub-optimal intake, for example Ca and P from the skeleton, Cu and Fe from the liver.

The concentrations of minerals required in formulated livestock rations are higher than those normally required in pasture. This may be due to source-specific differences in availability of minerals. For example, Se extracted from perennial ryegrass is metabolized more readily if present in herbage than in inorganic salts (MacNaeidhe, 1992; Verkleij and MacNaeidhe, 1992). Mineral concentrations in animal feedstuffs are affected by many factors, but certain feedstuffs contain higher minerals than others, and judicious compounding can lead to a well-balanced ration (Scott, 1977; MacNaeidhe, 1989).

The ranges of minerals within well-balanced soil and herbage are given in Table 15.1.

Management factors affecting the content of minerals in herbage

Organic standards require ruminant diets to be based primarily on forage. It follows that the mineral and trace element supply of organic ruminants is directly influenced by forage composition and intake. Whilst mineral supplementation is permitted, the farmer is expected to manage the grassland to optimize the supply of minerals from the herbage produced on the farm. The following are some of the ways that management methods affect that supply.

Management effects on major elements

An excessive supply of K in the soil leads to the phenomenon of luxury uptake, in which K ions are taken up by the plants in preference to elements such as Mg and Ca. Deficiencies of these two minerals can result, causing hypomagnesaemia and milk fever, respectively. Herbage Na content is also reduced by excessive K and may lead to Na deficiency in livestock. The risk of luxury uptake of K is normally greatest in spring, when soil K levels are at their highest. Ammonium ions can have the

Table 15.1. Ranges of major and trace elements in a well-balanced soil[a].

Nutrient	Soil concentration (ppm)	Herbage concentration (dry matter basis, %)
Major elements		
P	4–9	0.25–0.50
K	70–150	2.5–4.0
Mg	80–200	0.11–0.20
Ca		0.50–1.45
S	25–40	0.17–0.30
Na		0.15–0.65
Trace elements	(ppm)	(ppm)
Fe		50–150
Mn	50–150	25–250
Cu	5–8	7–9
Zn	2–10	20–60
Mo	1.0–2.0	2.0–5.0
Se		0.03–0.30
Co	3.0–10.0	0.05–0.15
I		0.08–0.30
B	0.8–1.5	2.0–10.0

[a]The soil nutrient values are given as available levels (Byrne, 1979); herbage values are given as total concentrations.

same effect indirectly because they displace K ions from the clay mineral fraction in the soil, causing them to move into soil solution and thus become available for luxury uptake. Thus, the application of cattle slurry to grazed grass should be avoided, especially in spring, since it contains high levels of both K and ammonium-N.

A good Na supply can counteract the effects of low Mg, and hypomagnesaemia is less likely when the Na supply in the pasture is adequate (Rogers and Gately, 1992). The Na supply in soil and herbage is likely to be satisfactory in western coastal areas of NW Europe, but is likely to be deficient in sheltered eastern or inland areas. The Na supply can be boosted by applications of agricultural salt or by calcified seaweed.

An adequate supply of P is required in the diet to maintain the cow's fertility. Surveys have shown that soil P is often deficient in organic situations (Newton and Stopes, 1995). Also, the P content of herbage decreases as it matures. Therefore, it is important to monitor soil P regularly. If hay or silage is regularly cut late, the P content of the herbage also should be checked regularly.

Effect of N supply on dilution of trace elements

High N concentration in the soil causes a dilution of Cu, Zn and Se in the herbage (especially when these are at low concentrations in the soil), and thereby can harm the animals' health (Fleming, 1977; MacNaeidhe, 1991). Therefore, heavy applications of fresh farmyard manure and a high clover content in the pasture should be avoided, especially when the rate of soil N mineralization is high (May to July). Composted farmyard manure has a lower N concentration than fresh farmyard manure and should be preferred when heavy applications are required to restore the nutrient balance following the harvesting of a heavy nutrient-demanding crop such as silage. In contrast to conventional farmers, who can adopt a relatively short-term crop nutrient strategy using soluble mineral fertilizers and have no restrictions on nutrient supplementation in livestock diets, the manuring strategy on organic farms must be based on a holistic long-term approach, employing biological N-fixation by legumes, mineralization of nutrients from the soil and efficient internal nutrient cycling within the farm. These elements of the strategy must take account not only of nutrient dynamics but also of animal health.

Effect of soil pH, drainage, poaching and soil ingestion on availability of trace elements

Soil pH can have a major influence on the uptake of minerals by plants. On soils with inherently low levels of minerals, care should be taken to avoid over-liming. Increasing soil pH (e.g. by application of lime) decreases the uptake of Fe, Zn and Co, and increases the uptake of Se and Mo and, to a smaller extent, S. Induced trace element deficiencies can occur in livestock as a result of lower absorbability in the rumen, caused by increased intake of other minerals. For example, the absorbability of Cu in the rumen is reduced in the presence of raised levels of Mo and S on high pH soils.

Drainage conditions also can significantly affect the uptake of trace elements. Poor drainage increases the Mo concentration in pasture and hence the risk of induced Cu deficiency in livestock (Fleming, 1977). The concentration of Fe and Mn is increased in grass and clover in poorly drained soils (Mitchell, 1960; Walsh and Fleming, 1978). High concentrations of these two elements in the herbage can cause Cu and Co deficiency in livestock.

Herbage is poorly anchored in poorly drained soil. When grazed, the whole plant and the soil adhering to the root may be eaten and digested by the animal. Poaching by livestock also increases soil ingestion. Soil ingestion is an important cause of mineral imbalance in livestock nutrition. Soils sometimes contain much higher concentrations of some minerals than do plants. Ingestion of soils with high Fe may induce Cu and Zn deficiency, whilst soils with high Zn may cause Cu and Fe deficiency. Similarly, soils with high Mo can induce Cu deficiency, whilst soils with high Mn can induce Co deficiency (Suttle *et al.*, 1975).

Overstocking and overgrazing of pasture also leads to increased soil ingestion, especially at the end of the grazing season. At a stocking rate of 30 lambs per hectare, two to six times more soil may be ingested than at a stocking rate of 20 lambs per hectare during September and October (McGrath *et al.*, 1982).

Effect of forage herbs on mineral content of herbage

Traditionally, organic farmers have looked favourably upon secondary or indigenous grass species and forage herbs, whether in unsown pastures and field boundaries or sown in complex seed mixtures. Improved livestock nutrition and health are the reasons normally given. Several species of herbs have deep tap roots and can reach deeper soil layers, exploiting a greater volume of soil than conventional herbage species and potentially enhancing the supply of nutrients (Wilman and Derrick, 1994). These potentially higher levels of minerals in herbage translate into higher mineral status in animals (e.g. Younie *et al.*, 2001). The presence of secondary plant products (e.g. aucubin, a natural antibiotic, in *Plantago lanceolata*) may improve livestock health in organic farming, but be harmful for other processes, such as silage fermentation (Isselstein and Daniel, 1996). Table 15.2 summarizes the differences in mineral content among perennial ryegrass (*Lolium perenne*), white clover (*Trifolium repens*), and a range of weeds and forage herbs.

Table 15.2. Mineral content of some common grassland herb species in comparison with perennial ryegrass (Swift *et al.*, 1990).

	P	K	Ca	Mg	Na	Cu	Co
Perennial ryegrass							
g/kg dry matter (DM)	3.0	20.0	5.0	2.0	1.8		
mg/kg DM						8.5	1.1
Broad-leaved dock	**	**	*	***	*	*	*
Chickweed	***	***	*	**	***	*	*
Stinging nettle	***	***	***	***	*	***	*
Creeping buttercup	***	***	***	*	***	**	*
Creeping thistle	***	***	***	**	*	***	–
Dandelion	*	**	***	**	***	***	*
Chicory	***	***	***	***	***	**	**
Ribwort plantain	*	**	***	**	**	**	*
Burnet	*	*	***	***	–	*	*
Sheep's parsley	**	***	***	***	**	NA	NA
Yarrow	**	***	***	**	–	**	–
White clover	*	*	***	**	**	***	**

– less than perennial ryegrass;
* similar to perennial ryegrass;
** more than perennial ryegrass;
*** much higher than perennial ryegrass;
NA, not available.

Despite these potential advantages, forage herbs have largely been overlooked in practice, even in organic farming (Foster, 1988). This is changing to some extent, at least in New Zealand and Australia, where a breeding effort has led to the commercial release of varieties of chicory and ribwort (Moloney and Milne, 1993). Perhaps the main reason for the low use of forage herbs is their generally low contribution to sward biomass, largely resulting from low seed rates and their poor establishment and persistence. New thinking is required regarding the companion species and mixtures of forage herbs most appropriate to the intended grazing management regime. For example, Umrani (1998) has shown that perennial ryegrass, with its vigorous, densely tillered growth habit, is too competitive. The upright and less well-tillered growth habit of timothy makes it less competitive, and coupled with the N-fixing ability of white clover makes for a more suitable mixture, perhaps sown in strips or as separate swards, with the main bulk of the grassland on the farm sown to a ryegrass-based mixture.

Dairy Cows

From a nutritionist's point of view, various issues need to be addressed if the feeding of organic dairy cows is to be optimized:

- Level of production.
- Temporal variability of nutrient supply compared with nutrient requirements.
- Nutrient balance.
- Maintenance of normal ruminant function.

Obviously, the probability of nutrition-related problems increases with increasing level of production, decreasing forage quality and a lack of home-grown feedstuffs with a high nutrient concentration. Generally, the risk of a severe undersupply of nutrients will be at a maximum during the first 2 months of lactation, when the daily milk yield increases considerably faster than feed intake. During this time, the cow is balancing the energy deficit by mobilizing adipose tissue. When lipid mobilization exceeds physiologically defined limits, which are associated with an energy deficit of about 15%, the risk for metabolic disorders increases markedly (INRA, 1989; Knaus *et al.*, 2001), an area that the NAHWOA Recommendations call for more research on.

Problems may occur not only as insufficient nutrient intake *per se*, but also in connection with an unbalanced supply of different nutrients. In organic farming, the protein intake will be in excess relative to the energy intake during certain phases of the vegetation period. The magnitude of this imbalance will depend, among other things, on the soil type, the proportion of legumes in the diet, and the availability of feedstuffs with a relatively high energy content (Baars, 1998; Van Eekeren, 2000). However, in several experiments, overfeeding cattle with protein did not result in acute metabolic diseases or in an immune suppressive effect (Sundrum, 1997). Not quite clear, on the other hand, are the health effects of protein deficiencies, which can be observed, for instance, on farms that focus on cheese production and therefore practise a strictly hay-based feeding regimen.

Milk fever can contribute significantly to health problems in organic dairy herds (Table 15.3; Krutzinna and Boehncke, 1997). Among other causes, a surplus of dietary cations during the dry period is thought to play a major role in causing this metabolic disorder (Horst *et al.*, 1971; Henriksen, 2000). Because of the lower use of inorganic fertilizers and greater diversity of plants on organic grassland, it can be expected that the dietary cation–anion balance (DCAB; $Na^+ + K^+ - Cl^-$) will be systematically different between organic and conventional farms. Nevertheless, it is not yet known to what extent the incidence of milk fever will be affected by changes in DCAB through organic farming (Henriksen, 2000). Nevertheless, the possibility of increased K levels (4–5% on a dry matter basis) in roughage harvested during the conversion period (Krutzinna and Boehncke, 1989) should be considered a risk factor.

Table 15.3. Incidence of certain diseases on 15 organic farms (1991 to 1995, *n*=1547 calvings (Krutzinna and Boehncke, 1997).

Disease	No. of cases	Overall incidence[a] (%)	Incidence by farm Minimum (%)	Incidence by farm Maximum (%)
Milk fever	68	4.4	0.8	13.6
Ketosis	17	1.1	0.0	5.9
Retentio placentae	129	8.3	2.8	14.2
Endometritis	146	9.4	2.1	19.7

[a] Number of cases/total number of calvings.

In general, the natural variation in feed production caused by external factors may have a specific effect on organic farms, which are limited with regard to feed imports (Olesen *et al.*, 1999). Typically, organic dairy farmers will focus on forage and pasture systems and use less concentrates and mineral supplements than comparable conventional farmers. Under these circumstances, nutrition will probably limit milk production and eventually affect the milk's nutrient content (Krutzinna *et al.*, 1996; Kristensen and Kristensen, 1998; Schwarzenbacher, 2001). The question remains whether, or by which factors, the incidence of metabolic disorders will be systematically influenced after conversion to organic farming (Krutzinna *et al.*, 1996; Henriksen, 2000; Schwarzenbacher, 2001).

Strategies that may help to overcome (seasonal) nutrient shortages include:

- Increasing forage nutrient content.
- Adaptation of the crop rotation.
- Optimizing management measures.
- Temporary import of specific supplementary feedstuffs.

Increasing forage nutrient content

A field survey conducted by Schwarzenbacher (2001) shows that both the milk yield and the milk protein content may be increased when the forage is harvested earlier. The most important reason for this is an increase in digestibility and hence in energy content, which also results in an increase in feed intake when plants are cut earlier in the vegetation period (Van Soest, 1994; Gruber *et al.*, 2000). A model calculation presented by Knaus *et al.* (2001) shows that in organic farms that rely on permanent grassland, yearly milk yield per cow will decrease by about 1000 kg if the energy concentration in forage dry matter decreases by 0.5 MJ net energy of lactation (NEL). This variation in feedstuff quality also affects the cow's protein supply because of the correlation between forage protein and energy content and because the ruminal microorganisms' protein synthesis rate depends on the energy supply (Van Soest, 1994). Therefore, diets based on high-quality forage allow high milk yield with a limited use of concentrates (Fig. 15.2).

Nevertheless, it must be emphasized that in organic farming systems that rely on permanent grassland, the need to maintain a stable pattern of species among the plant communities will pose a clear limit on increasing the number of harvests and the digestibility of organic matter by cutting earlier (Dietl, 1986). Intensifying the use of permanent grassland above a

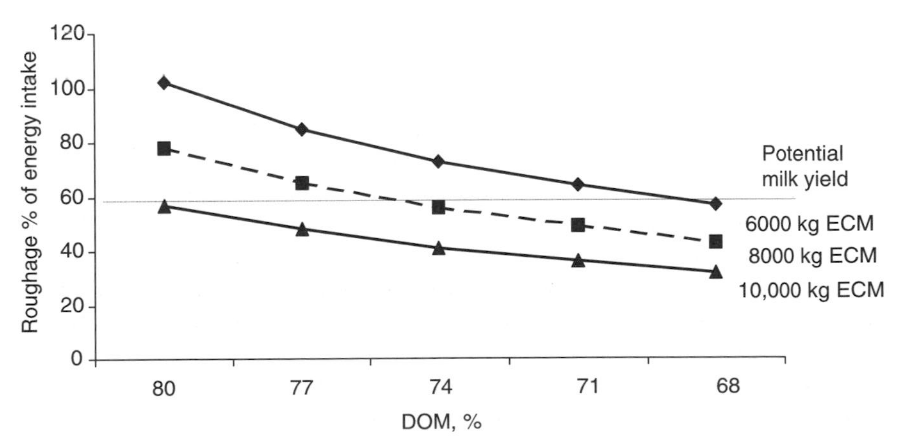

Fig. 15.2. Effects of milk yield (ECM; energy corrected milk) and digestibility of organic matter (DOM) on the amount of roughage in the diet of dairy cows.

certain limit will significantly reduce biodiversity and thereby endanger the sustainability of land use. Climate and soil-related factors impose among the most important limits on grassland use.

Adaptation of crop rotation

Feedstuffs such as fodder beets, whole-plant grain silage and potatoes will help to increase feed intake and optimize the energy supply of high-yielding dairy cows. The versatility of dietary components will be higher as the proportion of arable land available on-farm increases (Krutzinna *et al.*, 1996). However, the option of adapting the crop rotation to increase the supply of certain feedstuffs will be limited: one must consider the economic superiority of crops produced for human nutrition and the competition of other livestock species for these feedstuffs.

Optimizing management measures

Overcoming seasonal nutrient shortages can be helped by prolonged use of pasture, together with improved pasture management, appropriate timing of calving and a flatter curve of milk yield over the weeks of lactation (which would help to avoid short-term peaks in nutrient requirement and thereby reduce the probability of severe nutrient deficiencies), through an altered feeding strategy (Kristensen and Kristensen, 1998; Olesen *et al.*, 1999). Basically, the way the forage is used (pasture, stall-feeding, silage conservation) together with the specific timing of the harvest can help to improve the balance between energy and protein intake from specific forages (Van Eekeren, 2000). However, the farmer must evaluate the potential harm that these measures may exert on the whole system.

Temporary use of specific supplementary feedstuffs

Because organic farms rely on home-grown feedstuffs much more than conventional farms, low feedstuff quality, which may temporarily occur because of unfavourable conditions, will be important for the nutrition of dairy cows (Olesen *et al.*, 1999). Temporary nutrient deficits can be covered by using supplementary feedstuffs such as concentrates or different mineral sources.

For organic farms based on grass–clover mixtures, the typical oversupply of protein relative to energy that occurs as the proportion of clover increases during the vegetation period can be reduced by supplementing the diet with energy-rich feeds such as wilted silages with a lower clover content from an earlier cut, maize silage, fodder beets or concentrates. The urea content of the bulk milk can serve as an important tool for the correct timing of supplementary feeding (Van Eekeren, 2000), because together with the percentage of milk protein it indicates a relative surplus of protein over energy in the diet.

If the overall forage quality is low, the protein supply of ruminants will be negatively affected apart from low energy intake. Depending on the climate, such a situation is likely in permanent grassland systems in which hay is a predominant forage. Cobs or pellets produced from lucerne–grass or clover–grass mixtures can serve as a high-quality supplementary feedstuff under these circumstances (Krutzinna *et al.*, 1996; Thurner, 2001), which may also help to improve the intake of vitamins and minerals. Nevertheless, the question of overall energy efficiency and the import of energy (especially from non-renewable sources) into the production system must be kept in mind when choosing supplementary feedstuffs for organic dairy cows: system modelling has shown that in terms of energy costs, grains may be more favourable than grass pellets (Refsgaard *et al.*, 1998).

Depending on the type of soil and the available feedstuffs, the mineral supply of dairy cows may be deficient. Baars and Opdam (1998) showed that Se deficiency of heifers fed on a grass–red clover–lucerne mixture was among the causes of increased individual somatic cell counts. High contents of N and K in the diet can increase the risk of hypomagnesaemia in livestock turned out to pasture in spring. In such cases, administering trace element preparations obviously can improve the situation. However, alternative approaches to this problem still need to be analysed and developed, including measures that support the animal's self-regulating abilities.

Beef Cattle

Although beef cattle production can be economically favourable on organic farms (MacNaeidhe and Fingleton, 1997), very little information about specific aspects of beef cattle husbandry in organic farming has been published so far. In general, pasture will be the preferred feeding system for beef cattle. Pasture management, optimum stocking density, and the right choice of supplementary feedstuffs for meeting the animals' needs will be crucial for achieving the full potential of pasturage systems (Frankow-Lindberg and Danielsson, 1997).

The details of organic beef production systems vary among different European regions, as seen in Chapter 3, but in general suckler cows will play an important role. Because the growth rate of beef cattle in organic systems will be lower than in conventional farming, nutritional limits and related health problems are less likely to occur, as was found in a survey in Germany (Tenhagen *et al.*, 1998). An exception may be the surplus of protein relative to energy, as was already discussed regarding dairy cows, and a lack of minerals in some regions (Boehncke, 1997; Tenhagen *et al.*, 1998). However, the oversupply of protein will not necessarily harm the animals' health (Sundrum, 1997).

Significant correlations between the Se status of beef cattle and the mineral content of forage (Groce *et al.*, 1995) clearly indicate that a deeper understanding of the complexity of mineral availability is necessary for organic solutions to these problems. Until this has been achieved, farmers will have to stay with available measures of supplying adequate amounts of minerals for beef cattle (Bruce *et al.*, 1998).

Sheep

Sheep are kept in very varying systems throughout Europe, examples of which were described in Chapter 3. Therefore, sheep production will require different management and nutrition strategies, depending on the specific conditions on the farms in the region. In general, the level of production correlates with the availability of grassland of high productivity, as sheep are grazed mainly on pasture, and pasture herbage is by far the most important source of nutrition for these animals. However, depending on the production level, concentrates may be given to sheep under certain conditions, mainly as an energy supplement. The protein requirement of the animals will be mostly covered by forage, although there are situations (in late pregnancy and early lactation), in which the feeding of supplementary protein can have a significant effect on animal health by reducing ewes' peri-parturient faecal egg count of GI worms (see Chapter 14).

Organic sheep production systems can roughly be grouped into three different levels of productivity:

- Extensive systems mainly for meat production: these systems can be found in regions where large but less productive grassland areas are available; both the production costs and the productivity of the sheep are relatively low, but will be increased if the forage ration is supplemented with concentrates. Concentrates will be given mainly around the time of ovulation and early lactation, with the amount depending on the forage quality. As an example, pregnant ewes from lowland flocks are given concentrates when introduced to housing before lambing. Concentrates are given at 0.1 kg per day approximately 7 weeks before lambing, and are increased to 0.7 kg in the final week (MacNaeidhe *et al.*, 1996). During suckling, the feed intake of the ewe is not sufficient to replace the energy lost in milk production, and this supplementation is necessary to ensure that body condition is built up so as to maintain the health and welfare of lambs and sheep. Due attention should be given to the mineral supply, which may impair animal health and welfare if the mineral content of the forage is insufficient because of soil conditions.

- Moderately intensive systems: sheep produce either meat or relatively low amounts of milk. Their nutrition is based on more intensely managed grassland (paddock systems) or higher amounts of concentrates given to ewes, lambs and dairy sheep. Concentrates not only increase growth of lambs, but also affect the product quality. If lambs reach their slaughter weight earlier and their carcasses contain an optimum amount of fat, the taste of the meat will be relatively mild. This effect may be of special importance for certain markets, such as where consumers dislike the typical taste of mutton.
- Intensive systems for milk production: since sheep are better than cattle in coping with high amounts of concentrates (Van Soest *et al.*, 1994), their diets may contain 40% and more of concentrates, especially in early lactation (first 100 days). One of the most urgent problems involves the feeding of lambs. As the milk has a high market value, farmers are not interested in raising lambs on maternal milk unless they are needed as replacements. Feeding strategies need to be designed that guarantee that the lambs' needs are met and that are economically advantageous.

Out-of-season lamb production is difficult in organic flocks. Conventional methods such as administration of hormones to artificially induce and hence synchronize ovulation are not permitted under the organic standards. As a result, the supply of lamb during the period January to May is scarce. However, the use of breeds with an extended or non-seasonal period of reproduction, together with well-planned reproduction management, will help to extend the market supply. This approach could be supported by overwintering lambs in housing or outdoors, and giving additional feed so as to supply the market during January to May (Cole, 1966). However, these methods of production may entail high costs.

Pigs and Poultry

In accordance with the principles of organic farming, pigs and poultry are unlikely to play a role similar to their role in conventional farming. Among the reasons for this are that:

- Most monogastric animals use high-fibre feedstuffs very inefficiently.
- Therefore, they require feedstuffs with highly digestible organic matter, which usually are not available in sufficient amounts on organic farms, or which are otherwise sold for human consumption.

- Most plant proteins are deficient in one or several amino acids that are essential for these omnivorous species. With the very limited availability of feedstuffs of animal origin (an issue that the NAHWOA Recommendations call for further consideration of), nutrient supplementation, especially for amino acids, from home-grown feedstuffs will probably limit the performance of modern pigs and poultry.

Compared with ruminants, the nutritional status of pigs and poultry may be more homogeneous across Europe because there is a smaller variation in the range of potential feedstuffs for monogastric animals on organic farms. Currently, deficiencies in organic feeding strategies for pigs and poultry may be frequent because of a lack of information concerning the following aspects:

- Nutrient content of typical organic feedstuffs (species, varieties) for monogastric animals.
- Actual feed intake of pigs and poultry on the specific farms.

Balancing protein intake

Especially regarding amino acids, information about the actual feed intake of organic pigs and poultry is needed to optimize the diet. Currently, conventional recommendations are used for formulating diets. Because of the shortcomings mentioned above, this procedure will not give a realistic picture of the actual nutritional status of the animals.

In pig and poultry nutrition, the concept of an 'ideal protein' (i.e. one with the recommended ratios of lysine and other essential amino acids) is used for defining the optimal amino acid intake (Rose, 1997; NRC, 1998). If these amino acid ratios are compared with the amino acid contents of the most important organic protein feedstuffs (peas, faba beans, lupins, soybeans), it is obvious that there is a lack of organic protein sources with sufficient content of several essential amino acids, especially methionine and cystine. Organic feedstuffs that may be helpful in balancing the amino acid supply of organic pigs and poultry (e.g. potato protein, maize gluten, oilseed cakes, yeast, dried milk products) are not available in sufficient amounts and are therefore relatively expensive. A feasibility study for Austria revealed that even with a well-developed organic market, the availability of those feedstuffs from organic origin cannot be expected to increase dramatically in the near future (Wlcek, 2002). Therefore, including a certain percentage of conventional high-protein feedstuffs will probably be necessary if supplementation of certain amino acids is to be achieved. Besides the technical problem of avoiding nutrient imbalances by including certain feedstuffs in the diet, it will be necessary to develop organic strategies for pigs and poultry in regions where feedstuffs that are rich in protein typically cannot be grown because of environmental conditions, as in parts of Scandinavia.

The basic consequences of the consumption of highly unbalanced protein are well known: amino acids that are in surplus relative to the first limiting amino acid (probably lysine in pigs and methionine in poultry) are broken down and urea and uric acid are synthesized and excreted. Obviously, the performance of the animal will be affected as protein synthesis and feed intake are depressed after ingestion of diets with an unbalanced amino acid pattern (Forbes, 1986; D'Mello, 1994). Currently, it can be only speculated whether these processes pose a risk to animal health in organic farming, where the protein supply can be expected to be somewhat unbalanced.

Despite these uncertainties, an intensive discussion is currently taking place about the potential supplementation of diets of organic monogastric livestock with synthetic amino acids. In view of the general shortage of organic concentrate feedstuffs that are rich in protein, synthetic amino acids would undoubtedly simplify diet formulation. However, long-term consequences have to be taken into account. A key element in the discussion is the trend towards more intensive animal production systems, which will probably be fostered by the widespread use of isolated nutrients in general in organic diets, and specifically of synthetic amino acids; because of this concern, the NAHWOA Recommendations favour continuing the exclusion of synthetic amino acids. Obviously, farmers see advantages evolving from increasing growth and egg production and from the reduction of feed costs. Nevertheless, the pace of change towards more intensive production systems will be increased, since the use of synthetic amino acids would allow the introduction of genotypes with an even higher genetic potential for growth and egg production.

At the same time, the aim of organic farming to prohibit substances produced by processes involving genetically modified organisms will probably restrict the use of synthetic amino acids to DL-methionine. In most situations, this means that the practical relevance of this measure will be limited to the feeding of laying hens. For most other types of monogastric livestock, lysine rather than methionine will be the first limiting amino acid.

Nutrition of Organic Pigs

Because of their low nutrient requirements and their relatively high feed-intake capacity, formulating diets for non-lactating sows is not a serious problem for organic farmers. Different kinds of forage with a moderate fibre content contribute to the nutrient intake, allow the sows to perform their natural behaviour, and hence contribute positively to their health status.

Sows and piglets

The nutrition of the lactating sow and growing pigs in the earlier stages is much more critical because of their limited feed-intake capacity and relatively high nutrient requirements. The nutrient supply of the sows is of special importance, as nutrient deficiencies lower milk yield and therefore impair the development of the litter (King, 1998). It can be expected that the nutrient demands of lactating sows can be met more easily and that forage can be incorporated into the daily diets to a greater extent if the sows have a higher feed-intake capacity. As shown by Kongsted *et al.* (1999), there is a considerable individual variation in the consumption of whole-crop silage by lactating sows that have limited access to concentrates.

In the nutrition of sows and piglets, phase feeding (that is, adjusting dietary nutrient densities to changing nutrient requirements during different physiological stages) may help to solve some of these problems, because under the current regulations feedstuffs with a high feeding value may be used during the most critical phases (European Communities, 1999). A recent survey showed that fewer diets for phase-fed lactating sows were below the minimum acceptable lysine:energy ratio than when the sows got the same diet in gestation and lactation (Zollitsch *et al.*, 2000). In the absence of information about the actual feed intake, ratios of essential nutrients to energy seem to be relevant indicators for balanced diets, as dietary energy content is an important regulative factor for feed intake in sows (Koketsu *et al.*, 1996; Kemm *et al.*, 1997). In general, it can be expected that the problem of insufficient amino acid intake can be most prevalent in piglets, because feed intake of the litter may not be sufficient unless all factors associated with feeding are close to optimal. One strategy may be to allow a slightly prolonged suckling period: the body condition of piglets from farms where the litters were not weaned until they are 7–8 weeks old was better than that of piglets weaned at about 6 weeks (Andersen *et al.*, 1999; Leeb, 2000). The NAHWOA Recommendations cite optimal weaning age under various conditions as an area for additional research, including the identification of breeds of sows capable of suckling for the optimal period.

Growing/finishing pigs

Similar nutritional problems as were discussed for piglets can be expected at least for the growing phase. The following shortcomings in the nutrition of organic growing/finishing pigs can be expected to be relevant (Thielen and Kienzle, 1995; Wurzinger, 1999; Jakobsen and Jensen, 2000):

- Insufficient amino acid supply.

- Lack of phase feeding.
- Insufficient supply of major and trace elements.
- Unsatisfactory microbial status of the feed.

The amino acid supply of growing/finishing pigs can be optimized by importing protein sources high in the limiting amino acid, or it can be based solely on plant protein from home-grown feedstuffs. The first strategy is currently used by most organic pig producers. In contrast, the latter will be moderately detrimental for growth in an earlier stage and for lean meat percentage, while the intramuscular fat content, an important quality characteristic, clearly will be improved, as found by Sundrum *et al.* (2000). That study emphasized the need to evaluate critically the indicators used in defining nutrient requirements, and to examine their relevance for organic animal husbandry.

The hygienic status of feed may be a specific concern in organic farming. There is a risk that grains produced on farm for feeding purposes may not be properly cleaned and dried before storage. This will lead to microbial growth and deterioration of feedstuffs (Thielen and Kienzle, 1995; Wurzinger, 1999).

Because of their digestive systems, growing pigs cannot use significant amounts of forage, which are elements of the crop rotation on organic farms. Therefore the contribution of forage to the nutrient supply of growing pigs will be marginal (Danielsen *et al.*, 1999). However, these feedstuffs may have positive effects on animal welfare and can be used for an animal-friendly restriction of energy intake in finishing pigs (Bellof *et al.*, 1998; Sundrum *et al.*, 1999). However, the potentially harmful effects of forage on the composition, consistency, and oxidative stability of the animals' adipose tissue must be taken into consideration (Fischer and Lindner, 1998).

Nutrition of Organic Poultry

Before nutritional strategies for organic poultry are analysed, the genetic potential of the birds used in organic agriculture must be discussed. Typically, broiler hybrids used in organic farming have a much lower genetic potential for protein growth than conventional broilers. Therefore, the production of organic meat-type chickens can be considered to be in agreement with the demands formulated in the EU regulation (European Communities, 1999). On the other hand, truly 'alternative' breeds of laying hens are not available at the moment, and most European organic farmers rely on conventional hybrids. The extremely high genetic potential for egg production, together with the limitations of organic farming, will probably affect the health and welfare of the hens. For turkeys, these problems will be even more pronounced in most European countries.

Several specific aspects of the nutrition of organic poultry need to be addressed:

- The high requirements for protein and especially for sulphur-containing amino acids are a specific limitation in organic agriculture, as the most important organic protein feedstuffs (legumes) are low in these.
- Organic farms that focus on poultry production will probably have to import feedstuffs with a high nutrient density.
- Because of this and because of the importance of optimizing feed structure (colour, size and shape of particles), organic poultry diets will mostly have to be produced in commercial feed mills rather than on the farm.
- Except for water fowl and older turkeys, poultry do not use high-fibre feedstuffs well.
- Because of the relatively low feed intake of poultry, shortcomings in diet formulation usually depress their performance significantly and will probably contribute to health and welfare problems such as feather pecking and cannibalism, as discussed at length in Chapter 8.

As with other species, poultry's ability to adapt to a sub-optimal nutrient supply is not well known. Growing poultry will probably respond to variations in amino acid intake by changing protein accretion, and unless their protein intake is extremely low, their health and welfare are not likely to deteriorate. For laying hens, which usually have an extremely high genetic potential, protein deficiencies may pose a much higher risk to health and welfare. Insufficient provision of amino acids will act as a metabolic stress factor and has therefore been considered as impairing animal welfare (Hadorn *et al.*, 2000). The genetic 'improvement' of laying hens during the last decade was paralleled by a decrease in body weight and feed intake (Preisinger and Flock, 1998). In field surveys, laying hens had an average daily feed intake of 120 to more than 160 g (Wiener, 1996; Strobel *et al.*, 1998; Ingensand and Hörning, 1999; Staack *et al.*, 1999). Low feed intake is especially significant in organic farming, because feedstuffs with high nutrient and energy density are not available in great amounts and the need to formulate diets with a high nutrient density greatly reduces the flexibility needed to incorporate the variety of feedstuffs produced in organic farming.

Furthermore, feed structure is extremely important for nutrient intake and the feeding behaviour of laying hens (Trei *et al.*, 1997; Walser, 1997). This will be another limitation for diet formulation, and every measure must to be taken to optimize this feed characteristic.

Currently, feedstuffs such as potato protein and maize gluten are used to optimize the amino acid pattern of the dietary protein. However, these (conventional) feedstuffs may not be allowed in the future. Therefore, alternative feeding strategies based on organic feedstuffs for alternative poultry breeds need to be developed.

Conclusions and Prospects

Nutritional status, among other factors, will affect the health and welfare of animals in any farming system. In most systems, however, nutrition is mainly viewed in connection with productivity and efforts to combat diseases.

According to the principles of organic farming, however, livestock in general and especially their feed sources must be seen as part of the agroecological system if their health and welfare are to be optimized. The feeding strategy must therefore be considered in connection with the whole farming system and hence with several factors such as the productivity and the potential of the farmland, the genetic background of the animals, the husbandry system, and herd management.

Unlike in other farming systems, external inputs that could help to buffer internal shortcomings will clearly be limited in organic farming. Any deficit should primarily be understood as an indicator of imbalances in the whole system. This will require greater efforts to compensate for the deficits, but in the long term will contribute to a more stable system with a high buffering capacity.

Organic feeding strategies for any species should be based on maximum and efficient usage of home-grown feed. Especially for animals with a high genetic potential, all measures must be used that contribute to a high feed and nutrient intake, such as optimal crop rotations, pasture management, harvesting and conservation technology, and feeding techniques.

Despite this holistic approach, the temporary use of external inputs such as supplements of certain critical major and trace elements currently seems to be necessary to avoid harming the animals' health and welfare. To optimize the farm-based nutrition of organic livestock, more research is needed that is directed towards understanding better the internal nutrient flows that affect the animals' nutrient supply.

If the direction and pace of current activities in animal breeding are maintained, the increased genetic potential of livestock for productive traits will present an even greater challenge for organic systems from a nutritionist's point of view. Unless restrictions on importing synthetic nutrients into the system are loosened, organic strategies must be developed that offer a solution for this basic problem. New combinations of certain elements of organic systems should be an integral part of such a strategy, such as different livestock species within a system to lower the pressure on limited resources. In this connection, a key research area identified in the NAHWOA Recommendations was natural sources and appropriate feeding regimes to provide adequate amounts of vitamins and amino acids.

Despite organic farming's emphasis on home-grown feeds, the dynamic development of organic markets currently seen throughout

Europe may aggravate a situation in which organic producers try to solve problems by importing limiting nutrients into the system. Developing guidelines for this situation will need joint efforts by organic farmers, their organizations, extension personnel, scientists and consumers, always keeping in mind the basic concepts of organic farming.

These aspects ultimately lead to the question of the strategic development of organic animal husbandry. The current tendency of intensification of production on organic farms will eventually lead to a division of organic producers: traditional organic farmers focusing on balanced systems may be faced with another group who are intensifying production and who therefore will compete successfully in the market. Strategies to cope with this scenario are essential for the future development of organic farming.

The role of nutrition in a broader sense (including feed quality, nutrient supply, feeding techniques and other feed-related factors) in affecting animal health and welfare is still poorly understood. Current investigations on how nutrition affects the immune system are a step in the right direction. Ultimately, a holistic analysis of the whole complex will combine with an understanding of the specific quality of organic food from animal origin.

References

Ahmad, H.A., Bryant, M.M., Kucuktas, S. and Roland, D.A., Sr (1997) Econometric feeding and management for first cycle phase two DeKalb Delta hens. *Poultry Science* 76, 1256–1263.

Andersen, L., Jensen, K.K., Jensen, K.H., Dybkjaer, L. and Andersen, B.H. (1999) Weaning age in organic pig production. In: Hermansen, J.E., Lund, V. and Thuen, E. (eds) *Ecological Animal Husbandry in the Nordic Countries. Proceedings from NJF-seminar No. 303.* DARCOF (Danish Research Centre for Organic Agriculture), Tjele, pp. 119–123.

ARC (1988) Agricultural Research Council. *The Nutrient Requirements of Ruminant Livestock.* Commonwealth Agricultural Bureaux, Slough, UK.

Baars, T. (1998) Future systems in organic farming: grassland and fodder production. In: Isart, J. and Llerena, J.J. (eds) *The Future of Organic Farming Systems. Proceedings of the Fourth ENOF Workshop.* The European Network for Scientific Research Coordination in Organic Farming, Edinburgh, UK, pp. 39–48.

Baars, T. and Opdam, A. (1998) *Intake of Trace Elements in Relation to Individual Somatic Cell Counts in Heifers. A Participatory Study to Improve Animal Health in Organic Farming.* Louis Bolk Institute, Driebergen, The Netherlands.

Bellof, G., Gaul, C., Fischer, K. and Lindermayer, H. (1998) Der Einsatz von Grassilage in der Schweinemast. *Züchtungskunde* 70, 372–388.

Boehncke, E. (1997) Rindfleischproduktion unter den Bedingungen des ökologischen Landbaus. In: *Proceedings of the Satellite Symposium I in connection with the 48th Meeting of the EAAP, Beef Production and Special Aspects of Beef Quality, Vienna, Austria.* ZAR, Vienna, pp. 50–59.

Bruce, L.B., Torell, R.C. and Kvasnicka, W.G. (1998) Maintaining adequate blood selenium levels in deficient beef cows. *Large Animal Practice* 19, 18–20.

Byrne, E. (1979) *Chemical Analysis of Agricultural* Materials. An Foras Taluntais, Dublin.

Cole, H.H. (1966) *Introduction to Livestock Production.* W.H. Freeman and Company, San Francisco, California.

Corah, L. (1996) Trace mineral requirements of grazing cattle. *Animal Feed Science and Technology* 59, 61–70.

Danielsen, V., Hansen, L.L., Møller, F., Bejerholm, C. and Nielsen, S. (1999) Production results and sensory meat quality of pigs fed different amounts of concentrate and ad lib. clover grass or clover grass silage. In: Hermansen, J.E., Lund, V. and Thuen, E. (eds) *Ecological Animal Husbandry in the Nordic Countries. Proceedings from NJF-seminar No. 303.* Danish Research Centre for Organic Agriculture, Tjele, Denmark, pp. 79–86.

Dietl, W. (1986) Pflanzenbestand, Bewirtschaftungsintensität und Ertragspotential von Dauerwiesen. *Schweizerische landwirtschaftliche Monatshefte* 64, pp. 241–262.

D'Mello, J.P.F. (1994) Amino acid imbalances, antagonisms and toxicities. In: D'Mello, J.P.F. (ed.) *Amino Acids in Farm Animal Nutrition.* CAB International, Wallingford, UK, pp. 63–98.

Emmans, G.C. (1981) A model of the growth and feed intake of ad libitum fed animals, particularly poultry. In: Hillyer, G.M., Whittemore, C.T. and Gunn, R.G. (eds) *Computers in Animal Production.* Occasional Publication No. 5. British Society of Animal Production, Scottish Agricultural College, Edinburgh, pp. 103–110.

European Communities (1999) Council Regulation (EC) No. 1804/1999 of 19 July 1999 supplementing Regulation (EEC) No. 2092/91 on organic production of agricultural products and indications referring thereto on agricultural products and foodstuffs to include livestock production. *Official Journal of the European Communities 24.98.1999.* Brussels. L 222 1–28.

Fischer, K. and Lindner, P.J. (1998) Einzelaspekte der Fütterung nach Richtlinien des ökologischen Landbaus im Hinblick auf die Fleischqualität. In: Braun, C. (ed.) *Proceedings 110. VDLUFA-Kongress 'Einfluss von Erzeugung und Verarbeitung auf die Qualität landwirtschaftlicher Produkte'.* VDLUFA, Darmstadt, Germany, p. 58.

Fleming, G.A. (1977) Mineral disorders associated with grassland farming. In: *Proceedings, International Meeting on Animal Production from Temperate Grassland, Dublin.* Irish Grassland and Animal Production Association/An Foras Taluntais, Dublin, pp. 88–95.

Forbes, J.M. (1986) *The Voluntary Food Intake of Farm Animals.* Butterworths, London.

Foster L. (1988) Herbs in pastures. Development and research in Britain, 1850–1984. *Biological Agriculture and Horticulture* 5, 97–133.

Frankow-Lindberg, B.E. and Danielsson, D.-A. (1997) Energy output and animal production from grazed grass/clover pastures in Sweden. *Biological Agriculture and Horticulture* 14, 279–290.

Groce, A.W., Taylor, C.E., Pettry, D.E. and Kerr, L.A. (1995) Levels of selenium and other essential minerals in forages, bovine blood and serum relative to soil type. *Veterinary Clinical Nutrition* 2, 146–152.

Gruber, L., Steinwidder, A., Guggenberger, T., Schauer, A., Haeusler, J., Steinwender, R. and Steiner, B. (2000) Einfluss der Grünlandbewirtschaftung auf Ertrag, Futterwert, Milcherzeugung und Nährstoffausscheidung. In: *Proceedings 27. Viehwirtschaftliche Fachtagung*. BAL Gumpenstein, Irdning, Austria, pp. 41–88.

Hadorn, R., Gloor, A. and Wiedmer, H. (2000) Einfluss des Ausschlusses von reinen Aminosäuren bzw. potentiellen GVO-Eiweißträgern aus Legehennenfutter auf pflanzlicher Basis. *Archiv für Geflügelkunde* 64, 75–81.

Henriksen, B.I.F. (2000) Effect of organic fodder on prevention of milk fever – a step further. In: Hovi, M. and Bouilhol, M. (eds) *Proceedings of the 3rd NAHWOA Workshop, Clermont- Ferrand, 21–24 October 2000*. University of Reading, Reading, UK, pp. 152–154.

Horst, R.L., Goff, J.P., Reinhardt, T.A. and Buxton, D.R. (1971) Strategies for preventing milk fever in dairy cattle. *Journal of Dairy Science* 80, 1269–1280.

Ingensand, T. and Hörning, B. (1999) Zur Wirtschaftlichkeit der ökologischen Legehennenhaltung. In: Hoffmann, H. and Müller, S. (eds) *Beiträge zur 5. Wissenschaftstagung zum Ökologischen Landbau*. Verlag Köster, Berlin, pp. 287–290.

INRA (Institut National de la Recherche Agronomique) (1989) *Ruminant Nutrition. Recommended Allowances and Feed Tables*. Jarrige, R. (ed.), John Libbey Eurotext, Paris.

Isselstein, J. and Daniel, P. (1996) The ensilability of grassland herbs. In: Parente, G., Frame, J. and Orsi, S. (eds) *Grassland and Land Use Systems, 16th meeting of the European Grassland Federation, Grado, Italy, September 1996*, European Grassland Federation, Reading, UK, pp. 451–455.

Jakobsen, K. and Jensen, S.K. (2000) Zur Vitamin- und Mineralstoffversorgung in der ökologischen Schweinemast. 1. Produktionsergebnisse sowie Nährstoff- und Fettsäuren-zusammensetzung des Fleisches nach Verabreichung von ökologischem Futter mit oder ohne Zusatz von Vitaminen und Mineralstoffen und mit oder ohne Zugang zu Gras im Vergleich zu konventioneller Haltung. In: Pallauf, J. (ed.) *Proceedings of the Society of Nutritional Physiology 9*. DLG-Verlag, Frankfurt/Main, p. 79.

Kemm, E.H., Malan, D.D., vanVuuren, C.J.J, Loots, L.P., Siebrits, F.K., Moroeng, J. and Mashiane, J. (1997) What feeding method is the best for your lactating sows? *Porcus* 12, 24–26.

King, R.H. (1998) Dietary amino acids and milk production. In: Verstegen, M.W.A., Moughan, P.J. and Schrame, J.W. (eds) *The Lactating Sow*. Wageningen Press, Wageningen, The Netherlands.

Knaus, W., Steinwidder, A. and Zollitsch, W. (2001) Energy and protein balance in organic dairy cow nutrition – model calculations based on EU regulations. In: Hovi, M. and Baars, T. (eds) *Breeding and Feeding for Animal Health and Welfare in Organic Livestock Systems. Proceedings of the Fourth NAHWOA Workshop, 24–27 March 2001, Wageningen, The Netherlands*. VEERU, University of Reading, Reading, UK.

Koketsu, Y., Dial, G.D., Pettigrew, J.E., Marsh, W.E. and King, V.L. (1996) Characterization of feed intake patterns during lactation in commercial swine herds. *Journal of Animal Science* 74, 1202–1210.

Kongsted, A.G., Larcher, J. and Larsen, V.A. (1999) Silage for outdoor lactating sows. In: Hermansen, J.E., Lund, V. and Thuen, E. (eds) *Ecological Animal*

Husbandry in the Nordic Countries. Proceedings from NJF-seminar No. 303. DARCOF (Danish Research Centre for Organic Agriculture), Tjele, Denmark, pp. 125–129.

Kristensen, T. and Kristensen, E.S. (1998) Analysis and simulation modelling of the production in Danish organic and conventional dairy herds. *Livestock Production Science* 54, 55–65.

Krutzinna, C. and Boehncke, E. (1989) Zur Problematik der Natriumversorgung von Milchkühen in der Umstellungsperiode. *Bio-Land* 16, 29–30.

Krutzinna, C. and Boehncke, E. (1997) Ketose. In: Lünzer, I. and Vogtmann, H. (eds) *Ökologische Landwirtschaft (Loseblattsammlung), Sektion 04, Kapitel 4.08, Teil 8*. Springer Verlag, Berlin, pp. 1–10.

Krutzinna, C., Boehncke, E. and Herrmann, H.J. (1996) Organic milk production in Germany. *Biological Agriculture and Horticulture* 13, 351–358.

Leeb, T. (2000) Unpublished data. University of Veterinary Medicine, Vienna, Austria.

MacNaeidhe, F. (1989) Trace elements in cereal grain – their influence on diet. *Farm and Food Research* 20, 28–30.

MacNaeidhe, F.S. (1991) The effect of zinc and nitrogen on pasture growth in a zinc deficient soil. *Proceedings of the 17th Annual Research Meeting of the Irish Grassland and Animal Production Association*. University College Dublin, pp. 1–2.

MacNaeidhe, F.S. (1992) The uptake of selenium by *Lolium perenne* from inorganic biological sources. *Proceedings of the 18th Annual Research Meeting of the Irish Grassland and Animal Production Association*. University College Dublin, pp. 5–6.

MacNaeidhe, F.S. and Fingleton, W. (1997) Economic investigations of organic calf to beef and sheep production. *European Newsletter on Organic Farming* 6, 20–22.

MacNaeidhe, F.S., Lynch, M., Murphy, W., Fingleton, W. and Codd, F. (1996) *A Report on the Organic Farm, Johnstown Castle, Production and Economics*. Teagasc, Johnstown Castle Research Centre, Wexford.

McGrath, D., Poole, D.B.R., Fleming, G.A. and Sinnott, J. (1982) Soil ingestion by grazing sheep. *Irish Journal of Agricultural Research* 21, 135–145.

Mitchell, R.L., (1960) Trace elements in Scottish soils. *Proceedings of the Nutrition Society* 19, 148–154.

Moloney, S.C. and Milne, G.D. (1993) Establishment and management of grasslands Puna chicory used as a specialist, high quality forage herb. *Proceedings of the New Zealand Grassland Association* 55, 113–118.

Newton, J. and Stopes, C. (1995) Grassland productivity on organic farms, 1992–1994. *Elm Farm Research Centre Bulletin*, No. 18, 2–6.

NRC (1994) National Research Council. *Nutrient Requirement of Poultry*, 9th edn. National Academy Press, Washington, DC.

NRC (1998) National Research Council. *Nutrient Requirement of Swine*, 10th edn. National Academy Press, Washington, DC.

Olesen, I., Lindhard, E. and Ebbesvik, M. (1999) Effects of calving season and sire's breeding value in a dairy herd during conversion to ecological milk production. *Livestock Production Science* 61, 201–211.

Preisinger, R. and Flock, D.K. (1998) Changes in genetic parameters and genetic progress in layers during tow decades of intensive selection. In: van

Askham Bryan College
LIBRARY BOOK

Arendonk, J.A.M. (ed.) *Proceedings of the 49th Annual Meeting of the European Association of Animal Production*. Wageningen Press, Wageningen, The Netherlands, p. 38.

Refsgaard, K., Halberg, N. and Kristensen, E.S. (1998) Energy utilization in crop and dairy productions in organic and conventional livestock production systems. *Agricultural Systems* 57, 599–630.

Rogers, P.A.M. and Gately, T.F. (1992) *Control of Mineral Imbalances in Cattle and Sheep*. Teagasc, Dublin.

Rogers, P.A.M., Fleming, G.A. and Gately, T.F. (1989) *Mineral Imbalances and Mineral Supplementation in Cattle*. Teagasc, Agriculture and Food Development Authority, Dublin.

Rose, S.P. (1997) *Principles of Poultry Science*. CAB International, Wallingford, UK.

Schwarzenbacher, H. (2001) Vergleich von biologischen mit konventionellen Milchviehbetrieben in Niederösterreich. Diploma thesis, BOKU, University of Natural Resources and Applied Life Sciences, Vienna.

Scott, M.L. (1977) Trace elements in animal nutrition. In: Mortvedt, J.J., Giordano, E. and Linsay, D. (eds) *Micronutrients in Agriculture*. Soil Science Society of America, Inc. Madison, Wisconsin, pp. 555–587.

Staack, M., Trei, G., Hörning, B. and Fölsch, D.W. (1999) Ausgewählte Ergebnisse des Modellvorhabens, Artgereche Legehennenhaltung in Hessen' unter besonderer Berücksichtigung des Arbeitszeitaufwandes. In: Hoffmann, H. and Müller, S. (eds) *Beiträge zur 5. Wissenschaftstagung zum Ökologischen Landbau*. Verlag Köster, Berlin, pp. 283–286.

Strobel, E., Peganova, S., Dänicke, S. and Jeroch, H. (1998) Zum Einfluß von unter den Bedingungen des ökologischen Landbaus konzipierten Futtermischungen mit unterschiedlichem Rohproteingehalt auf die Eiqualität bei zwei Legehennenherkünften. In: Braun, C. (ed.) *Proceedings 110. VDLUFA-Kongress 'Einfluss von Erzeugung und Verarbeitung auf die Qualität landwirtschaftlicher Produkte'*. VDLUFA, Darmstadt, Germany, p. 80.

Sundrum, A. (1997) *Beurteilung der Auswirkungen überhöhter Rohproteinversorgung beim Rind mit biochemischen und immunologischen Blutparametern*. Schriftenreihe Institut für Organischen Landbau. Rheinische Friedrichs-Wilhelms-Universität Bonn, Germany.

Sundrum, A., Valle Zarate, A., Roeb, S., Rubelowski, I., Schoone, U. and Weber, R. (1999) *Auswirkungen von Grundfutter in der Schweinemast auf Tiergesundheit, Verhalten, Leistung und Produktionskosten unter den Prämissen des Organischen Landbaus*. Forschungsberichte Umweltverträgliche und Standortgerechte Landwirtschaft, Heft 71. Rheinische Friedrichs-Wilhelms-Universität Bonn, Germany.

Sundrum, A., Bütfering, L., Henning, M. and Hoppenbrock, K.H. (2000) Effects of on-farm diets for organic pig production on performance and carcass quality. *Journal of Animal Science* 78, 1199–1205.

Suttle, N.F., Alloway, B.J. and Thornton, I. (1975) An effect of soil ingestion on the utilization of dietary copper by sheep. *Journal of Agricultural Science, Cambridge* 84, 249–254.

Swift, G., Davies, D.H.K., Tiley, G.E.D. and Younie, D. (1990) *The Nutritive Value of Broad-leaved Weeds and Forage Herbs in Grassland*. Technical Note No. 223. Scottish Agricultural College, Aberdeen, UK.

Tenhagen, B.-A., Hoffmann, A. and Heuwieser, W. (1998) Mutterkuhhaltung im ökologischen Landbau in Brandenburg – Tiergesundheit und Wirtschaftlichkeit. *Tierärztliche Umschau* 53, 678–685.

Thielen, C. and Kienzle, E. (1995) Die Fütterung des 'Bioschweins' – eine Feldstudie. In: Giesecke, D. (ed.) *Proceedings of the Society of Nutrition Physiology*. DLG-Verlag, Frankfurt/Main, Germany, p. 91.

Thurner, J. (2001) Einflüsse der Verfütterung von Luzerne-Pellets an Milchkühe auf Bio-Betrieben. Diploma thesis, BOKU, University of Natural Resources and Applied Life Sciences, Vienna, Austria.

Trei, G., Kuhn, M., Fölsch, D.W. and Djadjaglo, D. (1997) Akzeptanz und Futterwert nach ökologischen Richtlinien erzeugten Futters für Legehennen in Volieren- und Bodenhaltung. In: Köpke, U. and Eisele, J.A. (eds) *Beiträge zur 4. Wissenschaftstagung zum Ökologischen Landbau*. Verlag Köster, Berlin, pp. 610–614.

Umrani, A.P. (1998) Sustainability in contrasting farming systems: Part 1 Utilisation of herbs and grass in temperate organic (sustainable) farming systems; Part 2 Sustainable management of semi-arid ranges in two locations in Pakistan. PhD Thesis, University of Aberdeen, Aberdeen, UK.

Van Eekeren, N. (2000) Balancing summer rations of dairy cows by means of urea in bulk tank milk. In: Soegaard, K., Ohlsson, C., Sehested, J., Hutchings, N.J. and Kristensen, T. (eds) *Grassland Farming, Balancing Environmental and Economic Demands. 18th General Meeting of the European Grassland Federation*. Aalborg, Denmark. European Grassland Federation, Reading, UK, pp. 555–557.

Van Soest, P.J. (1994) *Nutritional Ecology of the Ruminant*. Cornell University Press, Ithaca, New York.

Van Soest, P.J., McCammon-Feldman, B. and Cannas, A. (1994) The feeding and nutrition of small ruminants: application of the Cornell Discount System to the feeding of dairy goats and sheep. In: *Proceedings of the 1994 Cornell Nutrition Conference for Feed Manufacturers*. Rochester, New York, pp. 95–104.

Verkleij, F.N. and MacNaeidhe, F.S. (1992) Foliar application and uptake of selenium extracted from ryegrass. *Journal of Plant Nutrition* 15, 1227–1234.

Walser, P.T. (1997) Einfluss unterschiedlicher Futterzusammensetzung und –aufarbeitung auf das Auftreten von Federpicken, das Nahrungsaufnahmeverhalten, die Leistung und den Gesamtstoffwechsel bei verschiedenen Legehennenhybriden. Doctorate thesis, Eidgenössische Technische Hochschule, Zürich, Switzerland.

Walsh, T. and Fleming, G.A. (1978). *Trace Elements in Agriculture – An Overview. Symposium on Trace Elements – Soil, Plant, Animal Interface*. University of Aberdeen, Aberdeen, UK.

Weiss, W.P. (1998) Requirement of fat-soluble vitamins for dairy cows: a review. *Journal of Dairy Science* 81, 2493–2501.

Wiener, B. (1996) Vergleich zwischen Legehennen- und Milchkuhfütterung auf biologisch wirtschaftenden Betrieben in der Steiermark. Diploma thesis, BOKU, University of Natural Ressources and Applied Life Sciences, Vienna, Austria.

Wilman, D. and Derrick, R.W. (1994) Concentration and availability to sheep of N, P, K, Ca, Mg and Na in chickweed, dandelion, dock, ribwort and spurrey, compared with perennial ryegrass. *Journal of Agricultural Science, Cambridge* 122, 217–223.

Wlcek, S. (2002) Die systemkompatible Ernährung von Schweinen im biologischen Landbau – Untersuchungen zum Aufkommen und Futterwert von Nebenprodukten aus der Verarbeitung biologisch erzeugter Lebensmittel. Doctorate thesis, BOKU, University of Natural Resources and Applied Life Sciences, Vienna, Austria.

Wurzinger, M. (1999) Erhebung der Fütterungspraxis bei Mastschweinen of biologisch wirt-schaftenden Betrieben in Niederösterreich. Diploma thesis, BOKU, University of Natural Resources and Applied Life Sciences, Vienna, Austria.

Younie D., Umrani A.P., Gray D. and Coutts, M. (2001) Effect of chicory or ryegrass diets on mineral status of lambs. In: Isselstein, J. and Spatz, G. (eds) *Organic Grassland Farming. Volume 5, Grassland Science in Europe*. European Grassland Federation, pp. 561–563.

Zollitsch, W., Wlcek, S., Leeb, T. and Baumgartner, J. (2000) Aspekte der Schweine- und Geflügelfütterung im biologisch wirtschaftenden Betrieb. In: *Proceedings 27. Viehwirtschaftliche Fachtagung*. BAL Gumpenstein, Irdning, Austria, pp. 155–162.

16

Breeding Strategies for Organic Livestock

Jennie E. Pryce,[1] Jo Conington,[2] Paul Sørensen,[3] Hillary R.C. Kelly[4] and Lotta Rydhmer[5]

[1]Livestock Improvement Corporation Ltd, Private Bag 3016, Hamilton, New Zealand; [2]Scottish Agricultural College, West Mains Road, Edinburgh EH9 3JG, UK; [3]Danish Institute of Agricultural Sciences, Research Centre Foulum, PO Box 50, DK-8830 Tjele, Denmark; [4]School of Agriculture, Food and Rural Development, University of Newcastle, Newcastle NE1 7RU, UK; [5]Department of Animal Breeding and Genetics, Swedish University of Agricultural Sciences, Funbo-Lövsta, S-755 97 Uppsala, Sweden

Editors' comments

Organic animal husbandry favours animals with certain characteristics, such as the ability to thrive under loose housing and outdoor conditions, to be good mothers, to resist diseases, to search for food, and to sustain themselves and produce well on locally produced diets. Some of these abilities are supported through the housing system, the way young animals are kept, feeding and human contact, but the genetic predisposition of the animals is still the key to solving many problems in the organic herd. This raises many difficult questions for the authors to address: which heritable characters are desirable in organic animals and should be considered in organic breeding goals? Can these goals be reached through conventional breeding methods and breeds? Are family breeding and very local breeds suitable alternatives? The rapid development of technology raises additional questions. The focus on naturalness and natural behaviour in organic farming may argue in favour of natural mating instead of artificial insemination. Moreover, the organic ban on genetic modification raises the question of whether conventional bulls can be used at all by organic farmers. Finally, breeding goals are crucial to the question of what organic farming in Europe should look like: what production levels and what growth rates should we aim for in future organic herds?

© CAB International 2004. *Animal Health and Welfare in Organic Agriculture* (eds M. Vaarst, S. Roderick, V. Lund and W. Lockeretz)

Introduction

This chapter examines whether different strains or breeds are needed for organic livestock production. There are two reasons this might be so. The first is 'genotype–environment interaction' (or 'G × E'), the term used when animals being selected for performance traits in one environment rank very differently in another environment. Second, additional traits may appropriately be selected for in organic production beyond those used in conventional production, which have focused on production level.

It is open to debate whether organic production systems are sufficiently different from conventional systems to make switching to an alternative breed or strain worthwhile. The EU Regulation on Organic Livestock Production states that 'in the choice of breeds or strains, account must be taken of the capacity of animals to local conditions, their vitality and their resistance to disease' (Council Regulation [EC] No 1804/1999, B 3.1). According to this regulation, breeding decisions should also give preference to indigenous breeds and strains. Many organic farmers have already decided to change breeds, perhaps to indigenous breeds, although the reasons behind this decision may be connected to marketing opportunities associated with an unusual product rather than the suitability of the breed to the system.

In recent years, more precise genetic evaluations on a wider number of traits have become available, and this is continuing with advances in statistical methods and improvements in computing capability. For the organic movement to be able to make full use of advances in technology, clear guidelines on the traits to select for are required (i.e. the selection objectives). If the traits included in existing genetic evaluations do not include all the traits of value to organic farming, then either these need to be evaluated, or other traits need to be used as predictors. The latter is feasible only if the predictor is genetically correlated with the trait of interest. If current breeding programmes do not fulfil these requirements, then either alternative breeds should be considered, or existing breeding programmes adapted, or even new 'tailor-made' ones initiated.

Selection for specific genes or QTLs, either directly or using genetic markers, is already a commercial reality for some species. There have also been substantial advances in the area of reproductive technologies in conventional agriculture, although there are major ethical concerns already prohibiting the use of some of these in organic agriculture. Thus the NAHWOA Recommendations call for development of breeding strategies suitable for organic systems 'before the use of genetic manipulation makes conventional breeding systems inaccessible for organic farmers'. Later we discuss the role of new and existing technologies as tools and strategies for breeders of organic livestock.

Since many of the principles of genetic theory are equally applicable to all species, we have deliberately kept this chapter broad. We have also avoided giving a detailed explanation of some of the principles underpinning genetic improvement, as there are textbooks that do this already, e.g. Simm (1998). Instead, our purpose is to discuss some of the issues of selection that are specifically applicable to organic production.

Breeding Objectives

For most livestock species, breeding has been focused mainly on traits related to output, such as lean tissue growth rate in meat animals and weight of milk solids in dairy cows. Of course, any farm, whether conventional or organic, must still be economically viable, which inevitably means that some selection emphasis must be paid to traits important for production. Therefore, existing breeding programmes set up for conventional livestock farming can still be useful for organic production.

However, the philosophies and ethical values behind organic farming (as discussed in Chapter 5) will inevitably lead to differences either in the breeding objectives themselves, or in their relative importance. Nevertheless, there currently are few, if any, breeding programmes that produce animals bred specifically for organic production. Furthermore, little guidance is available on breeding animals for organic farms.

Regardless of species, the primary breeding objectives for organic farming are likely to include disease resistance and longevity. Another area of importance is increased reliance on forage in ruminant diets, rather than the trend in recent years to feed diets that are high in concentrates, especially for cattle. Good mothering ability is also an important selection criterion in pigs and sheep, as is the ability to seek forage outdoors. Production is seen as secondary to these objectives by many involved in organic farming, because in conventional farming, significant advances have already been made for production traits in most species. In a survey of priorities for dairy cattle breeding aimed at all sectors of the UK organic dairy industry, production-related traits were of low priority (Pryce *et al.*, 2001). There are several possible reasons for this: (i) that the level of milk production was already felt to be high enough; (ii) that selection for production is already efficient and that some emphasis should now be placed on traits other than production; and (iii) that selection for production has led to a deterioration in health and fertility (e.g. Rauw *et al.*, 1998).

For most breeding programmes, the favoured method of selecting individuals is to use a selection index, where breeding values on a range of traits contributing to the overall breeding goal are weighted by their economic importance. This works well, because most conventional farmers are interested primarily in improving profitability, and a selection index identifies individuals that most closely match this objective. However, with organic systems, a different approach may be necessary if there are multiple objectives. For example, maintaining genetic diversity may be an additional objective that is difficult to incorporate into a selection index.

Breeding for Disease Resistance

Production diseases caused by infectious agents are part of the usual farming environment. Conventional farmers control these challenges using management practices and prophylactic treatments such as antibiotics, vaccination, anthelmintic drenches and anti-fungal agents. The motivation for prophylactic treatments is often animal welfare. However, the decision to use such treatments is often an economic one, which depends on the actual or likely level of infestation in the herd, together with an assessment of risk (based on previous herd history and weather conditions (particularly for grazing stock). Raising animals without using such drugs is a primary aim of organic farming, for various reasons, including the human, animal and environmental health benefits of not having chemical residues in the animal products. However, as a substitute, homoeopathic and other alternative plant-based remedies are frequently used in some countries to control animal diseases in organic farming, as described in Chapter 13.

As there is growing evidence to show that selection for improvements in production has led to a decline in health and reproductive performance (Rauw *et al.*, 1998), any advances in breeding for disease resistance will be particularly welcome; there is a clear economic incentive to breed for improvements in health and fertility, for both conventional and organic systems. Beilharz *et al.* (1993) and subsequently Rauw *et al.* (1998) and Bakken *et al.* (1998) used resource allocation theory to explain the negative genetic correlations that many, but not all, researchers have observed between performance traits and those important for fitness and health. This theory means that if an animal is driven to allocate resources to one function, it may not have enough resources to perform another required function, so that overall fitness will be reduced. (For a comprehensive treatment of breeding for disease resistance in all farm animals, see Axford *et al.*, 2000.)

Breeding for disease resistance is new in most species. Morris (2000) noted that 10 years ago, the possibility of selecting sheep for resistance to internal parasite infection was being seriously questioned. Yet now it is generally accepted as feasible, although not all breeders have taken it up. Furthermore, genetic evaluations for somatic cell count (SCC) in dairy cattle have become available only in the last decade in most countries. As SCC is genetically correlated with mastitis, selection for reduced SCC will lower the incidence of mastitis. At present, several countries now include SCC as part of their dairy cattle genetic evaluation programme (Mark *et al.*, 2002, list 12 such countries).

In cattle and sheep, considerable effort is being put into research and consequently selection for disease resistance (Axford *et al.*, 2000). In growing pigs, recent results from Denmark suggest that the number of leukocytes has a moderate heritability and may be used as a selection criterion when breeding for resistance to clinical and sub-clinical diseases (Henryon *et al.*, 2002). In poultry, environment and management have been and probably will continue to be the main way to control disease. Thorp and Luiting (2000) stated that 'disease resistance will not be selected for if the cost in a loss of genetic improvements in other traits is too great and there are other effective methods of disease control'. But they go on to argue that 'although decisions in the poultry industry are largely and increasingly driven by economic considerations, the psychological impact of flock morbidity and mortality on the farmer and society cannot be ignored.' Thus, it is likely that breeding objectives already reflect values that are not purely economic.

Levels of mortality in livestock, greater reliance on management practices to reduce disease (such as housing or type of feed), and consumer intolerance to residues in animal products may lead to changes in conventional livestock systems as well. The speed of change may depend on whether legislation imposes penalties for production systems with unacceptably high levels of disease or mortality. If this happens, much greater emphasis will be placed on health and fertility traits in conventional farming. However, the pressure to breed for disease resistance is much greater in organic livestock production for health, economic and ethical reasons.

There are three practical options for implementing genetic improvement for these traits: (i) use breeds or strains that already demonstrate high levels of disease resistance ('breed substitution'); (ii) cross existing stock with another breed ('cross-breeding'), because even if disease resistance is not exhibited in the pure-bred, it may exist in the cross; and (iii) select for disease resistance within a breed ('within-breed selection'). More detailed information on the principles of these three methods is given below (see also Simm, 1998).

Selection Objectives Specific to Different Livestock

Dairy cattle

Progeny testing in dairy cattle relies on accurate phenotypic records. Milk volume and constituent components are routinely collected by recording agencies in most countries. Disease resistance, longevity and fertility are likely to be the most important traits to select for in organic farming. Records used to produce genetic evaluations on these are either estimated using measurements on the trait itself (such as records of clinical disease), or traits known to be closely genetically related to the trait of interest (such as SCC to select for mastitis resistance). Selection will be enhanced further by the use of genetic markers, or selection for specific genes.

Disease traits

The main limitations to genetic progress in disease traits in most countries have been:

- the low heritability estimates of such traits
- the lack of reliable phenotypic records
- the unfavourable genetic relationship between production and fitness traits
- the number of offspring required to get an accurate breeding value (about four to ten times as many as for production traits).

However, the Scandinavian countries have a strong tradition of recording health and fertility records, and these records make it possible to estimate breeding values for disease resistance. Data are collected by veterinarians on treatment of sick animals. If alternative treatments not administered by veterinarians, such as homoeopathy, become popular, this may pose a problem for data collection.

The emphasis placed on breeding for mastitis resistance is likely to be an area where organic and conventional dairy breeding differ. Currently, the country with the greatest emphasis on mastitis in its national selection index is Norway, with a relative weight of 22% on mastitis and 23% on protein yield (Arne Ola Refsdal, GENO, Norway, 2003, personal communication). The relative weight on milk yield compared with mastitis dropped from 4.5 in 1980 to 1.6 in 1990 to 1 in 1999 (Svendsen, 1999). Breeding values for mastitis frequency showed negligible change for bulls born between 1971 and 1989, whereas for bulls born between 1990 and 1993 there was an improvement of 0.43% per year (Svendsen, 1999), with genetic progress still being made in milk yield. Mastitis resistance has low heritability, so large daughter groups together with a high index weight are essential for reliable and effective selection. Norway tests about 125 bulls annually and aims to get 250–300 daughters per sire (Svendsen, 1999).

Condition score

Another trait that might be more important in organic than conventional dairy cattle breeding is condition score, a visual and tactile appraisal of the amount of fat cover in the rump and back area. Condition score has commonly been used to indirectly monitor feeding levels, but several studies in recent years have demonstrated that it has a substantial genetic component as well. Heritability estimates are around 0.3 (e.g. Koenen and Veerkamp, 1998; Pryce *et al.*, 2000; Harris, 2002).

A major reason for interest in condition score in breeding programmes is its potential as a proxy for fertility. Genetic correlations with fertility are reasonably high, and some countries are already using it in combination with fertility data in indexes to improve fertility. However, its value in future breeding programmes, especially for organic livestock, may go further. Condition score typically shows a depletion of reserves in early to peak lactation to fuel milk production, followed by a period of recovery to the next lactation (Coffey *et al.*, 2001). Genetic differences in the shape of these profiles across lactation (and possibly across environments) may help to identify animals most suitable for organic production. Also, a flatter lactation curve may be a way of avoiding short-term nutrient deficits in organic dairy herds, as discussed in Chapter 15. With the advent of test-day models, breeding values related to persistency and lactation curve shape are now available in some countries. At present only a few countries use a test-day model for genetic evaluations of production traits, but several others are working on implementing it in their national evaluations (Interbull, 2001).

Ecological indexes

Bapst (2001) and Baumung *et al.* (2001) have discussed the merits of ecological indexes, designed with organic producers in mind. The Swiss ecological index has a relative weighting of 24% for performance traits and 76% for functional traits, whilst corresponding conventional weightings are 57% and 43% respectively (Bapst, 2001). Baumung *et al.* (2001) proposed an adaptation of the existing total merit index for dual-purpose Austrian Simmentals. Economic weights were derived by varying the assumptions on feed costs and composition of diet. As feed costs increased, the relative economic importance of milk and beef decreased, while the importance of fitness traits increased.

Sheep

Breeding goals

Conventional breeding goals for sheep in the UK include litter size, lamb growth rate and, more recently, carcass composition traits such as the amount of fat and muscle, as determined by on-farm ultrasound measurements (Simm and Dingwall, 1989). The traits included as breeding goals vary according to the breed type. For instance, 'terminal sire' or meat-type sheep such as the Suffolk and Texel breeds only have growth rate, fat and muscle depths as their breeding goals. Maternal breed types have traits that are important for maternal performance, such as litter size and milking ability.

Organic sheep production requires broad breeding goals that address issues of sustainability (e.g. improving ewe longevity), health, fitness and flock efficiency (such as reducing lamb losses). Such breeding goals are likely to appeal to the ethical concerns of organic livestock producers as well as ensuring that organic production is a viable and competitive alternative to conventional production. A detailed discussion of breeding goals suitable for organic sheep production is given by Conington *et al.* (2002).

Some of these traits have recently been incorporated into 'new age' selection indexes for hill sheep breeds in Britain. These selection indexes are a first attempt to use appropriate breeding goals that are tailored to suit different categories of hill environments. Ewe longevity and lamb survival are included in the selection objective alongside more conventional traits to improve lamb carcass production in tandem with maternal characteristics. The indexes are currently being tested in Scotland at the Scottish Agricultural College (Conington *et al.*, 2001), and their use in breeding programmes is predicted to improve flock performance without detriment to ewe or lamb survival.

Disease resistance

Selection of sheep with improved resistance to disease potentially increases animal health, welfare and productivity. It is particularly pertinent for organic sheep production to consider including resistance to diseases as breeding goals because non-genetic control measures for many diseases (e.g. nematode parasites) are expensive and not fully effective, and are becoming less effective because of nematode resistance to chemotherapy. In New Zealand and Australia, established breeding programmes that include faecal egg count as a breeding objective are currently being implemented. In the UK, breeding values for faecal egg count were available for the first time in 2001 although they are not yet integrated into the overall breeding objective. As resistance to one species confers a degree of cross-genera resistance (McEwan *et al.*, 1992), selection against one species also results in enhanced immunity against another, although little is known about the genetic relationships with other common diseases such as footrot and mastitis. The primary requirement for including disease resistance in breeding programmes is knowledge of the extent of genetic variation in resistance and the genetic correlations between resistance and all other important traits. Therefore, considerable research effort is being expended to estimate these quantities (Bishop *et al.*, 1996; Morris, 1998; Bishop and Stear, 2000; Raadsma, 2000).

Breeding for resistance to other major sheep diseases has shown that there is genetic variation for flystrike, facial eczema and footrot (Raadsma, 2000), and for mastitis in dairy sheep (Mavrogenis *et al.*, 1995). Diseases that are difficult to measure are ideally suited to selection using molecular information. In the UK, the National Scrapie Plan is based on selective breeding only from non-susceptible genotypes in an attempt to eliminate the disease (DEFRA, 2002). In New Zealand, efforts are underway using microsatellite markers to identify regions of the genome that influence particular phenotypic characteristics related to parasite resistance (McEwan and Kerr, 1998). Such QTLs are expected to be soon available for both conventional and organic breeding programmes for several other diseases (Conington *et al.*, 2002).

Behaviour

Behavioural characteristics are difficult and expensive to measure, but potentially, could be ideal candidates for selection using QTLs in the future. Genetic variation in maternal behaviour has been quantified both between (Dwyer and Lawrence, 1998) and within breeds of sheep (Lambe *et al.*, 2001). Other behavioural characteristics that are important for lamb survival, such as the time from birth to stand, and the time to suckle successfully has been shown to be influenced considerably by the sire of the lamb (C. Dwyer, Scottish Agricultural College, UK, 2003, personal communication). Integrating such traits into organic breeding indexes for some lowland sheep breeds would set them apart from conventional sheep production.

Different genotypes of sheep and cattle have different preferences for grazing particular grass species or land areas (see review by Hohenboken, 1986). This could also be an important breeding goal, particularly for environmental pasture management, and could be useful attributes for an organic sheep system. With more sheep being tended per shepherd than ever before, breeding strategies that include traits conferring greater ability to survive and adapt to extensive management practices should be adopted by organic sheep producers.

Pigs

Breeding goals

All conventional pig production is based on cross-breeding, where the pigs reared for slaughter are three- or four-breed crosses. The goal of most breeding companies is to provide the mass market with a better quality product, i.e. less fat in the carcass. During the last decade, the annual genetic progress in the pure-bred lines in most European pig breeding programmes has been +20 g for daily gain, +0.5% for lean meat percentage, and 0.2 for piglets per litter (Merks, 2000). Although the organic pig sector requires significant numbers of boars and sows, there are no specific breeding programmes for organic pig production. Among the production traits, feed conversion rate is probably as important in organic as in conventional pig production. A high feed efficiency reduces nutrient wastage and thus leads to less environmental pollution. Utilization of forages is not regarded as important in conventional pig production, but may be valuable in organic production.

Regardless of whether there are important G×E interactions, different traits may have different economic weights in organic compared with conventional production. Thus, traits like body constitution and mothering ability may be given higher weights and growth rate a lower weight in breeding evaluations for organic production. Sows farrowing outdoors should have good mothering abilities, good temperament, high milk production, and strong legs and feet (Edwards, 1995). In a Danish field study, leg weakness was the most common culling reason for sows in outdoor piglet production (Larsen and Kongsted, 2001). Osteochondrosis is one reason for leg weakness in young pigs. Osteochondrosis is heritable, and it is related to life span of sows (Yazdi *et al.*, 2000). Osteochondrosis and leg weakness are already included in the breeding evaluation in some countries. Strong legs and good abilities to move are probably even more important when sows are held outdoors, for example on rocky pastures.

Outdoor pigs

Some pig breeding companies have special 'outdoor dam lines'. Such lines are usually based on a combination of Landrace–Large White and Duroc, where the Duroc is supposed to increase hardiness.

According to van der Steen (1994), there is a big difference between dam lines bred for outdoor and indoor piglet production. However, there are no published studies that compare reproduction traits in different lines of sows in both indoor and outdoor environments. Reproduction results for sows in organic production have not been reported in the literature either, but there are some comparisons between outdoor and indoor production systems. In general, the number of stillborn piglets is lower outdoors than indoors (ITP, 2001; Edwards, 2001; Thodberg and Kongsted, 2001), but the outdoor piglet mortality from birth to weaning tends to be either higher (ITP, 2001) or about the same as indoors (MLC, 2000).

Behaviour

The newborn piglet, which has almost no fat, is prone to cooling down. Thus, if the sow is farrowing outdoors or in simple buildings, her ability to build a good nest at farrowing is important for piglet survival. Nest-building behaviour is also related to the calmness of sows during and after farrowing and consequently to the risk of crushing piglets

(Thodberg, 2001). (See Chapter 7 for a discussion of both thermoregulation and nest building in sows.) It is not known whether the variation between sows in nest-building behaviour has a genetic background. When piglet survival was compared for sows of different breeds in different types of farrowing huts, no interaction between breed and type of hut was found (McGlone and Hicks, 2000).

When the sow is loose housed, a good temperament is very important. Aggressive loose sows are dangerous for the farmer. Swedish and Norwegian studies in conventional nucleus herds have shown that aggression towards the farmer is a heritable trait in sows (Vangen *et al.*, 2002; Grandinson *et al.*, 2003). These studies also showed that fear of humans is a heritable trait. Furthermore, there is a genetic correlation between fear of humans, recorded as the sows' avoidance of the farmer, and piglet mortality (Grandinson *et al.*, 2003). Sows with a genetic disposition for fearfulness have a lower genetic ability for piglet survival. Sow behaviour can be recorded in the field with help of questionnaires, and such records on aggression and fearfulness could be included in the breeding evaluation of pigs for organic production.

Lactation and fertility

It is difficult to formulate an organic feed for piglets because of their high protein demands. Furthermore, weaning is much later in organic than in conventional farming. Nursing behaviour and milk production of sows is therefore more important in organic than conventional production. Sows nurse their piglets approximately every hour, but there is variation in nursing frequency between sows, and nursing frequency is related to piglet growth (Valros, 2003). Sows also vary in their motivation to nurse piglets (Thodberg and Jensen, 1999). This trait is probably even more important in organic production, where sows are held loose on pasture or in large pens where the sow can move away from her piglets.

Normally, sows do not show oestrus during lactation, since nursing inhibits ovulation. After weaning, the ovarian activity starts quickly and most sows ovulate within a week. When sows and piglets are held in groups, e.g. on deep straw, some sows ovulate during lactation (Hultén, 1997). These sows often fall out of the mating scheme and decrease the possibility of keep them in farrowing batches. Another problem is that when sows are in groups, some will not get enough food because of competition, resulting in loss of weight and body fat (Brouns and Edwards, 1994); this in turn increases the risk of delayed oestrus after weaning (Sterning *et al.*, 1990). (How to deal with competition for feed is discussed in Chapter 7.) 'No detected oestrus' or 'Not pregnant' are among the most common culling reasons reported for sows in outdoor production (Larsen and Kongsted, 2001). Furthermore, the interval from weaning to next successful insemination seems to be longer outdoors than indoors (ITP,

2001). In conventional piglet production, hormonal treatments are routine in many countries, for example to synchronize first oestrus, to induce farrowing and to stimulate oestrus after weaning. In organic production, such use of hormonal treatments is prohibited. Thus, good reproduction ability without hormonal treatments should have a high priority in breeding pigs for organic production.

Poultry

Breeding goals

Conventional breeding goals for broiler production have focused on rapid growth, increased feed efficiency and increased processing yield. Not surprisingly, the NAHWOA Recommendations give special emphasis to poultry in discussing breeding goals. Genotypes that have been bred to suit intensive, indoor management systems are not generally suitable for organic systems, for two main reasons. First, the high productivity of these strains depends on high levels of feed, health treatments and other inputs. If these are not available, or are not allowed, for example under organic rearing systems, these birds may be prone to ill-thrift and disease. The second reason that most genotypes from large breeding companies are not suitable for organic systems is that they often are bred to suit a specific market destination. For example, large, well-fleshed birds are required for the processing and ready-meal markets, whereas birds of 2–3 kg are required for the roasting market. As these birds often are only 5–6 weeks old, this does not meet the current EU requirement for organic poultry to be at least 12 weeks. Because layer and broiler strains have become so different, they are not appropriate for dual-purpose organic systems. For example, with layers, cockerel chicks are killed after hatching because feeding them to 10–12 weeks will not produce marketable chickens. Similarly, an adult hen at end of lay is not suitable for chicken meat.

Another incompatibility between conventional and organic principles for egg production is the use of hens for only one laying period in conventional production, although in practice, more are possible. There is at least one breeding company that supplies hens (in the USA) that are not discarded as soon as they start to moult, and produce eggs for several laying seasons. Using these hybrids may be appropriate for organic laying units, not just because of the high price of new layers (Deerberg, 2001) but also because it is more in keeping with organic principles.

For these reasons, the development of dual-purpose breeds could overcome the apparent disparity between separate layer and broiler strains, to combine both egg and meat production. Hirt *et al.* (2000) showed that dual-purpose chickens can producing 240 eggs per hen and that cockerels can reach 1.4–2.5 kg body weight in 10–12 weeks. They concluded, though, that dual-purpose chickens are still not profitable enough to be commercially useful. However, they could be a start for organic breeding, although to be feasible this approach would require considerable price increases for meat and eggs.

Health problems

There is considerable evidence that genetic selection for high productivity in conventional strains of broilers and layers has led to deterioration in animal health. Rauw *et al.* (1998) classified the problems into three categories: reproductive difficulties, disease and immuno-competence, and metabolic. Reproductive problems include more defective eggs, 'erratic' ovulations (Siegel and Dunnington, 1985; Anthony *et al.*, 1989) and chromosomal abnormalities (Siegel and Dunnington, 1985).

Health problems include:

- Reduced immune performance (Maatman *et al.*, 1993; Rauw *et al.*, 1998; Yunis *et al.*, 2000);
- Higher incidence of tibial dyschondroplasia (Whitehead and Wilson, 1992);
- Ascites (Thorp and Luiting, 2000; Moghadam *et al.*, 2001; Navarro *et al.*, 2002);
- Pectoral myopathy (Henckel, 2002; Nielsen *et al.*, 2003);
- Heart failure (Chambers, 1986; Rauw *et al.*, 1998);
- Skeletal deformities (Whitehead *et al.*, 2003);
- And leg and bone problems (Mercer and Hill, 1984; Sørensen, 1992; Whitehead and Wilson, 1992; Bishop *et al.*, 1996; Bakken *et al.*, 1998).

Metabolic problems include:

- Damaged satiety mechanisms (hypothalamic failure to diminish the hunger drive leading to over-consumption). Selected chickens consume feed above their metabolic requirements until they reach a limit set by gastrointestinal capacity (Burkhart *et al.*, 1983; Barbato *et al.*, 1984; Siegel and Dunnington, 1985).
- Enhanced susceptibility to stress linked to larger muscle fibres (Henckel, 2002).
- Hypoxia leading to myopathies, particularly deep pectoral myopathy ('Oregon disease') (Siller *et al.*, 1978; Henckel, 2002).
- Metabolic 'exhaustion' leading to high pH and a high number of glycolytic muscle fibres (Henckel, 2002).
- Hypothyroidism (Scheele *et al.*, 1991; Buys *et al.*, 1999, cited by Henckel, 2002).

Many of these problems have been recognized by conventional breeders, some of whom have taken positive steps to prevent a further deterioration in fitness. Including some traits important for health and welfare alongside more conventional traits in poultry breeding indexes can capture a market opportunity for breeding birds that are more robust than conventional strains, and that are specifically tailored to the organic egg or poultry meat market. In this way, 'modern' levels of production are maintained, but the associated problems of poor health and fertility are halted.

Behaviour

Behavioural characteristics such as feather pecking, cannibalism and lack of interest for nests at oviposition can be a problem with caged as well as non-caged birds. As these characteristics are heritable, they can be overcome through genetic selection (Sørensen, 2001). Feather pecking and cannibalism have heritability estimates that justify consideration in an organic breeding programme for laying stock (Kjær, 1995; Kjær and Sørensen, 1997; Kjær *et al.*, 2001). Cannibalism, causing an unacceptably high level of mortality, was shown by Muir and co-workers to be markedly reduced by family selection over a few generations (Muir, 1996). Other characteristics, such as the ability to use roughage and reaction to predators could potentially be worthy organic breeding goals. Therefore, to include such behaviour into breeding programmes would be a relevant and feasible goal for organic poultry systems.

Selection techniques

As health traits such as ascites are only partially correlated genetically with high production, it is possible to select animals that not only perform well, but that also have good health. (For a discussion of the main diseases affecting poultry and the opportunities for overcoming them by breeding, see Thorp and Luiting, 2000.) There is considerable evidence of genetic variation in susceptibility, tolerance, resistance and immune response to many of the main diseases or conditions that affect poultry. More losses are attributable to metabolic disorders than infectious diseases, with cardiovascular and musculoskeletal disorders being the main problems, which in turn are strongly influenced by nutrition and environment but also have a genetic component.

Conventional breeding (selecting on phenotypic traits) is the easiest strategy for the genetic improvement of disease resistance, and has already been applied to Marek's disease and Newcastle disease (Friars *et al.*, 1972, and Gordon *et al.*, 1971, respectively). Indirect selection via genetic markers has great potential and has already been used for control of Marek's disease (Thorp and Luiting, 2000). Particular research emphasis has focused on typing the MHC (major histocompatibility complex) genes, which exert major genetic control over host resistance to disease. Although there is potential for transgenesis to become available for the control of avian disease, it is very unlikely that this would be considered acceptable in either conventional or organic farming because of public opinion.

Achieving the Chosen Breeding Objectives

Selection between breeds

The only objective way to choose among breeds is to use across-breed evaluations. For example, across-breed evaluations have been routine in the New Zealand dairy population since 1996 (Harris *et al.*, 1996), so that the best animals can be identified regardless of the breed (or cross). Two large projects are underway in the southern Australian beef and sheep industries to improve genetic evaluation by developing Multi Breed Estimated Breeding Values (EBVs). It is expected that they will soon be incorporated into the Australian genetic evaluation tools BREEDPLAN and LAMBPLAN (Brien *et al.*, 2000).

Cross-breeding

Cross-breeding is often a good way to increase the overall productivity of animals by using the resulting heterosis (or 'hybrid vigour'). Heterosis is defined as the deviation in performance from the mid-parent average. The amount of heterosis is a function of the gene frequency of the trait of interest in the two pure-bred populations. If the gene frequency is similar, there will be little heterosis. Traits such as health, fertility and fitness appear to benefit most from heterosis (Falconer and Mackay, 1996) The additive genetic component of these traits is small, and therefore difficult to improve through selection, yet the dominance component is relatively large. Heterosis for survival from first to fifth lactation in first cross Jersey × Holstein–Friesian dairy cattle was between 9.6% and 18.3%; for proportion in calf at first AI, heterosis was between 6.8% and 10.1% (Harris and Kolver, 2001). Cross-breeding may be an attractive option for organic farmers, as fitness traits gain most.

Cross-breeding is common in pigs and poultry, with hybrids or lines the product of most breeding companies. It also is widely used in the sheep and beef industries. Cross-breeding is not as common in dairy cattle in most countries, with the exception of New Zealand, where approximately 30% of cattle are cross-bred, predominantly Jersey crossed with Holstein–Friesian. This is because in New Zealand, the best Jersey and Holstein–Friesian sires are similar in overall genetic merit (Breeding Worth).

Cross-bred animals can be profitable where conditions are difficult. This is why cross-breeding is common in cattle and sheep in New Zealand

and Australia, where animals are kept in continuous grazing, low-cost environments. Also, cross-breeding is common in the tropics because local breeds are more adapted to the environment than the high-yielding European breeds. By cross-breeding, good characteristics of both breeds can be used. Organic farming often faces tougher conditions than other conventional systems because of the restrictions and standards placed on the use of medicines and preventative disease treatments, levels of feeding, grazing, and use of fertilizers and manure.

The first cross between two breeds (F1) shows 100% heterosis. There is a choice of what to do after the first cross. One strategy is continuous production of F1s, which involves maintaining a proportion of the population or herd as pure-bred and mating the best ones within each breed to maintain the pure-bred population, while mating the rest of the purebreds to the desired male of the other breed to provide replacements for the cross-bred part of the population or herd. Alternatively, continuous rotational cross-breeding strategies can be used. A two-breed rotational cross maintains 67% of the direct heterosis, while three- and four-breed crosses maintain 86% and 94% of the direct heterosis respectively. The problem is to find several breeds of suitable merit such that the cross-bred population is better than the pure-bred population.

Selection indexes

Hazel (1943) developed a method, known as selection index methodology, of combining information from several sources to measure genetic merit. Selection indexes are generally viewed as the method of choice to make genetic progress within a breed when there are multiple selection objectives (e.g. Simm, 1998). Across-breed selection is possible where across-breed evaluations are done (as previously discussed). Applying existing selection indexes to organic production depends on: (i) whether the right traits are recorded and available; (ii) whether weightings applied to each of the traits are appropriate for organic circumstances; and (iii) whether the genetic evaluations gained from conventional systems are appropriate for organic farms.

Well-designed breeding programmes usually have an overall breeding objective or set of objectives that encompass the vision of the future market either for a single customer type or a range of customer types. The ultimate breeding objective has traditionally been profitability in conventional production, but it could encompass an alternative objective, such as animal welfare, especially in organic systems. The selection criteria (traits that are measured) are not necessarily the same as the breeding objective, as in the example of using somatic cell count where the breeding objective is mastitis resistance. Breeding values or predicted transmitting abilities are multiplied by appropriate economic weights or relative importance and summed. The result is a single score that gives an animal's ranking according to the population with which it was evaluated. This can be either within individual herds or flocks, or across flocks and herds that are genetically linked.

Economic and other values

Normally the weights applied to traits are derived through optimizing their monetary value to the producer (e.g. Weller, 1994), using economic models that give the expected return from a 1% change in genetic merit in the trait in question. A trait such as mastitis resistance would probably be more valuable and therefore be given a greater economic weight on an organic than a conventional farm.

Evaluating the direct economic impact of disease may not cover all the perceived value in improving a disease trait, as there also is a welfare benefit to improving disease resistance. Therefore we need to consider what alternative approach could replace or enhance the conventional economic framework for estimating weights. One approach would be to pre-determine the rate of genetic progress, such as desired gains indexes, or restricted indexes (Brascamp, 1984). To avoid further deterioration in a trait, a restricted index would be used, and to improve a trait at a given rate, a desired gains index would be used. An example would be selecting for production while pre-determining genetic change in fitness traits. The advantage with these methods is that derivation of economic weights is entirely avoided.

An alternative strategy is to consider each trait to have both a market and a non-market value; individual traits may differ in whether they have only one or both types of value. Research in this area is new, although several authors have already given it thought (e.g. Olesen *et al.*, 2000). This approach raises the issue of the determination of the non-market (or ethical) values for traits. Research is needed to develop this area further and determine how ethics, welfare and social considerations can be included into breeding goals for organic livestock. Potentially, this is where organic livestock breeding could differentiate itself from conventional breeding.

Customized indexes

As breeding has become more sophisticated with the aid of new statistical tools and computer capability, the number of selection criteria and the complexity of the breeding goals have expanded. Selection has become more precise and based on more traits simultaneously. Yet the majority of customers are conventional farmers, and breeding programmes cater to their requirements. Although organic farms have become much more numerous in Europe in recent years, they still are overshadowed by conventional farms.

National breeding objectives are geared towards the 'average' farm. This is reasonable, as selection indexes are insensitive to modest changes in economic values. For example, Baumung *et al.* (2001) found a correlation of close to 1 between the economic efficiencies of the existing Austrian Simmental selection index and an index where the economic weights of functional traits were increased by 50%. The lower response in milk traits is nearly offset by higher responses in functional traits. Weller (1994) warns that if no single trait dominates the index, then altering index weights will change expected responses to selection by more. This is important, because health and fertility traits are likely to have greater importance relative to production in organic breeding goals than in conventional breeding goals. Also, as genetic evaluations become available on more traits, farmers may wish to select which traits to include and which to exclude in the breeding objective. These are just some of the reasons why customized selection indexes are gaining ground as an alternative way of selecting animals.

A customized selection index works the same as a standard selection index, except the assumptions used to derive economic weights rely on data specific to the farm or type of farms, provided there is no G×E and the appropriate traits are being evaluated. The Internet provides a convenient way of allowing this technology to become reality. The breeder would answer several questions related to sale prices and costs of inputs; from these a set of economic weights could be calculated. The output could be a list of sires tailored exactly to the requirements specified in the inputs. Before an index customized to organic farming can be developed, a clear understanding of requirements is needed. An example of such a computerized index customization is one in Australia for sheep producers, called OBJECT (Atkins *et al.*, 1994).

Alternative selection methods

Alternative ways of selecting animals are based on lifetime production, line-breeding or on-farm 'kin breeding' (Baars, 1990). Most of the followers of breeding on lifetime production are from Germany, Austria and Switzerland. The assumption is that cows that survive have all the desirable traits for sustainable production over many lactations. With family or kin breeding, breeding is entirely based on the farm herd, as this is believed to avoid G×E. It relies on the farmer knowing which animals exhibit the traits that suit his or her breeding goals.

This system is based on 'low-scale' inbreeding, where most animals are related, but the same ancestor appears more than once in an animal's pedigree four or more generations back. To run the scheme successfully, inbreeding needs to be monitored to ensure it does not get too high. Several farmers are already following this sort of breeding scheme successfully. However, making genetic progress in health and fertility traits would be very difficult in this type of scheme, as it is realistic only with large-scale progeny testing because of the low heritability estimates of these traits.

Genetic Evaluations and Environments

When animals for use in conventional systems are selected under a narrow range of conditions, is it realistic to expect the same ranking in organic systems? This depends to a degree on the species. Breeding pigs and poultry are kept and tested in environments that are very different from those in which organic animals are kept. In contrast, progeny testing of cows and sheep may take place in environments that range from similar to organic to very intensive systems, especially for dairy cattle.

As noted, companies that provide the parents for the next generation are likely to heed the requirements of the majority of their customers. Major poultry breeders, for example, breed specific lines to suit the different types of commercial customers, depending on whether they are mainly producing for the processing, ready cooked or fresh markets. Some also breed strains of poultry tolerant to heat stress for their worldwide customers. Some companies use sub-indexes that include elements of health and fitness (such as leg strength) that are more suited to the needs of organic producers. If the demand were great enough, there would be considerable scope for breeding companies to supply animals that meet the needs of the organic producer.

Obtaining genetic evaluations using just organic livestock is unrealistic because there are so few of them compared with conventional livestock. To date, no articles appear to have been published on genetic correlations between organic and conventional systems. The reason is that to get robust estimates requires thousands of observations and good genetic ties between individuals in different environments. However, there is information on genetic correlations of the same trait measured in systems that rely heavily on concentrates versus extensive systems. Studies on different conventional environments show G×E interactions ranging from 'not of significant, practical importance' to 'meaningful'.

For example, estimated genetic correlations between countries for protein yield in Holsteins range from 0.95 to 0.75 (Interbull, 2003a). Genetic correlations between countries for somatic cell count (udder health) are generally similar to production traits, ranging from 0.63 to 0.97 (Interbull, 2003b). Correlations between countries for longevity are lower: the average genetic correlation for Holsteins was 0.56 (van der Lindhe and de Jong, 2003). Some of the differences are due to weak genetic ties and the variety of methods used to produce genetic evaluations for longevity, such as in Israel, which had an average correlation of 0.21 with other countries. However, even for countries that use very similar models and have adequate genetic ties, correlations appear to be low. For example, the genetic correlation between France and Switzerland for longevity was 0.66, although both use a method known as the 'survival kit' (Ducrocq and Sölkner, 1994). Such a genetic correlation suggests that in different environments there will be extensive re-ranking of animals by breeding values.

One of the most difficult parts of organic pig production is getting adequate feed composition. Growing pigs need feed with a high-quality protein, such as soybean. An aim of organic farming is to use locally produced feed, but in many northern countries soybean will never be locally produced. Furthermore, a large part of the soybean available at the market is genetically modified, but genetically modified organisms are prohibited under organic standards. During the 1980s, pigs were selected for high lean-tissue growth rate in two lines at the Swedish University of Agricultural Sciences (Stern, 1984). One line was fed a low-protein feed during the rearing and fattening period, the other a high-protein feed. Pigs with a high genetic capacity for lean-tissue growth rate at the high-protein feed also performed better when fed the low-protein feed. Thus, no G×E interactions were found. However, conventional and organic production differ in more than feed composition. In another study at the Swedish University of Agricultural Sciences, growth rate during the fattening period showed an interaction between sire breed (Hampshire and Swedish Landrace) and rearing environment (indoors or outdoors) (Lundeheim *et al.*, 1995).

According to a recent Dutch study, pig reproduction traits may be governed partly by different genes in different environments with different infection pressure. The genetic correlation between piglet survival in conventional farms and in SPF farms was only 0.6. The corresponding correlation for farrowing interval was 0.8 (Bergsma *et al.*, 2001). However, the rate of stillborn piglets seemed to be influenced by the same genes in both environments: the genetic correlation between those traits was close to 1.

The importance of G×E depends on the difference between organic and conventional systems. If the difference is small and no G×E is expected, then provided genetic evaluations of all the traits required are available, a customized organic index may be the most appropriate strategy. If the difference is large, then a solution worth considering is genetic evaluations specifically for organic herds. This may be a worthwhile and economical solution if there are enough farms to make the evaluation feasible. Across-country organic evaluations may be a realistic alternative to ensure that sufficient numbers of animals are included in the evaluation.

This could also open opportunities for across-country progeny testing specifically for organic farms. Widespread use of AI makes this technically feasible. The problem with this solution is the great expense of running a progeny-testing scheme.

To be able to establish whether existing breeding programmes fulfil the requirements of organic livestock, estimates of GxE interactions would be useful. Without this information, existing genetic evaluations could be used, assuming there is no re-ranking of animals. If there is good reason to believe that significant re-ranking would occur, then an alternative strategy, such as cross-breeding, might be a good option.

The Use of Molecular Genetics

In many species, considerable genetic progress has been made in commercially important traits using phenotypic information, that is, measurements made on individual animals. Selection of traits such as disease resistance, fertility and meat quality have been limited, because these traits either are difficult to measure or have low heritabilities (or both).

This is where recent advances in molecular biology may be of particular benefit. Until recently, it was thought that most traits were controlled by a large number of genes. However, research has shown that a large part of the variation in these quantitative traits are controlled by a moderate number of loci. For example, even in dairy cattle, where selection for production has been intensive, QTLs explaining substantial variation are still being found, which gives encouragement for under-explored species, such as sheep (Haley, 2001). With this identification of single loci that have major effects, one can apply marker-assisted selection if a QTL and genetic markers associated with it have been identified, or gene-assisted selection if the exact position of the gene is known.

There are two main approaches to identifying trait loci: candidate genes and genome scans (Haley, 2001). In both the aim is to identify a marker that has two or more DNA variants (i.e. alleles) present in the population that are associated with variation in the trait of interest. Both approaches have been successful in identifying genes with appreciable effects.

Once suitable DNA markers or the gene itself are identified, the next questions are what to do with the information and whether this technology fits in with organic principles. Transgenesis (transfer of genes from one organism to another), while possible, is unlikely to be accepted even in conventional livestock breeding because of consumer opinions. The biggest benefits of molecular technologies are likely to be in improving the accuracy of selection of individuals and helping to estimate the genetic merit of an animal for a trait in the absence of observations, e.g. early on in life. The technology could also be used to introgress or cross in desirable genes from one population to another. An example of this is introducing the polled gene to a horned population of dairy cows.

The EU Regulation on Organic Livestock Production allows dehorning and disbudding on animal welfare and safety grounds. Currently, all the biodynamic organizations (in German-speaking countries) prohibit dehorning and disbudding, but in most other European countries the interpretation of this regulation has been that disbudding is acceptable (see Chapter 8). However, the polled gene exists in some cattle populations and has been located on chromosome 1 (Brenneman *et al.*, 1996). Georges *et al.* (1993) have mapped suitable markers for this gene. Introducing the polled gene to non-polled populations of cattle through crossing (or introgression) would mean that calves would not have to be dehorned or disbudded, thereby removing a stressful procedure. The polled gene is dominant to the horned gene, so that carriers of the gene will be polled. Assuming there is no use of reproductive technologies other than AI, crossing the polled gene from Friesians into the Holstein population would take about 16 years (4 years per generation, three backcross generations plus inter-cross) and would come at a substantial cost to genetic progress in traits such as production (Pryce *et al.*, 2001). Since the mid-1980s, polledness has been included in the Norwegian dairy breeding objective (without the use of gene technology), and naturally polled cattle are common among Norwegian Red cattle (NRF). Today about 15% of the population is polled, and in the breeding programme, preference is given to polled sires and 'sire mothers' (Arne Ola Refsdal, GENO, Norway, 2003, personal communication , 2003)

Molecular techniques are already being applied to animal breeding, e.g. to eliminate scrapie in sheep. Pig breeders have long known about the lesion of the CRC gene that causes poor meat quality (Fujii *et al.*, 1991), and cattle and chickens are being tested for complex vertebral malformation and Newcastle disease, respectively. Marker-assisted selection is already being used by some of the larger pig and dairy cattle breeding organizations. More widespread use seems likely, as efforts to identify QTLs continue. From an organic perspective, the technology may offer even more, as markers can also be used to explore and maintain genetic diversity in small populations. The extent to which these new technologies should be used is an urgent issue for the organic farming movement to discuss, taking into consideration the pros and cons in relation to the overall aims of organic farming.

Reproductive Technologies

Breeding technologies can be used to accelerate genetic progress, but some may not fit with the values and principles of organic farming. Whether to allow their use is a matter to be decided by the organic movement, as represented by IFOAM and embodied in the EU organic standards, among others.

Appreciable genetic progress in production traits has already been made in poultry and pig breeding, with little use of reproductive technologies. This is because the generation interval is short and litters large. Genetic progress in sheep and cattle has generally been slower, because the generation interval is long and only one or two offspring are born per year. Also, with dairy cattle, the traits of interest are expressed later in life and are sex-limited.

The dairy cattle industry, in particular, has used reproductive technologies extensively to accelerate the genetic progress possible through selection. There are two important pathways: through the nucleus population (controlled by breeding companies), and through dissemination of the genetic gain to commercial populations. The genetic difference between the two populations is termed 'genetic lag'. AI is the main way that genetic progress in the nucleus population gets disseminated to the commercial population; its main advantage is in reducing genetic lag.

In the past, genetic gain has been monitored primarily through progress in milk production. However, more emphasis is now being put on health and fertility traits, a trend that can be expected to continue, especially in organic production. Regardless of the breeding goal, AI is the most effective way of disseminating progress in desired traits, and in practice is the only way to improve traits with a low heritability, such as health and fertility. As these are traits that organic farmers are interested in, this is a strong argument against using natural mating. The drawback, however, is that organic systems are supposed to allow animals to act out their natural behaviour, which AI obviously prevents. Furthermore, inbreeding is now becoming a serious issue in some countries, as the best bulls are used worldwide through AI. The average inbreeding in Holstein cows in the USA increased by about 4% between 1982 and 2002 (USDA, 2003).

Reducing the effective population size (the result of inbreeding) reduces biodiversity, although maintaining or increasing biodiversity is a goal of organic farming. The rate of inbreeding can be reduced both in the nucleus population and in the commercial population, and judicious planning can encourage, rather than suppress genetic diversity. On balance, using AI technology should be encouraged, both to ensure that the best animals for organic production can be used and to minimize the risk of an outbreak of a sexually transmitted disease. In addition, further considerations of human safety issues associated with keeping bulls on farm tilt the argument towards the continued use of AI for the organic dairy sector.

Sexed semen is now commercially available. Semen can be sorted either physically or chemically, or a combination of physical and chemical separation. Physical separation is currently allowed under EU rules.

However, commercially viable separation methods currently rely on a combination of physical and chemical separation. The main benefit of using sexed semen is to reduce the number of unwanted male dairy-type calves in the dairy herd by inseminating cows with sexed semen to produce replacements and breeding the remainder of the herd to beef bulls.

Embryo transfer (ET) is not used widely in commercial populations, and is currently prohibited by most organic certifying bodies. However, it is used very widely in the dairy cow nucleus population. Of the most popular bulls currently available worldwide, few have no ET ancestors. Cloning is very unlikely to be acceptable to the organic movement.

The question of what reproductive technologies are acceptable for organic farming presents an ethical dilemma, particularly regarding the nucleus populations. If we exclude ET from organic herds, but allow it for ET sires, do we have a double standard? If we exclude all ET sires (or sires with ET ancestors), we exclude most AI bulls. A pragmatic view is to accept treatments that are not applied on the organic farm itself, as the germplasm from ET bulls, e.g., is one step removed from the ET process.

Indigenous Livestock and Genetic Diversity

Although maintaining genetic diversity is a major objective of organic livestock production, in The Netherlands 63% of organic dairy herds are Holsteins (Nauta, 2001), and a similar situation holds in the UK (Roderick and Hovi, 1999). The issue of inbreeding in dairy breeds, such as Holsteins, has already been mentioned, and there also is the serious question of whether such breeds, selected to produce high yields from high levels of purchased inputs, are appropriate for organic systems.

Attitudes regarding the role of indigenous breeds in organic farming have varied from country to country. For example, in The Netherlands, some farmers have said that they were not satisfied with the Holstein breed, and a few use more traditional Dutch breeds, such as the Dutch Friesian, MRIJ and Groninger Blaarkop (Nauta, 2001). Norwegian organic farmers have deliberately tried to use indigenous breeds, whilst other countries, such as Sweden, have placed much less emphasis on them. One reason could be that economic conditions have been much tougher in Sweden than in Norway.

In organic pig production, there are niche markets for traditional breeds. For example, the Rare Breeds Survival Trust in the UK has developed a Meat Marketing Scheme with independent butchers to encourage consumers to appreciate and demand traditional breeds, such as the Gloucester Old Spot. A survey of UK organic herds in 1999 found nine different traditional breeds in use, the most numerous being the British Saddleback (Kelly *et al.*, 2001). In Spain, particularly in the extensive grazing areas, there is a great diversity of breeds, and although many of the indigenous breeds are in decline, others, such as the Iberian, are increasing.

The future place of indigenous breeds in organic farming will be determined by many considerations: market forces; the suitability of conventionally bred livestock for organic farms; the adaptability of indigenous livestock; and the economic reality of keeping animals that are less productive than the alternatives. If existing breeding programmes focus on traits that are of less value to organic farmers or fail to include traits of interest, such as health and fertility, we can expect indigenous breeds to become more popular, provided they have characteristics that are valuable in organic farming. Of course, using pure-bred indigenous breeds is only one option; crossing with conventionally bred livestock may be an attractive alternative. In conjunction with molecular technologies, there is potentially more to be gained through introgression (crossing in) of desirable genes from indigenous donor to modern recipient populations.

Conclusions

The choice of strategy to guide the breeding decisions of organic farmers is a complex matter, and the best solution for one farmer may be wrong for another. Therefore it is important to consider each case independently. The following considerations enter the breeding strategy:

- The first major step is to identify the most appropriate breed or cross. Consideration should be given to the environment and source of replacements, and to how much existing breeding programmes focus on organic objectives.
- Regardless of species, the most important objectives are likely to include disease resistance, fertility, longevity, high feed efficiency, foraging ability, temperament and moderate production levels.
- As soon as the breeding objectives have been identified, it is necessary to obtain suitable genetic evaluations on traits that would enable selection for the chosen selection objectives.
- If genetic evaluations for the desired traits are available, the obvious solution is to create a customized selection index geared specifically to organic farmers' requirements, if the variation in the existing population is big enough to provide suitable selection candidates.
- If the G×E interactions are substantial or key selection criteria are absent, then ideally organic breeding schemes could be set up. This may be feasible and economically sensible if there are enough farms in each country to warrant this task, or if it could be implemented across countries.
- Indigenous livestock breeds could be used as an alternative to

existing selected populations, as long as they possess characteristics that would make them commercially viable. Again, ideally an organic breeding scheme would be set up, so that genetic progress could be made. However, the populations of many indigenous or rare breeds is too small to make efficient genetic progress. Still, they are important for preserving genetic diversity. Alternatively, introgression (crossing in) of valuable genes from indigenous breeds into another population could be the best way forward. Either way, it is necessary to preserve the indigenous populations to conserve genetic diversity.

References

Anthony, N.B., Dunnington, E.A. and Siegel, P.B. (1989) Egg production and egg composition of parental lines and F1 and F2 crosses of White Rock chickens selected for 56-day body weight. *Poultry Science* 68, 27–36.

Atkins, K.D., Semple, S.J. and Casey, A.E. (1994) OBJECT – personalised breeding objectives for Merinos. *5th World Congress on Genetics Applied to Livestock Production* 22, 79–80.

Axford, R.F.E., Bishop, S.C., Nicholas, F.W. and Owen, J.B. (eds) (2000) *Breeding for Disease Resistance in Farm Animals*, 2nd edn. CAB International, Wallingford, UK.

Baars, T. (1990) *Dirk Endendijk – 21 jaar familieteelt 1967–1988*. Louis Bolk Instituut, Driebergen, The Netherlands.

Bakken, M., Vangen, O. and Rauw, W.M. (1998) Biological limits to selection and animal welfare. *6th World Congress on Genetics Applied to Livestock Production, Armidale, Australia* 27, 381–388.

Bapst, B. (2001) Swiss experiences on practical cattle breeding strategies for organic dairy herds. In: Hovi, M. and Baars, T. (eds) *Breeding and Feeding for Animal Health and Welfare in Organic Livestock Systems. Proceedings of the Fourth NAHWOA Workshop, Wageningen, 24–27 March 2001*. University of Reading, Reading, UK, pp. 44–50.

Barbato, G.F., Siegel, P.B., Cherry, J.A. and Nir, L. (1984) Selection for body weight at 8 weeks of age. *Poultry Science* 63, 11–18.

Baumung, R., Sölkner, J., Gierzinger, E. and William, A. (2001) Ecological total merit index for an Austrian dual purpose cattle breed. In: Hovi, M. and Baars, T. (eds) *Breeding and Feeding for Animal Health and Welfare in Organic Livestock Systems. Proceedings of the Fourth NAHWOA Workshop, Wageningen, 24–27 March 2001*. University of Reading, Reading, UK, pp. 14–22.

Beilharz, R.G., Luxford, B.G., and Wilkinson, J.L. (1993) Quantitative genetics and evolution: is our understanding of genetics sufficient to explain evolution? *Journal of Animal Breeding and Genetics* 110, 161–170.

Bergsma, R., Knol, E.F., Merks, J.W.M. and van Groenland, G.J. (2001) Performance levels, genetic parameters and genotype–health interactions for production traits in pigs. ASAS, Indianapolis, July 2001.

Bishop S.C. and Stear, M.J. (2000) The use of a gamma-type function to assess the relationship between the number of adult *Teladorsagia circumcincta* and total egg output. *Parasitology* 121, 435–440.

Bishop S.C., Bairden, K., McKellar, Q.M. and Park, M. (1996) Genetic parameters for faecal egg count following mixed, natural, predominantly *Ostertagia circumcincta* infection and relationships with live weight in young lambs. *Animal Science* 63, 423–428.

Brascamp, E.W. (1984) Selection indices with constraints. *Animal Breeding Abstracts* 52, 645–654.

Brenneman, R.A., Davis, S.K., Sanders, J.O., Burns, B.M., Wheeler, T.C., Turner, J.W. and Taylor, J.F. (1996). The polled locus maps to BTA1 in a *Bos indicus* × *Bos taurus* cross. *Journal of Heredity* 87, 156–161.

Brien, F.D., Graham, J.F., Cummins, L.J., Clark, A.J. and Fogarty, N.M. (2000) Improving genetic evaluation in the beef and lamb industries in Southern Australia. In: *Proceedings of the 51st EAAP Annual Meeting, The Hague*, p. 36.

Brouns, F. and Edwards, S.A. (1994) Social rank and feeding behaviour of group-housed sows fed competitively or ad libitum. *Applied Animal Behaviour Science* 39, 225–235.

Burkhart, C.A., Cherry, J.A., Van Krey, H.P. and Siegel, P.B. (1983) Genetic selection for growth rate alters hypothalamic satiety mechanisms in chickens. *Behavioural Genetics* 13, 295–300.

Buys, N., Scheele, C.W., Kwakernaak, C., and Decuypere, E. (1999) Performance and physiological parameters in broiler chicken lines differing in susceptibility to the ascites syndrome: 2. Effect of ambient temperature on partial efficiencies of protein and fat retention and plasma hormone concentrations. *British Poultry Science* 40, 140–144.

Chambers, J.R. (1986) Heritability of crippling and acute death syndrome in sire and dam strains of broiler chickens. *Poultry Science* 65 (Suppl.), 23.

Coffey, M.P., Emmans, G.C. and Brotherstone, S. (2001) Genetic evaluation of dairy bulls for energy balance traits using random regression. *Animal Science* 73, 29–40.

Conington, J., Bishop, S.C., Grundy, B., Waterhouse, A. and Simm, G. (2001) Multi-trait selection indexes for sustainable UK hill sheep production. *Animal Science* 73, 413–423.

Conington, J., Lewis, R.M. and Simm, G. (2002) Breeding goals and strategies for organic sheep production. In: Kyriazakis, I. and Zervas, G. (eds) *Organic Meat and Milk from Ruminants. Proceedings of the joint Hellenic Society of Animal Production and British Society of Animal Science meeting, Athens, October 2001.* EAAP Publication No. 106.

Deerberg, F. (2001) Eine Legepause kann sich lohnen. *Bioland* 1, 14–15.

DEFRA (2002) *National Scrapie Plan for Great Britain*. Department for Environment, Food and Rural Affairs (UK). http://www.defra.gov.uk/corporate/regulat/forms/Ahealth/nsp/nsp1.pdf

Ducrocq, V. and Sölkner, J. (1994) The Survival Kit – a Fortran package for the analysis of survival data. *Proceedings of the 5th World Congress on Genetics Applied to Livestock Production* 22, 51–52.

Dwyer, C.M. and Lawrence, A.B. (1998) Variability in the expression of maternal behaviour in primiparous sheep: effects of genotype and litter size. *Applied Animal Behaviour Science* 58, 311–330.

Edwards, S.A. (1995) Sistemas extensivos en ganado porcino. In: *III Curso-simposium Internacional de Reproduccion e I.A. Porcina*. INIA, Madrid, Chapter 7.
Edwards, SA. (2001) Perinatal mortality in the pig: environmental or physiological solutions? Presented at the 52nd Annual Meeting of the EAAP, Budapest, Hungary, August 2001.
Falconer, D.S. and Mackay, T.F.C. (1996) *Introduction to Quantitative Genetics*, 4th edn. Longman, Harlow, UK.
Friars, G.W., Chambers, J.R., Kennedy, A., and Smith, A.D. (1972) Selection for resistance to Marek's disease in conjunction with other economic traits in chickens. *Avian Diseases* 16, 2–10.
Fujii, J., Otsu, K., Zorzato, F. De Leon, S., Khanna, V.K. Weiler, J.E., O'Brien, P.J. and Maclennan, D.H. (1991) Identification of a mutation in porcine ryanodine receptor associated with malignant hypothermia. *Science* 253, 448–51.
Georges, M., Drinkwater, R., King, T., Mishra, A., Moore, S.S., Nielsen, D., Sargeant, L.S., Sorensen, A., Steele, M.R. and Zhao, X. (1993) Microsatellite mapping of a gene affecting horn development in Bos taurus. *Nature Genetics* 4, 206–210.
Gordon, C.D., Beard, C.W., Hopkins, S.R. and Siegel, H.S. (1971) Chick mortality as a criterion of selection towards resistance or susceptibility to Newcastle disease. *Poultry Science* 50, 783–789.
Grandinson, K., Rydhmer, L., Strandberg, E. and Thodberg, K. (2003) Genetic analysis of on-farm tests of maternal behaviour in sows. *Livestock Production Science* (accepted for publication). www.sciencedirect.com/science/journal/03016226
Haley, C.S. (2001) Mapping genes for milk and meat quality. *Proceedings of BSAS Annual Meeting*, Scarborough, pp. 275–278.
Harris, B.L. (2002) Genetics of body condition score in New Zealand dairy cattle. In: *7th World Congress on Genetics Applied to Livestock Production*. CD-ROM Abstract.
Harris, B.L. and Kolver, E.S. (2001) Review of Holsteinization on intensive pastoral dairy farming in New Zealand. *Journal of Dairy Science* 84 (E. Suppl.), E56–E61.
Harris, B.L., Clark, J.M. and Jackson, R.G. (1996) Across breed evaluation of dairy cattle. *Proceedings of the New Zealand Society of Animal Production* 56, 12–15.
Hazel, L.N. (1943) The genetic basis of constructing selection indexes. *Genetics* 28, 476–490.
Henckel, P. (2002) Genetic improvement in meat quality: potentials and limitations. In: *7th World Congress on Genetics Applied to Livestock Production, August 19–23, Montpellier, France*. Communication no. 11–12.
Henryon, M., Juul-Madsen, H.R. and Berg, P. (2002) Genetic variation for total and differential numbers of leukocytes exists in growing pigs. In: *7th World Congress on Genetics Applied to Livestock Production, August 19–23, Montpellier, France*. Communication 13-02.
Hirt, H., Hördegen, P. and Zeltner, E. (2000) Laying hen husbandry: group size and use of hen-runs. In: Alföldi, T., Lockeretz, W. and Niggli, U. (eds) *IFOAM 2000: The World Grows Organic. Proceedings of the 13th International IFOAM Scientific Conference, Basel, Switzerland, 28 to 31 August, 2000*, p. 363.
Hohenboken, W.D. (1986) Inheritance of behavioural characteristics in livestock. A review. *Animal Breeding Abstracts* 54, 623–639.

Hultén, F. (1997) Group housing of lactating sows. Effects on sow health, reproduction and litter performance. PhD thesis, Veterinaria 27. Swedish University of Agricultural Sciences, Uppsala.

Interbull (2001) Interbull guidelines for national and international genetic systems in dairy cattle with focus on production traits. *Interbull Bulletin* 28.

Interbull (2003a) *Genetic Evaluations: Production*. International Bull Evaluation Service. http://www-interbull.slu.se/eval/framesida-prod.htm

Interbull (2003b) *Genetic Evaluations: Udder Health*. International Bull Evaluation Service. http://www-interbull.slu.se/udder/framesida-udder.htm

ITP (2001) Gestion technique des tropeaux de truies, résultats des élevages en plein air. ITP-GTTT. L'Institute Technique du Porc, Rennes, France.

Kelly, H.R.C., Browning, H.M., Martins, A.P., Pearce, G.P., Stopes, C. and Edwards, S.A. (2001) Breeding and feeding pigs for organic production. In: Hovi, M. and Baars, T. (eds) *Breeding and Feeding for Animal Health and Welfare in Organic Livestock Systems. Proceedings of the Fourth NAHWOA Workshop, Wageningen, 24–27 March 2001*. University of Reading, Reading, UK, pp. 86–93.

Kjær, J.B. (1995) Genetic variation in feather pecking behaviour in chickens. *Applied Animal Behaviour Science* 44, 266.

Kjær, J. and Sørensen, P. (1997) Feather pecking behaviour in White Leghorns: a genetic study. *British Poultry Science* 38, 335–343.

Kjær, J., Sørensen, P. and Su, G. (2001) Divergent selection on feather pecking behaviour in laying hens. *Applied Animal Behaviour Science* 71, 229–239.

Koenen, E.P.C. and Veerkamp, R.F. (1998) Genetic covariance functions for live weight, condition score and dry matter intake measured at different stages of Holstein Friesian heifers. *Livestock Production Science* 57, 67–77.

Lambe, N.R., Conington, J., Bishop, S.C., Waterhouse, A. and Simm, G. (2001) A genetic analysis of maternal behaviour score in Scottish Blackface sheep. *Animal Science* 72, 415–425.

Larsen, V.A. and Kongsted, A.G. (2001) Frilandssohold. Produktion, foderforbrug, udsaetningsårsager og graesdaekke. DJF rapport nr 30, husdyrbrug. Foulum, Denmark.

Lundeheim, N., Nyström, P.-E. and Andersson, K. (1995) Outdoor vs indoor raising of growing/finishing pigs. Does a genotype × environment interaction exist? Presented at the 46th Annual Meeting of the EAAP, Prague, Czech Republic, September 1995.

Maatman, R., Gross, W.B., Dunnington, E.A., Larsen, A.S. and Siegel, P.B. (1993) Growth, immune response and behavior of broiler and leghorn cockerels fed different methionine levels. *Archiv für Geflügelkunde* 57, 249–256.

Mark, T., Fikse, W.F., Emanuelson, U. and Philipsson, J. (2002) International genetic evaluations of Holstein sires for milk somatic cell and clinical mastitis. *Journal of Dairy Science* 85, 2384–2392.

Mavrogenis, A.P., Koumas, A., Kakoyiannis, C.K. and Taliotis, C.H. (1995) Use of somatic cell counts for the detection of sub-clinical mastitis in sheep. *Small Ruminant Research* 17, 79–84.

McEwan, J.C. and Kerr, R.J. (1998) Further evidence that major genes affect host resistance to nematode parasites in coopworth sheep. *6th World Congress on Genetics Applied to Livestock Production, Armidale, Australia* 27, 335–338.

McEwan, J.C., Mason, P., Baker, R.L., Clarke, J.N., Hickey, S.M. and Turner, K.

Askham Bryan College
LIBRARY BOOK

(1992) Effect of selection for productive traits on internal parasite resistance in sheep. *Proceedings of the New Zealand Society of Animal Production* 52, 53–56.

McGlone, J.J. and Hicks, T.A. (2000) Farrowing hut design and sow genotype (Camborough-15 vs 25% Meishan) effects on outdoor sow and litter productivity. *Journal of Animal Science* 78, 2832–2835.

Mercer, J.T. and Hill, W.G. (1984) Estimation of genetic parameters for skeletal defects in broiler chickens. *Heredity* 53, 193–203.

Merks, J.W.M. (2001) One century of genetic changes in pigs and the future needs. In: Hill, W.G., Bishop, S.C., McGuirk, B., McKay, J.C., Simm, G. and Webb, A.J. (eds) *The Challenge of Genetic Change in Animal Production*. British Society of Animal Science Occasional Publication No. 27, BSAS, Edinburgh.

MLC (2000) *Pig Yearbook*. Meat and Livestock Commission, Milton Keynes, UK.

Moghadam, H.K., McMillan, I., Chambers, J.R. and Julian, R.J. (2001) Estimation of genetic parameters for ascites syndrome in broiler chickens. *Poultry Science* 80, 844–848.

Morris, C.A. (1998) Responses to selection for disease resistance in sheep and cattle in New Zealand and Australia. *6th World Congress on Genetics Applied to Livestock Production, Armidale, Australia* 27, 295–302.

Morris, C.A. (2000) Genetics of susceptibility in cattle and sheep. In: Axford, R.F.E., Bishop, S.C., Nicholas, F.W. and Owen, J.B. (eds) *Breeding for Disease Resistance in Farm Animals*, 2nd edn. CAB International, Wallingford, UK.

Muir, W.M. (1996) Group selection for adaptation to multiple-hen cages: selection program and direct responses. *Poultry Science* 75, 447–458.

Nauta, W.J. (2001) Breeding strategies for organic animal production: an international discussion. In: Hovi, M. and Baars, T. (eds) *Breeding and Feeding for Animal Health and Welfare in Organic Livestock Systems. Proceedings of the Fourth NAHWOA Workshop, Wageningen, 24–27 March 2001*. University of Reading, Reading, UK, pp. 7–13.

Navarro, P., Koerhuis, A.N.M., Chatziplis, D., Visscher, P.M. and Haley, C.S. (2002) Genetic studies of ascites in a broiler population. In: *7th World Congress on Genetics Applied to Livestock Production, August 19–23, Montpellier, France*. Communication No. 04-19.

Nielsen, B.L., Thomsen, M.G., Sørensen, P. and Young, J.F. (2003) Feed and strain effects on the use of outdoor areas by broilers. *British Poultry Science* 44, 161–169.

Olesen, I., Gjerde, B. and Groen, A.F. (2000) Definition of animal breeding goals for sustainable production systems. *Journal of Animal Science* 78, 570–582.

Pryce, J.E., Coffey, M.P. and Brotherstone, S. (2000) The genetic relationship between calving interval, body condition score and linear type and management traits in registered Holsteins. *Journal of Dairy Science* 83, 2664–2671.

Pryce, J.E., Wall, E.E., Lawrence, A.B. and Simm, G. (2001) Breeding strategies for organic dairy cows. In: Hovi, M. and Baars, T. (eds) *Breeding and Feeding for Animal Health and Welfare in Organic Livestock Systems. Proceedings of the Fourth NAHWOA Workshop, Wageningen, 24–27 March 2001*. University of Reading, Reading, UK, pp. 23–34.

Raadsma, H.W. (2000) Genetic aspects of resistance to ovine footrot. In: Axford, R.F.E., Bishop, S.C., Nicholas, F.W. and Owen, J.B. (eds) *Breeding for Disease Resistance in Farm Animals*, 2nd edn. CAB International, Wallingford, UK.

Rauw, W.M., Kanis, E., Noorhuizen-Stassen, E.N. and Gommers, F.J. (1998)

Undesirable side effects of selection for high production efficiency in farm animals: a review. *Livestock Production Science* 56, 15–33.

Roderick, S. and Hovi, M. (1999) *Animal Health and Welfare in Organic Livestock Systems: Identification of Constraints and Priorities.* A report to the Ministry of Agriculture, Fisheries and Food. The University of Reading, Reading, UK.

Scheele, C.W., DeWitt, W., Frankenhuis, M.T. and Vereijken, P.F.G. (1991) Ascites in broilers. I. Experimental factors invoking symptoms related to ascites. *Poultry Science* 70, 1069–1083.

Siegel, P.B. and Dunnington, E.A. (1985) Reproductive complications associated with selection for broiler growth. In: Hill, W.G., Manson, J.M. and Hewitt, D. (eds) *Poultry Breeding and Genetics. Proceedings of 18th Poultry Science Symposium.* British Poultry Science – Longman, Harlow, UK, pp. 59–72.

Siller, W.G., Wright, P.A.L. and Martindale, L. (1978) Exercise induced deep pectoral myopathy in broiler fowls and turkeys. *Veterinary Science Communications* 2, 331–336.

Simm, G. (1998) *Genetic Improvement of Cattle and Sheep.* Farming Press, Ipswich, UK.

Simm, G. and Dingwall, W.S. (1989) Selection indices for lean meat production in sheep *Livestock Production Science* 21, 223–233.

Sørensen, P. (1992) The genetics of leg disorders. In: Whitehead, C.C. (ed.) *Bone Biology and Skeletal Disorders in Poultry.* Carfax Publishing, Abingdon, UK, pp. 213–229.

Sørensen, P. (2001) Breeding strategies in poultry for genetic adaptation to the organic environment. In: Hovi, M. and Baars, T. (eds) *Breeding and Feeding for Animal Health and Welfare in Organic Livestock Systems. Proceedings of the Fourth NAHWOA Workshop, Wageningen, 24–27 March 2001.* University of Reading, Reading, UK, pp. 51–61.

Stern, S. (1994) Lean growth in pigs: response to selection on high and low protein diets. Doctoral thesis, report no 108. Department of Animal Breeding and Genetics, Swedish Agricultural University, Uppsala, Sweden.

Sterning, M., Rydhmer, L., Eliasson, L., Einarsson, S. and Andersson, K. (1990) A study on primiparous sows of the ability to show standing oestrus and to ovulate after weaning. Influences of loss of body weight and backfat during lactation and of litter size, litter weight gain and season. *Acta Veterinaria Scandinavica* 31, 227–236.

Svendsen, M. (1999) A retrospective study of selection against clinical mastitis in the Norwegian dairy cow population. *Interbull Bulletin* 23, 99–106.

Thodberg, K. (2001) Individual variation in maternal behaviour of gilts and sows – effects of environment, experience and reactivity. PhD thesis, Department of Animal Science and Animal Health, The Royal Veterinary and Agricultural University, Copenhagen.

Thodberg, K. and Jensen, K.H. (1999) How to test sows' motivation to nurse their piglets. Presented at the 33th ISAE meeting, Lillehammer, Norway.

Thodberg, K. and Kongsted, A.G. (2001) Pattegrisedødelighed. Bilag til temamøde 'Økologisk og udendørs svineproduktion – Hvor står vi?'. Intern Rapport nr. 145. Danmarks JordbrugsForskning.

Thorp, B.H. and Luiting, E. (2000) Breeding for resistance to production diseases in poultry. In: Axford, R.F.E., Bishop, S.C., Nicholas, F.W. and Owen, J.B.

(eds) *Breeding for Disease Resistance in Farm Animals*, 2nd edn. CAB International, Wallingford, UK.

USDA (2003) *AIPL Inbreeding Coefficients for Holstein Cows, Calculated February, 2003.* Animal Improvement Programs Laboratory, United States Department of Agriculture. http://aipl.arsusda.gov/dynamic/inbrd/current/kindx.html.

Valros, A. (2003) Behaviour and physiology of lactating sows – associations with piglet performance and sow postweaning reproductive success. PhD thesis, University of Helsinki, Finland.

van der Lindhe and de Jong, G. (2002) Feasibility of MACE for longevity traits. *Interbull Bulletin* 29, 55–60.

van der Steen, H.A.M. (1994. Genotypes for outdoor production. *Pig News Info* 15, 129–130.

Vangen, O., Holm, B., Rossly, T., Vasbotten, M., Valros, A. and Rydhmer, L. (2002) Genetic variation in maternal behaviour of sows. In: *7th World Congress on Genetics Applied to Livestock Production, August 19–23, Montpellier, France.* Communication 14–12.

Weller, J.I. (1994) *Economic Aspects of Animal Breeding.* Chapman and Hall, London.

Whitehead, C.C. and Wilson S. (1992) Characteristics of osteoporosis in hens. In: Whitehead, C.C. (ed.) *Bone Biology and Skeletal Disorders in Poultry.* Carfax Publishing, Abingdon, UK, pp. 265–280.

Whitehead, C.C., Fleming, R.H., Julian, R.J. and Sørensen, P. (2003) Skeletal problems associated with selection for increased production. In: Muir, W.M. and Aggrey, S.E. (eds) *Poultry Genetics, Breeding and Biotechnology.* CAB International, Wallingford, UK, pp. 29–52.

Yazdi, M.H., Lundeheim. N., Rydhmer, L., Ringmar-Cederberg, E. and Johansson, K. (2000) Survival of Swedish Landrace and Yorkshire sows in relation to osteochondrosis: a genetic study. *Animal Science* 71, 1–9.

Yunis, R., Ben-David, A., Heller, E.D. and Cahaner, A. (2000) Immunocompetence and viability under commercial conditions of broiler groups differing in growth rate and in antibody response to *Escherichia coli* vaccine. *Poultry Science* 79, 810–816.

17

Organic Animal Husbandry: the Future Challenges

Mette Vaarst,[1] Stephen Roderick,[2] Vonne Lund,[3] Willie Lockeretz[4] and Malla Hovi[5]

[1]Research Centre Foulum, Danish Institute of Agricultural Sciences, PO Box 50, DK-8830 Tjele, Denmark; [2]Organic Studies Centre, Duchy College, Rosewarne, Camborne, Cornwall TR14 0AB, UK; [3]Department of Animal Welfare and Health, Swedish University of Agricultural Sciences, PO Box 234, SE-532 23 Skara, Sweden; [4]Friedman School of Nutrition Science and Policy, Tufts University, Boston, MA 02111, USA; [5]Veterinary Epidemiology and Economics Research Unit, University of Reading, PO Box 236, Reading RG6 6AT, UK

Editors' comments

A great many authors have drawn upon their knowledge and experience to offer a diversity of viewpoints throughout the book. In this final chapter, the editors identify challenging but unavoidable questions for future discussions and development of organic animal husbandry. The aim of this book was not only to report the status of organic animal husbandry. It also was to open questions that cannot be expected to be answered once and for all, because the solutions to some problems are closely linked to changing living and farming conditions, and also to the basic ethical view held by the problem-solver, which similarly can be expected to change. Therefore, this chapter, in opening questions for the future, is intended to support the overall aim of the book: to help in addressing the critical outstanding questions on the basis of the most current knowledge and understanding of animal health and welfare.

Introduction

In this chapter, we synthesize the various challenges and conclusions set out in this book, and discuss these as a framework for the development of

© CAB International 2004. *Animal Health and Welfare in Organic Agriculture* (eds M. Vaarst, S. Roderick, V. Lund and W. Lockeretz)

organic livestock farming. The content of the book has ranged from discussion of the broad philosophical concepts of animal welfare to practical considerations in the application of these concepts, as well as offering an insight into the technological developments and practices of keeping animals under organic farming conditions.

Creating sustainable agroecosystems is the main aim of organic farming. Although animals play a critical role, and their welfare considered important, this element of organic farming has not been a major focus of research. The research that has been conducted has tended to concentrate on the disease status of animals, and how this is influenced by standards. This tendency is perhaps understandable, given the measurable nature of disease and the less distinct definition of animal welfare. Also, the economic consequences of health and disease problems are what tend to trouble the converting farmer.

The philosophical framework of organic farming is the source of several dilemmas in relation to animal welfare that need to be confronted. The organic idea of animal production emphasizes the animal's chance to live a natural life as a precondition for welfare and health. However, how we can achieve this has not been well documented, and the practical implications of this concept of animal welfare have not been discussed sufficiently. This presents us with complex and demanding challenge.

Understanding the Advantages, Risks and Epidemiological Implications of Organic Practices

The standards and practices of organic farming directly influence the disease status of animals. The emphasis on natural living gives many welfare benefits whilst diminishing many of the problems with intensive indoor systems arising from crowded conditions and poor ventilation (including behavioural disturbances and bacterial infections). Nevertheless, the legislative restrictions on disease treatment and the 'free-range' nature of organic animal production have raised the question of whether the welfare of the animals is at risk at the same time. There is evidence that may justify these concerns for specific diseases, under some conditions, although ultimately, the organic farmer has resources, flexibility and support available to prevent the animal from suffering. The fact that the organic standards emphasize animal health and welfare, and not only the minimizing of suffering which is included in law by many European countries, is too often ignored by the critics. Much evidence shows that with correct feeding, stocking, breeding and care, disease risks for animals in organic herds need not be a major concern.

Epidemiological studies can greatly enhance our understanding of disease control requirements for organic herds and flocks. However, we must be cautious when using comparative studies to evaluate organic farming. The challenge is not just to understand organic farming in comparison with conventional; even more, it is to understand which ideas, standards and elements of the production system influence disease incidence and welfare, and how to deal with these in ways which are appropriate in the organic context. Patterns of disease in both humans and animals are influenced by biological, economic, cultural and environmental factors, and an understanding of these effects needs to be incorporated into the study of epidemiology and disease control. When making cross-national comparisons of research results, national and historical differences in organic standards and in the way organic farming is understood can be important. Furthermore, organic farming has multiple objectives, creating the challenge of developing multidisciplinary approaches in the context of many different systems.

Approaches to Health Management

Animals' own handling of disease challenges

Although organic farming aims for healthy animals, the general approach is to cooperate with nature and improve the animals' ability to deal with challenges such as parasites and pathogenic microorganisms, rather than eradicating them. The animals need to build up and stimulate their capacity to handle infections. The definition of 'tolerable disease levels' is open to debate, as is the question of whether outdoor and free-range systems enable a more balanced immune response or present a welfare problem. Disease risk in some cases is an inevitable consequence of allowing the animals a more natural life. The judgement of whether this is acceptable should be based on the nature of the disease, i.e. its zoonotic character, infectivity, resulting suffering, available treatment, control options and potential immunological response, as well as an understanding of its epidemiology. A great deal of knowledge must be taken into account to provide a basis for such a judgement.

The question of vaccination

The use of vaccination has been questioned in organic farming, partly because routine use can be viewed as a management tool that may be masking critical flaws in the production environment and system. Decisions regarding the use of vaccinations need to involve risk and impact assessment. For some diseases, such as the clostridial diseases affecting sheep, the incidence is inconsistent and not predictable, and decisions may be heavily influenced by the relative risk averseness of the farmer. Above all, in the context of contagious diseases, the organic farmer is part of a mixed industry of both conventional and organic farms. Within this community, the level of risk and the need for protection from or acceptance of exposure is directly linked to the specific epidemiological situation in a country or a region. Geographic conditions, statutory approaches to animal disease, and even cultural practices regarding biosecurity and disease control vary significantly among countries. Within Europe, the UK and Finland offer two extreme examples: in the UK, the sheep and cattle industries struggle with over 25 endemic diseases that the farmer must either live with or protect the animals against, without any statutory or communal control programmes. In Finland, most of these diseases have been eradicated or are carefully controlled by communal or statutory efforts, so that the risk of introducing one of them into a farm is very low. In these two widely different situations, the 'organic' approach to disease control and health promotion necessarily will differ significantly also.

The role of the veterinarian

For the producer, the challenge is often seen as developing a system of management that meets the legislative requirements. Increasingly, this involves superficial changes in the methods used before conversion. The real challenge is to develop innovative systems of production that allow 'optimum' disease risks and to adopt good health promotion and health care practices, whilst encompassing the other organic and economic objectives. In some cases, this will involve fundamental changes to the existing system. We discuss later in this chapter how farmers' knowledge, coupled with epidemiological and ethological understanding, are critical in this respect.

An iterative and responsive approach, where decisions and developments are based on analysis of herd- and farm-specific conditions, goals, possibilities, results of previous efforts and experience, are essential for the development of the entire organic animal sector. The veterinarian can be vital in this process. However, there is a general feeling, supported by various studies, that the veterinary profession does not accept or trust the approaches to animal health management inspired by the goals of organic farming; this apparent mistrust was one of several reasons behind this book.

There are several likely reasons for this mistrust. One reason may be that veterinarians have a different understanding of animal welfare, one that emphasizes health, and focuses on disease in particular. Another may be that they do not approve of the organic focus on alternative treatment methods. Yet another may be that organic farming's emphasis on health promotion without routine drug use – in some countries and some organizations with little drug use in general – causes veterinarians to feel that they cannot really provide the farmer with anything. Since organic farming now is firmly entrenched in Europe, it is imperative that veterinarians become more aware and understanding of the organic approach, and that they engage more actively with producers and advisers and become more proactive in the development of organic animal husbandry.

The knowledge base has shifted from the requirements for prescribing medicine to one of understanding the role of animals within a farming system, including their behavioural needs. This also changes the role of veterinarian or adviser from that of expert and specialist to that of partner. Veterinarians need to develop a greater understanding of the goals for organic farming if they are to be effective in working with organic farmers. This means a greater understanding of the farmer's objectives and the individual characteristics of the farm, and systematic use of epidemiological analysis and monitoring of the health and disease situation of the herd, with the aim of better supporting the decisions of the farmer. Communication with health professionals and support through health planning and disease prevention should partially replace the focus on treatment.

There also needs to be a greater openness to alternative or complementary treatments, practised on so many organic farms and encouraged by EU regulations. Veterinarians can offer support without themselves specializing in homoeopathy or herbal medicine; roles the veterinarian should and can take, by way of offering advice and assessment, include verifying that the therapy used is safe and humane, and helping the farmer monitor and judge the efficacy of new treatment modalities. In addition, research resources are needed to provide the practising veterinarians and advisers with sufficient background to judge appropriate ways of handling actual diseases in individual herds.

Standards and Development of Good Practice

Standards are becoming ever more stringent as various derogations are lifted. At the same time, there is growing pressure from the farming lobby and some retailers, particularly among new entrants, to be more flexible and lenient. Increasingly, it seems that being organic means fulfilling the legal requirements. This misses an important point: meeting them does not by itself guarantee that one is a good organic farmer.

The standards are built around a basic understanding that organic farms will integrate diverse enterprises – home-grown feeds, animals as a source of fertility building, crop rotations, etc. Yet, many organic farms have only or two enterprises, and it may be very difficult for them to establish a truly organic system, particularly in areas of intensive production. In the more marginal areas, such as mountainous regions, single-enterprise systems may be the only option. Although they have only one enterprise, they often function within a nearly natural environment, and the integration here is not agricultural but ecological. In other areas, should the aim not be to develop multi-species systems that better comply with the organic aim for diversity?

With this in mind, the focus in the development on organic farms moves from simple input substitution to redesign of the system. Current organic standards and certification procedures have been blamed for fostering the former at the expense of the latter, because they tend to be goal-orientated and inflexible and do not encourage a continuous improvement in animal welfare. Yet, continuous improvement is required to achieve real change.

Codes of practice adopted by many private organic certification bodies allow a more flexible approach, and could be used to achieve the underlying aims of organic principles. To foster this, the standards should not only be quantitative, but should also be sufficiently flexible to result in a system that the animal easily fits into. Although standards aim to achieve a desired level of health promoting actions by the farmer, they should also be motivational. Should farmers simply be instructed to comply with the standards, or should they be helped and encouraged through a process of participation involving advisers, certification bodies and researchers?

This is a major challenge, because current developments in Europe seem to be driving organic certification in the opposite direction, with legislative requirements of quality assurance demanding clearer demarcation between farms and their certification bodies that traditionally acted as development partners in organic farming. Also, by making the standards into EU legislation they not only face the challenge of having to fit all kinds of geographical, traditional and climatic contexts – they also become much more difficult to change (a fact that has been pointed out by IFOAM as a major problem for organic farming). As a result, there is a growing risk of organic farming becoming too bureaucratic and prescriptive, resulting in substantial loss of the innovation and creativity of the pioneer generation.

These developments emphasize the need for farmers to have a profound understanding of what the organic concepts and values are, and how they can be transferred into practice. The farmer is the most important human in organic production, and thus all concepts such as 'good practice' and animal health plans must fully include the farmer and be genuinely participatory. The same applies to research, where the ultimate aim is the practical application of problem-solving technology. The farmer not only needs to operate the system and to work with the animals, but also carries the most knowledge of all aspects of the farm. Planners, advisers and researchers have a role as facilitators, and need to apply their knowledge within a framework in which the farm must be considered as a whole.

Integration of animal welfare assessment into health planning

An important tool in the development of animal friendly systems is welfare assessment procedures. Welfare assessment tools relevant to organic systems are needed to support the farmer, and the farmer must see them in that way, or as an integrated part of a health plan. They should be flexible and enable a more coherent approach to improving the life of the animals, consistent with other possibly conflicting objectives. There is evidence from the development of environmental plans in agriculture that involvement and full participation of the farmer is crucial to success. Welfare assessment needs to be more closely linked to herd management, with the behavioural needs of the animal being central to the procedure and to the resulting interpretation and action.

There are very good examples of farms that have evolved systems to combine what can be perceived as a good quality of life under farming conditions with economic efficiency. In these systems, the health of the animals is generally good, providing evidence of the link between healthy animals and a natural environment. The key to their success is the farmer and the other people managing and taking care of the herd. The role of the humans has been identified in many studies as critical in creating a positive environment, respecting the needs of the animal, and providing care. Yet, we are likely to see much variation in the approaches to management and care. Formal training and the development of 'good practice' have been proposed to accompany organic standards. Ultimately it is the caretaker's attitude and respect for the animals that will determine whether or not the 'integrity' of the animals is fully considered and incorporated in the development and daily routine of an organic farm; an important aim of the training would be to assist in developing such attitudes.

The Farm, the Herd and the Animal

Animal welfare as an important goal for the organic farm

A great challenge for organic farming is to balance the requirements for high standards of animal welfare with high standards of environmental care and protection of human health, whilst ensuring economical viability. The task is made more difficult by the fact that many concepts of animal welfare remain merely philosophical notions that seem difficult to put directly into practice. This becomes obvious in speaking with the people directly involved in production, i.e. farmers. Their notions of animal welfare often appear to diverge from those held by ethologists, animal scientists, policy makers or consumers. This is not to say that farmers themselves are not concerned with the lives of the animals in their care. Their approach just has to be more pragmatic, often by necessity and partly determined by how much consumers are willing to pay for their products. The challenge therefore includes closing the gap between what is perceived as the ideal and the practical reality.

Farm animals make important contributions to most organic farming systems. Environmental models demonstrate clearly that for organic farms to be ecologically sound, the integration of animals at the farm level is vital. It remains a question as to whether today's more specialized production actually achieves this, or even attempts to achieve it.

The challenge of integrating monogastric animals into organic systems

Whilst there are often great similarities between organic and conventional ruminant production, the requirement to be largely outdoors makes organic pig and poultry production very different from conventional. The feed requirements also pose special challenges on organic monogastric animal husbandry. This is particularly true for poultry, as most conventional production in Europe is not landbased, but rather is done in very large units of intensively housed birds. The growth in demand for organic products has required the rapid development of commercial outdoor production units that are larger and more intensive than traditional 'backyard' outdoor production. Whilst there is growing experience with free-range conventional production, the evolution of organic systems is in many ways still in its infancy.

Integration of monogastric animals into organic farming systems presents a significant challenge, not only practical, but also philosophical. It is a matter of discussion to what extent monogastric animals can contribute to the agroecosystem, and thus to what extent they are justified in organic farming. In contrast to ruminants, with their special roles – among others as forage eaters and manure producers – monogastric animals compete with humans for high-quality protein, and in some cases do little to improve the soil. However, the opposite sometimes holds: monogastric animals can be an important part of the crop rotation, and on farms with different crops and different animal species, these may all fit very well together, and monogastric animals could also be important parts of integrated animal husbandry systems, where more animal species interact at the same farm.

Conventional animal production commonly uses artificial dietary inputs such as synthetic amino acids and mineral and vitamin supplements. The prohibition of synthetic amino acids in animal diets has provoked arguments based on animal welfare concerns, as well as on economic grounds, particularly concerning poultry. In large parts of Europe, organic poultry can be produced without using synthetic amino acids, but there is still a need to develop systems that work well in harsh climates where several important protein crops, such as soybeans, cannot be grown. Development of breeds that do not have excessively high requirements is an important part of this work. Production from these birds will be more expensive, and could reduce the market for organic eggs and chicken meat. A concerted research effort is required to identify feed sources that make artificial feed supplements unnecessary. Keeping poultry in orchards, or using crop residues, provided the birds are allowed full access to natural pasture, are viable options with positive benefits to crop production.

Such an approach in turn raises questions concerning the current structure of animal production. Rather than relying on a few big egg producers, for example, perhaps it would be advisable to keep more animal species in each farm, that is, for each organic farmer to have a flock of laying hens. Maybe this can be carried out in practice by collaborations between nearby farms. This may demand changes in the supply of different animal products, and consequently changes in consumption patterns also.

The challenge of self-sufficiency

Organic farming systems are supposed to be based on local resources, the aim being to create closed systems where nutrients are recycled. With regard to nutrition, the EU regulation as well as national standards are putting increasing pressures on farmers to feed animals from on-farm sources, or, if that is not entirely possible, to link to neighbouring farms and to purchase organic feeds. The ideal role of organic animals would be to consume forage directly from the fields, whilst recycling nutrients in manure. Where this is not possible (and in most countries all year round grazing is not feasible for all species), the animals should at least be ingesting and returning nutrients grown on the farm.

For some producers and in some areas with difficult climatic conditions, this is a difficult challenge, particularly for those keeping pigs and poultry and for dairy producers using animals with a high genetic yield merit. There is a real risk that systems not designed for self-sufficiency will result in poor animal welfare because of nutritional deficiencies. The organic solution is to raise species and breeds that are well suited to the production environment. Besides being largely forage based, organic animal production should start with the goal of being self-sufficient, with deviations from this goal occurring only when there are genuine problems with the feed supply. Even then, the alternatives ought to adhere as closely as possible to regional self-sufficiency within the constraints of the organic standards. In particular, linking farms within a region offers opportunities to share resources.

The Practical Realities of Naturalness

What is natural behaviour and how 'natural' can we be in the framework of an organic farm?

'Naturalness' and the expression of 'natural behaviour' for animals are often used to describe the conditions that we strive for in organic systems, but how achievable is this, given the unnatural environments we have to keep them in to raise them commercially? Another important question is the extent to which the aim of naturalness contributes to the welfare of the individual animal; this turns out to be one of the inherent dilemmas for organic farming.

Given this emphasis on satisfying animal behavioural needs, we need to develop a much better understanding of what natural behaviour really is, and how far we can go to fostering it under farming conditions. We also need a better understanding of what naturalness means for animals that have been bred for highly artificial conditions. We cannot expect a high-yielding Holstein or a hybrid hen to live a natural life if some of its breed characteristics interfere with its doing so. Using terms such as 'naturalness' can be contentious, not only because they are ambiguous and imprecise, but also because they can be seen as conflicting with other farming practices. Organic farming is basically a mixed farming system, and in a sense, the confinement of animals assists in maintaining such a system insofar as manure can be collected and farm-grown cultivated feeds can be fed. Enclosure within fields enables more effective management and integration in crop rotations. Moreover, animal products in many respects can be more efficiently produced with confinement.

In such situations, our farm animals are always going to have their freedom restricted. The organic aim, then, is to create systems that permit a maximum of freedom and expression of naturalness while still being agronomically and commercially viable, thereby replacing the notion of purely natural freedom. Still, the starting point for all farmers when developing production systems should be due consideration of the animals' 'natural' patterns of behaviour.

Some systems, such as extensive, grass-based production, largely offer natural conditions. An outdoor life is one key to a natural life – although even on extensive pasture, maintaining animals still usually requires careful management and intervention. This is particularly true on organic farms, where risks of parasite infestation may be high, and where grazing animals have to fit into the farm's crop rotation. Integrated health management is often necessary. Helminth control is a good example: knowledge of epidemiology, integration of more than one animal species, selection for disease resistance characteristics and the adoption of new techniques, such as on-farm faecal egg counting, together require a much more complex process of monitoring, evaluation and implementation than the conventional approach of strategic prophylaxis. At one time the conversion of mountainous hill farms was thought to be automatic, obvious and merely an administrative process. We now know that this is not the case, because these farms present health, welfare and environmental problems that are not easily dealt with through organic methods.

Naturalness in intensive farming systems?

However, the more intensive production systems are where the real challenges lie regarding how much 'naturalness' we can introduce without impinging on the realities of commercial farming. There is little that is natural about the controlled seasonal and artificial breeding, machine milking, and feeding of grain and conserved silage found on most dairy farms, even organic farms. Questions such as whether the cows should be dehorned may seem of little importance in relation to 'naturalness', considering these major constraints to living a 'natural' life. Yet the desire to provide animals with more freedom within the farm environment should remain an aim for organic farming. The organic standards certainly stress conditions such as outdoor access, diets adapted to animal physiology and avoidance of some mutilations or tethering. An important area for debate is how to meet the aim of allowing the animals to approach true 'naturalness' under farming conditions that can offer the farmer a livelihood.

Natural life, commercial pressure and economics

Innovation is not all that determines the conditions we provide for the animals in our care. There is also the matter of costs, which can be considerable. Although organic farming practices and principles should not be dictated by consumer demands, it is more expensive to provide enhanced living conditions for animals. Commercial pressures are limiting progress in this respect. When the price differential between organic and conventional products is eroding, the capital to provide enhanced production environments will not be readily available. A significant part of the costs of welfare improvement occur before an animal leaves the farm to be slaughtered, and substantial restructuring of the market chain is required for this to be reflected in the price the farmer receives. It is also necessary to maintain a dialogue with the consumers regarding animal welfare issues in organic farming.

Challenges connected with naturalness

Despite human caretaking, some risks are inevitable in natural production environments and free-range systems. For example, in organic poultry production – which clearly differs from conventional – the mortality risk from predation is higher, which partly can be solved by increased vigilance and more better designed systems. An area of particular concern with organic poultry is the exposure to parasites and zoonotic diseases, such as *Campylobacter* infection and salmonellosis.

Natural systems are not automatically achieved, nor do they automatically bring about animal welfare. If we face this challenge with openness and acceptance, we may help the development of organic farming systems in a dynamic process that seeks to reduce the negative and promote the positive elements of naturalness. Again we need innovation coupled with ethological, epidemiological and environmental knowledge.

Reconciling and Promoting Diversity

Some people see the diversity in organic farming practices and trends across Europe as a problem, and focus on the logistical issues of legislation, enforcement and interpretation raised by that diversity. The wide differences in climate, culture, geographical conditions and demand for different animal products underscore the need for individualized approaches to animal husbandry. Whilst we should accept the variations in conditions and approaches to organic farming, we still are faced with the task of ensuring that the basic principles are applied and the legislative requirements met. However, a pragmatic approach to interpretation and implementation is required if we are to embrace this diversity.

Despite the differences, there are lessons to be learned from all situations, particularly with regard to developing innovation in the way we farm. Communication also becomes paramount in ensuring that this diversity is viewed positively and constructively. Collaborative efforts of organic researchers across Europe are now both commonplace and fruitful, but despite international bodies such as IFOAM, extending this collaboration to incorporate producers and advisers is still at a very early stage. There is little history of international understanding and collaboration in farming generally, although many organic farmers traditionally have had concerns and contacts that cross regional and national borders. There may be greater commonalities among organic farmers in different countries than between some organic and conventional farmers in the same country. This speaks in favour of not only maintaining but also extending cooperation among organic farmers. Internet accessibility and the greater emphasis on information dissemination provides good opportunities to share and understand this diversity. In combination with the need to promote much more innovative development – such as of multi-species and multi-enterprise systems – this offers a major challenge for future organic animal husbandry.

Technological Developments and Technology Transfer

The pioneers of organic farming were partly driven by scepticism towards high-tech agriculture, and consequently preferred low-tech, low-investment systems. This attitude seems to have diminished as organic farming has become more mainstream. Still, some technological approaches create dilemmas. For example, should organic farming favour robotic milking that allows the animals a more natural behaviour (and perhaps the farmer more freedom)? Or, is it better to keep to those systems of manually operated machines and twice-daily milking which enable better grazing management, require less capital, and allow closer contact between human and animal?

Technological developments associated with breeding and reproduction present especially difficult moral and ethical problems throughout society, and the organic sector is no exception. However, it is possible to find some answers within the organic philosophy on how to handle these problems. The organic notions of integrity, naturalness and environmental care provide the basis for the response, which in part is to reject the use of genetic engineering. However, there are new and emerging technologies where the organic sector is less likely to reach consensus, such as genetically engineered vaccines, or the use of genetic mapping to select more suitable plant species for organic production. As another example, rejection of artificial insemination in favour of natural mating would also make it more difficult to attain many breeding goals that are central to organic farming, such as disease resistance and longevity.

A particularly difficult set of questions arises because in organic animal production there is an increasing requirement for parent animals to be organically managed, not just their offspring. However, as long as this does not rule out AI, it is likely that some unacceptable breeding technologies, for example embryo transfer, will indirectly be involved in organic farming because they will be used to produce the bulls whose semen will in turn be used in organic farming. However, we also need to discuss the consequences of separating organic production completely from conventional production. This discussion is closely related to the discussion of whether organic farming should be an alternative for most farmers or if it should be an exclusive niche sector. If AI cannot be used even for parent animals, organic farming is cut off from conventional breeding programmes, which eventually will cause a dramatic difference in genetic traits (such as yield or disease resistance) between conventional and organic cattle – for good or for bad may be discussed.

There is also the possibility of transferring the values and practices of organic farming to the broader agricultural world. For example, organic farming can offer models of how animals can be raised without reliance on antimicrobial agents and anthelmintics – an objective now being adopted throughout many livestock sectors, particularly in the intensive poultry industry as well as in the dairy industry to control production diseases such as mastitis. However, there is a risk when a single practice is isolated and transferred to conventional production without the fundamental system changes needed to make it effective. Organic poultry production systems in particular often are very different from large industrial units, and their health control practises involve more than just avoiding prophylactic antibiotics. Breed resistance qualities, feed practices and free-range environments are all intended to stimulate the animals' favourable response to disease.

Organic Animal Production in a Global Perspective

This book is focused on European agriculture. What about the relevance of European organic farming to the wider world, tropical regions in particular? Although climatically such regions are very different, the basic principles often apply, and often are more achievable than they are in parts of Europe. However, with respect to animal production there are significant differences, largely related to the difference in the nature of animal diseases and in attitudes towards animal welfare.

Many of the common animal diseases in Europe are what we term 'management diseases', whereas some of the most economically important in the developing world are largely environmental or vector-borne. Management diseases can, by definition, be controlled by changing how we manage the production environment. However, the vector-borne diseases prevalent in many parts of the world, such as the tsetse-borne trypanosomosis and the tick-borne theileriases group, are largely determined by climate, combined with social and economic factors. Such diseases can be devastating, not only because of how many animals die, but also because of the severity of the resulting economic losses. Non-chemical approaches show potential, such as the odour-baited trapping of tsetse flies and the use of trypanotolerant breeds of cattle, but chemoprophylaxis and chemotherapy still are the main control options. Because of the complexities of the epidemiology and control options for such diseases, we must treat with caution the notion that European approaches to health management are transferable to all situations. Local solutions to local problems, provided that the environmental and social risks are heeded, are the most appropriate in most instances.

Turning the matter around, much that has happened in the developing world may be transferable to European agriculture. The use of disease-resistant breeds and vector trapping has been mentioned, and the knowledge gathered during the development of these approaches should be embraced more enthusiastically. Furthermore, the broader approach to epidemiological studies that have been implemented in many African, Asian and Latin American countries, incorporating social, economic, environmental and production system factors, has resulted in a deeper understanding of disease control options. Whilst it may be argued that farming in Europe is very different, and that these approaches are less appropriate, organic farming in particular would greatly benefit from adopting them. The need for greater farmer participation in research has already been recognized as one area that requires more serious attention, and where tropical research can offer experience and knowledge.

In Conclusion

In this book, we have focused on the situation of organic animal husbandry in Europe, and chosen to approach animal welfare in the broadest sense, including perceptions of welfare and the conditions to which we subject animals. Although the overall framework for production has been developed through the IFOAM, EU and national standards, its application has varied significantly among farms, regions and nations. The last decade has seen rapid growth of the organic sector, which includes a broad range of interests, views and approaches, and ranges from the very large to the very small scale. The growth has brought tension and conflicts. Today there is considerable retail and consumer interest in organic products and increasing political acceptance of organic production systems. Considering all these elements, considerable challenges lie ahead to ensure that the highest standards of welfare are maintained and the integrity of the organic approach is upheld.

Appendix
A European Network for Animal Health and Welfare in Organic Agriculture (NAHWOA)

Malla Hovi

Veterinary Epidemiology and Economics Research Unit, University of Reading, PO Box 236, Reading RG6 6AT, UK

Presentation of the NAHWOA Network

In order to present the background for this book, the Network for Animal Health and Welfare in Organic Agriculture (NAHWOA) will be shortly described in the following, and the final conclusions and recommendations from this network will be presented. NAHWOA was a network of researchers from 17 different institutes in 13 European countries. In the course of a 3-year period, the Network included many activities, while EU funding has enabled the organization of a series of workshops. The workshops involved researchers, advisers, certification body representatives and organic farmers into identification and prioritization of issues that are important for research and development in organic livestock production in the EU countries. This has been particularly important and timely as the EU Regulation 1804, covering organic livestock production, was implemented in August 2000.

The Objectives of NAHWOA

1. To provide a joint platform for organizations and institutions involved in organic livestock production, particularly in animal health and welfare research.
2. To create a forum for an ongoing discussion on animal production and welfare and their interrelationship within the framework of organic livestock.
3. To provide back-up material for advisory bodies working within

© CAB International 2004. *Animal Health and Welfare in Organic Agriculture* (ed. M. Vaarst, S. Roderick, V. Lund and W. Lockeretz)

organic farming by producing a series of publications from the workshops.
4. To contribute towards the utilization of preventive veterinary medicine, alternative animal health and management practices, and animal behaviour and welfare sciences within livestock production systems, and towards teaching these issues in agricultural and veterinary education.
5. To increase information availability and sharing on animal health and welfare issues both in organic and conventional livestock production.
6. To promote collaboration between and to contribute towards other EU-funded projects on information sharing in organic agriculture research.

The Final Conclusions of NAHWOA

NAHWOA concluded its funded activities at the end of 2001 and publicized final conclusions and recommendations at its website (www.veeru.reading.ac.uk/organic), where proceedings from all the workshops can also be found. The main conclusions and recommendations are presented later in this appendix, but shortly summarized in the following:

1. It is important to formulate a philosophical definition and basis for animal welfare in organic farming and to seek to solve potential conflicts between animal welfare and other organic farming aims (environmental protection, sustainability, public health protection, etc.).
2. Animal health is considered a vital part of animal welfare. It is therefore of concern that a growing body of evidence suggests that animal health situation on organic farms is no better than that reported in conventional livestock production systems. Whilst some diseases and conditions appear to be more frequent in organic than in conventional systems, others are found more frequently in conventional systems than in organic ones. Overall, it appears that improvement of health situation has not been adequately in focus to guarantee a clear improvement in animal health and welfare when organic livestock systems have been developed.
3. Whilst the development of organic standards should be driven and informed by research that reflects the practice and experience of organic farmers, there is also a need to guarantee that policy makers seek advice from ethologists/other animal welfare experts and from public health and veterinary experts when developing the standards.
4. The central feature of EU Regulation 1804/1999, requiring selection of appropriate breeds and strains for particular farm or conditions, should be given more weight in the development of organic livestock systems. There is a need to carry out research that clarifies the suitability of differ-

ent breeds and breeding aims in organic systems. In particular, there is a need to develop both standards and breeding practices for organic poultry production, in order to avoid the inherent animal welfare problems that are prevalent in conventional poultry systems.
5. Animal health management on organic farms should be based on evidence-based and ongoing planning that should preferably be produced in written format and should be able to demonstrate a gradual improvement of health and welfare status and decreasing reliance on medicinal therapy and prophylaxis.
6. There is a need to develop organic livestock production systems that are fully integrated with other production systems on the farm and that are focused on providing animals access to natural behaviour as part of the system.
7. There are specific research needs in all the above-mentioned areas of development. These needs are detailed further on the website.
8. In regard to research direction and methodology, it was felt that an innovative approach to husbandry system development and a participatory approach to research were needed.

Standard Development under Widely Differing Animal Husbandry Systems

One of the major 'findings' of the networking has been that such activities are vital in the European situation where generic organic 'rules' are imposed on many countries with widely differing livestock production systems and climatic constraints. Networking has allowed the different partners to understand better how and where the common standards should be developed. Bringing stakeholders and researchers together has enabled cross-fertilization of ideas, collaboration in activities and exchange of ideas. Many of the NAHWOA partners will also be involved in another, wider network, financed by the EU from 2003 onwards: Sustaining Animal Health and Food Safety in Organic Farming (SAFO).

Activities of the NAHWOA Network

A series of five workshops were organized (see below), with a proceeding from each of them. By the end of the project, it was:

1. Production of a series of four thematic publications from the theme workshops.
2. Producing a book based on contributions from partners and participants in the workshops.

Askham Bryan College
LIBRARY BOOK

3. Production of a series of recommendations regarding common organic standards within the EU.
4. Setting up and maintaining an Internet site for the project.
5. Setting up and servicing a mailing list of target organizations.

Five *workshops* were organized in the 3 project years:

- *1st Workshop in Reading, UK, 4–6 June 1999* (Inaugural workshop). This workshop was not thematic and no technical papers were presented. Only representatives of the partner organizations were invited to the workshop that had the main aim to plan the other project activities, establish final themes for the workshops and set up the working groups. A total of 28 delegates from 13 partner organizations attended the workshop and three working groups (see below) were set up.
- *2nd Workshop in Cordoba, Spain, 8–11 January 2000*. The first theme of the second workshop was the diversity of organic livestock systems within the EU and the difficulty of developing uniform production standards. The second theme was animal welfare within organic livestock systems. A total of 71 delegates from 19 countries participated in the workshop.
- *3rd Workshop in Clermont-Ferrand, France, 21–24 October 2000*. The general theme of the third workshop was the human–animal relationship in organic livestock systems. The subject was approached from two directions: stockmanship and housing. A total of 47 delegates from 14 countries participated in the workshop.
- *4th Workshop in Wageningen, Holland, 24–27 March 2001*. The first theme of the fourth workshop was breeding and feeding for health and welfare in organic livestock systems. A total of 70 delegates from 16 countries participated in the workshop, in spite of the difficulties caused by the concurrent foot-and-mouth outbreaks in the UK and The Netherlands.
- *5th Workshop in Roddinge, Denmark, 11–13 November 2001*. The general theme of the fifth and last workshop was positive health management in organic livestock systems. The subject was approached from two directions: preventive measures and alternative strategies. A total of 50 delegates from 14 countries participated in the workshop.

Recommendations

The following recommendations and comments form the final output of the project and have been formulated and collated in the five workshops described above, which were run during the course of the project in 1999–2001. The recommendations are divided into three subject areas: development of standards, development of systems, training and tools and research needs. Those recommendations that address the development of common European organic livestock standards within the framework of Regulation 1804/99. These recommendations are considered particularly important as the Regulation came into force during the project period and contains several inbuilt review opportunities and needs further development.

1. Development of standards/regulations

Animal welfare

- Standards should be driven and informed by research that reflects the practice and experience of organic farmers.
- Policy makers need to put more emphasis on the advice and expertise of ethologists/other animal welfare experts when developing standards.
- It is important to formulate a philosophical definition and basis for animal welfare in organic farming.
- Conflict areas between animal welfare and other organic farming aims (environmental protection, sustainability, public health protection, etc.) need to be defined and answers need to be sought to these conflicts.
- Organic standards should be sufficiently flexible to allow continuous improvement of animal welfare and to meet the different needs in the different parts of the EU.

Breeding

- The central feature of EU Regulation 1804/1999, requiring selection of appropriate breeds and strains for particular farm or conditions, should be given more weight in the development of organic livestock systems. In particular, there is a need to develop both standards and breeding practices for organic poultry production, in order to avoid the inherent animal welfare problems that are prevalent in conventional poultry systems.
- For selection, the poultry breeding flocks, and not only the parent stocks, should be kept in free-range systems.
- Organic breeding standards, particularly for poultry, need to be developed and enforced in order to guarantee the development of a sustainable and socially and environmentally acceptable systems and to maintain high standards of health and welfare, and public health protection.
- Breeders who start producing organic stock before the derogation for using conventional animals runs out, should be supported by the

certification bodies that need to enforce sourcing from these breeders. The unavailability of suitable stock should be assessed on a similar basis as currently used for the availability of seeds, either by the inspection or certifying body providing an overview of the market situation for that particular year or by the individual producer who would be obliged to supply some proof that enquiries have been made.

Feeding

- Standards on the extent of buffer grazing/cut-and-carry practices should be set.
- Mineral deficiencies were not considered a particular problem as the EU Regulation 1804/99 allows routine supplementation in organic systems. There is, however, a need to emphasize an evidence-based approach on organic farms with appropriate soil, forage and blood analyses in order to avoid deficiencies.
- Minimum forage content of ruminant diets should be maintained at 60%.
- Synthetic amino acids should be continued to be excluded from monogastric diets to support land-based systems and to avoid intensification.
- The proportion of home-produced feed required is not stipulated and leaves room for interpretation. Not all regions of the EU can produce cereals for livestock feed locally, so that a continuing degree of derogation to this may be required. It is suggested that clearer guidelines are elaborated as to when such derogations apply.
- The rigid application of a positive feed list could limit further development in the feedstuffs available for organic systems. The network would welcome the development of criteria as to what feeds can be used on organic farms, in order to supplement the current list in Annex II C and assist certification bodies in producing positive lists adapted to national/regional conditions.
- Because of the benefit to animal health, the feeding of young mammals should be based on natural milk, preferably from the same species.
- Recommendation as to what should be done with milk withheld from sale following veterinary treatment should be developed.
- It is suggested that a limited range of feedstuffs from animal origin for monogastric species should be considered in future.

Veterinary management

- The role of bio-security, largely within a closed flock/herd, in maintaining a high health status should be further emphasized in the standards.
- EU regulation should include a requirement for written health/welfare plans.
- The following suggestions were made in regard to the use of alternative/complementary medicine:
  - The standards should require and implement similar recording for complementary medicine as for conventional medicine.
  - There should be a compulsory training in animal health management and the use of complementary medicine for converting farmers.
  - The use of complementary medicine should be presented more clearly in the context of preventive measures and health planning, in order to ensure that conversion changes in health management are not limited to a shift from conventional medicine to alternative/complementary medicine.
- The limit on the number of conventional treatments administered to an animal should be reconsidered. The emphasis should be on avoidance of suffering and disease. It was suggested that the number should be based on a maximum average number of treatments per animal in herd/flock.
- The acceptability of coccidial water-administered vaccines in organic poultry hatcheries should be clarified. In layer systems, serious coccidiosis is unlikely to be a problem in adult birds that have acquired immunity, as long as hygiene standards are good, stocking densities are not too high and site rotation is practised.

2. Development of systems, advise, training and tools

Animal welfare

- A tool to assess and improve animal welfare on organic farms is needed. The tool should be action orientated, aimed at communication and improvement, rather than assessment and ranking. Whilst the Austrian/German Animal Needs Index (ANI/TGI) was considered a useful starting point for the development of such a tool, several improvements and additions were suggested.
  - All tools should be based on sound epidemiological and ethological understanding.
  - Ecopathological (zoonotic) aspects also need to be considered.
  - The purpose of the tools is to offer practical solutions to identified problems on the farm (i.e. should lead to action).

- The 'tool pack' should be transparent (i.e. it should be clear *why* each aspect is measured).
- Development and inclusion of practical and accurate animal health parameters (or links to simple health monitoring tools such as SCC in milk, lameness and body condition scoring, etc.) into existing assessment tools.
- Development and better inclusion of ways to measure stockmanship/human–animal interaction on the farm.
- Development of ways to include transportation and slaughter conditions in the existing animal welfare indices.
- Development of separate indices for production systems with minimum housing (e.g. Scottish hill farming).

- It is important to develop links between real animal welfare and real animal health – it should not be acceptable to measure animal welfare against production parameters alone (e.g. good daily weight gain/milk yield does not equal good animal welfare, if the animal is simultaneously suffering from chronic production disease or an infectious disease).
- Consumer education is needed to demonstrate the links between welfare and product quality and the cost of improved animal welfare at farm level.

Stockmanship

- Working with a participatory approach, on the basis of knowledge networks and farmer-to-farmer interactions, was recognized as better than the traditional approach of expert consultation. It was also pointed out that the individuality of the stockpeople involved must be considered. The constraints in regard to infrastructural limitations must be recognized in each farm in order to avoid undue expectations on what 'good' stockmanship can do.
- Training and information for farmers, stockpeople, vets and other 'co-workers' on husbandry solutions is needed during the conversion process.
- It is important to raise awareness about the fact that observation of the animals and recording of the observations is an important task. Time spent on observation/recording is valuable work – both when carried out by the owner-farmer or by paid staff.
- Attention paid on the development of practical templates and novel methods to record data on farm may pay off in increased motivation. Analysis of recorded data and its presentation in ways that help practical work (e.g. action lists, graphs about trends in production levels or disease incidence) also increase motivation (i.e. 'somebody is interested in what I am doing and what the results of my actions are'). The use of recorded and analysed data as a decision making tool can also

empower paid staff if used in a creative way ('my data helped them/forced them to change the feeding system', etc.).
- A greater degree of communication and participation between the producer, the researcher and the adviser is required.

Breeding

- Develop suitable breeds and breeding strategies for organic systems before the use of genetic manipulation makes conventional breeding systems inaccessible for organic farmers. The focus should initially be placed on monogastric systems, where the existing breeds and breeding systems are often inappropriate for organic production aims.
- There is a need to establish a relationship/dialogue with breeding companies and societies across Europe. (In a fashion similar to that which has been happening on organic plant/seed production).
- It may be advantageous to draw upon the breeding experience/expertise within Eastern Europe, as many breeding programmes in that region have maintained local breeds and in so doing maintained a wider genetic pool.
- It is suggested that education and empowerment of farmers in the identification of breeding goals in organic systems may be important. This could be particularly important for establishing breeding goals that put more emphasis on female selection criteria and avoided the limitations of a 'total index system'.
- It was also suggested that breeding for clear and concise animal health targets should be practised where possible and where management has already been optimized. E.g. if breeding for worm resistance, select on a whole herd/flock basis rather than on an individual basis. It was, however, pointed out that care should be taken in using this strategy at the expense of other breeding targets and in situations where the benefits of genetic resistance are not well established.
- The following specific points were made in regard to poultry breeding:
  - There is a need for breeds that are flexible and can adapt to variability of surroundings:
  - The animals should show normal behaviour in large groups.
  - Fattening poultry should grow slowly (81 days).
  - Facilities common in organic farming, like free range and perches, should be used in breeding facilities.
  - Birds should be able to adapt to regional characteristics, like local climate.
  - Birds should be able to cope with changing food composition.
  - Birds should show resistance against diseases and parasites, even if selecting for a certain trait means poorer performance in another area.

  - It was concluded that, whilst it may not be an easy task to find suitable lines, pure lines and local breeds should be tested on organic farms and direct acceptance of existing commercial breeds should be avoided unless tested in free-range and organic conditions.

Feeding

- It was suggested that full utilization of existing know-how on grassland management and conservation of forage combined with 24-h access to forage on an *ad libitum*-basis should be the basis of organic milk production. Data from Austria and Sweden suggest that relatively high yields are possible on forage-based diets without any concentrates.
- Training and advisory materials on poultry ration formulation for home mixing should be developed. It was also suggested that poultry could be allowed to practice choice feeding where they choose their own rations.

Veterinary management

- Facilitate a greater and more rational use of alternative therapies and meet the legislative requirements for veterinary medicines use in some member states.
- Develop and implement animal health plans that guarantee maximal public and animal health standards in organic livestock production.
- The following suggestions in regard to the technical aspects of health management plans were made:
  - Target setting should be farm-specific and should be based on both short- and long-term targets. It was suggested that setting of 'national' or 'organic' target levels (e.g. for disease levels) would be counterproductive.
  - It was suggested that a step-wise establishment of health plans would be needed, particularly on converting farms. Existing data may not always be adequate to assess the existing disease situation on the farm and to allow target setting and targeted. In this context, the need to ensure the dynamic nature of health management plans was emphasized. Regular updating and reviewing of the plans should be compulsory.
  - The health management plan should be utilized to help to introduce alternative/complementary therapies on a farm in a way that allows feedback and assessment of the impact of these therapies.

Systems development

- The following suggestions were made to solve the problem of unwanted dairy bull calves: use of dual-purpose breeds for dairying, organic veal production, communal/cooperative fattening system for dairy farms, collaboration between organic arable farms and dairy farms.
- It was also felt that there was adequate information on the influence of castration on growth rates, meat quality and other aspects of beef production. It was considered important to weigh the animal welfare implications of castration against other welfare-related aspects of management on an individual farm basis (e.g. if castration would make finishing by grazing possible, whereas entire animals would need to be kept housed, castration might be considered a lesser breech of welfare than housing).
- It was suggested that organic beef production needs a higher profile as a sustainable part of organic systems. Simultaneously, environmental problems related to the use of marginal land for beef production on organic farms need to be solved in order to maintain customer confidence.

3. Research needs

General

- Research should reflect the practice and experience of organic farmers, i.e. participatory research and action-orientated research should be encouraged by the funding bodies.

Animal welfare

- There is a need to identify and clarify the problem areas in animal welfare specific to organic farming systems. This research should particularly address the areas where there is a potential conflict between animal welfare and other aims of organic production.
- There is a need to develop holistic quantification methods for animal welfare on organic farms and to quantify the impact of organic management on animal welfare at farm level in order to maintain consumer confidence and to help organic farmers to maintain high welfare standards on their farms.

Stockmanship and housing

- It was suggested that further research into what is and creates good stockmanship/husbandry is needed. It was also recognized that, in this area, there was likely to be very little difference between conventionally managed and organic systems, i.e. research in conventional

systems could benefit organic systems and vice versa. The following questions need to be answered:
- What is a definition of good stockmanship?
- Which and whose values are represented in definitions?
- Is there a need for a special definition of good stockmanship in organic farming?
- How can the learning of stockmanship be best studied? Do we need social science methodology, and which of the methodologies are most appropriate?

- The impact of changing housing and husbandry systems on stockmanship, e.g. the welfare implications of automated or free-range systems with less human–animal interaction, needs to be studied.
- There is a need to quantify the benefits of good stockmanship to the farmers, e.g. financial benefits, reduced disease incidence, etc.
- Research is required into the development and application of 'codes of practice' for organic livestock producers.

Breeding

- Research is needed to define organic breeding goals. This research should be based on the experience and expertise of existing organic farmers.
- There is a need to carry out research into poultry breeding in order to set sustainable standards for poultry breeding systems. Again, this research should draw on the experience of existing organic breeders and they should actively be involved into this research.
- There is a need to correlate existing data, in the first instance from the dairy sector, to see how well organic cows perform in comparison to their genetic potential in order to help to set organic breeding targets.
- There is a need to document what breeding strategies are used and adopted by organic breeders in different European countries.
- There is a need to carry out specific research into organic pig production with regard to meat quality (proportion of subcutaneous and intramuscular fat), boar taint/need for castration, leg problems, mothering ability, etc., in order to establish to what extent breeding can address these problems.
- Identification of suitable crosses for 'production' of desirable dairy bull calves. This research would need to take into consideration the varying conditions in different production systems and individual farms.

Feeding

- Identify natural sources and feeding regimes in order to replace synthetic vitamins and amino acids in the rations, particularly for monogastric animals and synthetic vitamins for organic dairy cows in the northern parts of the EU.

- Research into the long-term effect of early lactation energy deficit in organic dairy cows: health and welfare monitoring, including post-mortem findings.
- Research into the biological efficiency and environmental impact of feeding systems.
- Research into innovative feeding and rearing systems for young stock in all organic livestock systems.
- Development of advice based on existing information on clover management, grassland management and forage conservation adapted to organic system.
- Development of advice based on existing knowledge on the formulation of feeding rations adapted to organic systems.
- Development of systems that allow young animals to stay with their dams as long as possible.
- Identification of optimal weaning age for the health and welfare of piglets under specific farm conditions, and identification of sow breeds that are capable of suckling to this period and development of optimized system at weaning: adequate control of parasite burdens, management of salmonella, etc.

Veterinary management

- It was suggested that there is little need for more knowledge on the general epidemiology of endoparasites in ruminants. However, understanding of the local situation at farm level in regard to parasite dynamics is vital for effective control.
- There is a need to establish a better understanding of internal parasite epidemiology in monogastric animals and how these parasites affect animal welfare under free-range conditions.
- It was suggested that collaborative European projects on parasite epidemiology and control in pigs and poultry were needed to ensure adequate and representative data.
- There is still a great need to research and identify alternative therapies and systems level strategies for parasite control in all species.
- It was also suggested that there is a need for controlled field trials in parasite control to produce data for simulation models that could eventually be used as decision support tools.
- There is a need to further develop health and welfare indices and to assess their usefulness to all stakeholders.
- Animal health planning research should be innovative and participative: demonstration farms, study groups, socio-psychological studies (e.g. to establish what makes farmer willing to accept and adopt advice).
- There is a need to further explore the research needs and methodologies in complementary/alternative medicine.

Systems development

- Identification of constraints to collaborative approach to fattening of store cattle (including dairy bull calves) and development of systems that have minimal impact on health and welfare (e.g. minimal travelling time and distances, herd health safety procedures).
- Research into the practicalities of producing non-castrated bulls for organic beef.
- Research into fully integrated (at farm level), land-based poultry systems and combination of monogastric livestock systems with orchards/agroforestry.
- Research into the management and husbandry of dairy goat and sheep systems under organic standards.
- There is a need to develop livestock production systems that are fully integrated with the other production systems of the farm (e.g. poultry combined with horticulture, forestry or dairy production) and where animal behaviour (such as foraging) becomes an integral part of the system.

Index